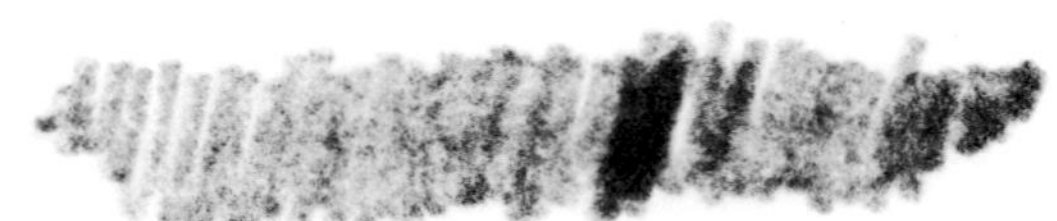

Developments in Petroleum Science, 7

BITUMENS, ASPHALTS and TAR SANDS

FURTHER TITLES IN THIS SERIES

1 A. GENE COLLINS
GEOCHEMISTRY OF OILFIELD WATERS

2 W.H. FERTL
ABNORMAL FORMATION PRESSURES

3 A.P. SZILAS
PRODUCTION AND TRANSPORT OF OIL AND GAS

4 C.E.B. CONYBEARE
GEOMORPHOLOGY OF OIL AND GAS FIELDS IN SANDSTONE BODIES

5 T.F. YEN and G.V. CHILINGARIAN
OIL SHALE

6 D.W. PEACEMAN
FUNDAMENTALS OF NUMERICAL RESERVOIR SIMULATION

Developments in Petroleum Science, 7

BITUMENS, ASPHALTS and TAR SANDS

Edited by

G.V. CHILINGARIAN

Professor, His Imperial Majesty Shahanshah Arya Mehr Pahlavi Chair, Department of Petroleum Engineering, University of Southern California, Los Angeles, Calif., U.S.A., and Abadan Institute of Technology, Abadan, Iran

and

T.F. YEN

Associate Professor, Departments of Chemical Engineering, Environmental Engineering and Medicine (Biochemistry), University of Southern California, Los Angeles, Calif., U.S.A.

ELSEVIER SCIENTIFIC PUBLISHING COMPANY
AMSTERDAM — OXFORD — NEW YORK 1978

ELSEVIER SCIENTIFIC PUBLISHING COMPANY
335 Jan van Galenstraat
P.O. Box 211, 1000 AE Amsterdam, The Netherlands

Distributors for the United States and Canada:

ELSEVIER NORTH-HOLLAND INC.
52 Vanderbilt Avenue
New York, N.Y. 10017

Library of Congress Cataloging in Publication Data
Main entry under title:

Bitumens, asphalts and tar sands.

(Developments in petroleum science, 7)
Includes bibliographical references and index.
1. Bitumen. 2. Asphalt. 3. Oil sands.
I. Chilingarian, George V., 1929- II. Yen, Teh Fu, 1927- III. Series.
TN850.B55 553'.27 78-3660
ISBN 0-444-41619-6

ISBN 0-444-41619-6 (Vol.7)
ISBN 0-444-41625-0 (Series)

Printed in The Netherlands

We dedicate this book to Dr. Zohrab A. Kaprielian, Dean of the School of Engineering and Executive Vice President of the University of Southern California, for his foresight in furthering energy research.

"The history of man is witness to his ingenuity in exploiting the earth's natural resources in building civilizations. In the 20th century, man has succeeded in this pursuit so well that he has transformed oil — his chief source of energy in this modern age — from what was once an inexpensive and seemingly limitless resource into one that is becoming increasingly costly and in danger of early exhaustion. It is comforting to see in this volume, which is one in a coming series, how man is applying his ingenuity and the power of modern technology to develop alternative energy sources so that the use of oil can eventually be relegated to only the most essential purposes and its reserves preserved for future generations."

Z.A. KAPRIELIAN

"Every other source of energy should be found to replace wherever and as much as possible this precious oil which should in my opinion, in the future — not too distant, I hope — be used mainly for the noble purpose of the petrochemical industry."

SHAHANSHAH ARYA MEHR MOHAMMED REZA PAHLAVI OF IRAN

FOREWORD

As the fact of depletion of the major fossil-fuel sources is gradually being unveiled on a worldwide scale, the exploration and exploitation of the earth's other natural fuels assume ever-increasing importance. Because of practically global availability of bitumens, asphalts, and tar sands, these natural deposits could offer a significant supplement to the waning oil and gas reserves. The interest which was generated by these potential fuels many years ago, however, was rather cursory and has not been followed by systematic scientific investigation and analysis. Although oil companies and some public concerns have invested heavily in the research on these potential fuels, the results of the investigations are not readily available. The small volume of published material on the subject tends to inhibit further the development of bitumens, asphalts and tar sands into viable fuels.

Guided by the purpose of focusing interest on these fuels, the editors invited a number of internationally known experts to summarize the current status of knowledge on origin and exploitation of bitumens, asphalts and tar sands. The editors hope that this book will be useful to the scientists and engineers who are already engaged in research in this field, as well as provide the impetus to others to devote themselves to the research on and development of these fuels. Regardless of its magnitude, any contribution to the encouragement of research work on these fuels, which may be stimulated by this book, would totally grafity the editors and contributors.

The editors wish to extend their gratitude to the contributors who worked hard to produce their chapters and, then, found the patience to wait out the span of time it has taken for the full realization of this book. The editors would also like to thank C.S. Wen, Michelle Hsi, Valerie Alexander, Mary A. Smith, and Keith Manasco for their efforts in the preparation of this manuscript.

GEORGE V. CHILINGARIAN | TEH FU YEN

CONTRIBUTORS

J.G. BURGER	French Petroleum Institute, Rueil-Malmaison, France
G.V. CHILINGARIAN	University of Southern California, Los Angeles, California, U.S.A.
J. CONNAN	Société Nationale des Pétroles d'Aquitaine, Pau, France
T.M. DOSCHER	University of Southern California, Los Angeles, California, U.S.A.
W.H. FERTL	Dresser Atlas Division, Dresser Industries, Inc., Houston, Texas, U.S.A.
W.K.T. GLEIM	University of Southern California, Los Angeles, California, U.S.A.
G. KAPO	Universidad Metropolitana, Caracas, Venezuela
P.H. PHIZACKERLEY	British Petroleum Exploration Co. (Colombia), Bogota, Colombia
L.O. SCOTT	British Petroleum Exploration Co. (Colombia), Bogota, Colombia
S.R. SILVERMAN	Chevron Oil Field Research Co., La Habra, California, U.S.A.
J.G. SPEIGHT	Research Council of Alberta, Edmonton, Alberta, Canada
F.K. SPRAGINS	Syncrude Canada, Ltd., Edmonton, Alberta, Canada
C.S. WEN	University of Southern California, Los Angeles, California, U.S.A.
T.F. YEN	University of Southern California, Los Angeles, California, U.S.A.

CONTENTS

Chapter 1

INTRODUCTION: ORGANIC MATTER AND ORIGIN OF OIL AND TAR SANDS

GEORGE V. CHILINGARIAN and T.F. YEN

Origin of oil and gas

Most of the scientists working on the origin of oil rule out all theories on inorganic origin, mainly because recent research work supplied overwhelming evidence for the biogenetic origin of oil. The main evidence lies in the similarity between petroleum (the composition and the nature of its constituents) and the basic structural subunits of components occurring in organisms. The principal model of petroleum formation can be presented as follows:

Organic matter
(mainly lower forms of plants — plankton)
↓
Hydrocarbons dispersed in sedimentary rocks
↓
Microoil
↓
Petroleum deposits $\xrightarrow[\text{e.g., due to uplift}]{\text{oxidation}}$ Solid and semisolid naphtides

The composition of living matter is presented in Table 1-I. Simple decarboxylation of lipids, which are present in large concentrations in both bacteria and spores, could yield hydrocarbons. A schematic diagram of the origin and maturation of petroleum is presented in Fig. 1-1.

The residual organic matter and hydrocarbons in some carbonate rocks and shales are presented in Table 1-II. There are about 30—150 times as much disseminated hydrocarbons in the source beds (both shales and carbonates) than in the petroleum reservoirs.

The soluble organic matter (in various organic solvents, such as benzene or ether) can be separated into (1) hydrocarbons and (2) nonhydrocarbons (asphalts) by column chromatography [2,3]. The asphalts contain sulfur, nitrogen, and oxygen, in addition to carbon and hydrogen. The residual organic matter (kerogen) has been described in detail by numerous authors (e.g., see [4]).

Representative elemental composition of natural gas, oil, asphalt, and

TABLE 1-I

Composition of living matter and petroleum

Part A: Living matter

Substances	Major constituents (wt%)		
	lipids	proteins	carbohydrates
Green plants	2	7	75
Humus	6	10	77
Phytoplankton	11	15	66
Zooplankton	15	53	5
Bacteria (veg.)	20	60	20
Spores	50	8	42

Part B: Petroleum

Substances	Elemental composition (wt%)				
	C	H	S	N	O
Lipids	80	10			10
Proteins	53	7	2	16	22
Carbohydrates	44	6			50
Lignin	63	5	0.1	0.3	31
Petroleum	82—87	12—15	0.1—5	0.1—5	0.1—2

kerogen is presented in Table 1-III. As shown in this table, the H/C ratio decreases in the following order: natural gas → oil → asphalt → kerogen. On the average, crude oil consists of 30% gasoline (C_4—C_{10}), 10% kerosene (C_{11}—C_{12}), 15% gas oil (C_{13}—C_{20}), 20% lube oil (C_{21}—C_{40}), and 25% asphaltic bitumens ($>C_{40}$). In terms of hydrocarbon types, oil contains 30% paraffins, 50% naphthenes, 15% aromatics, and 5% asphalts. Composition and boiling-point ranges of major petroleum fractions are presented in Fig. 1-2.

Asphalts constitute either straight-run residues of distilled crude oil or are obtained from the bottom of the pot with the "visbreaker" thermal-cracking process. They also result from the oxidation of the crude residues by the "blown"-air oxidation process. Asphalt is a composite material consisting of asphaltenes, waxes, and resins, and may be converted to asphaltites or asphaltoids by weathering and metamorphism [6]. As indicated in Fig. 1-2, vanadium and nickel trace-element concentrations are characteristic of asphaltic bitumens.

Kerogen is a solid, insoluble organic material of biological origin, which is incorporated in sediments. It is a cross-linked multipolymer [4,7], composed

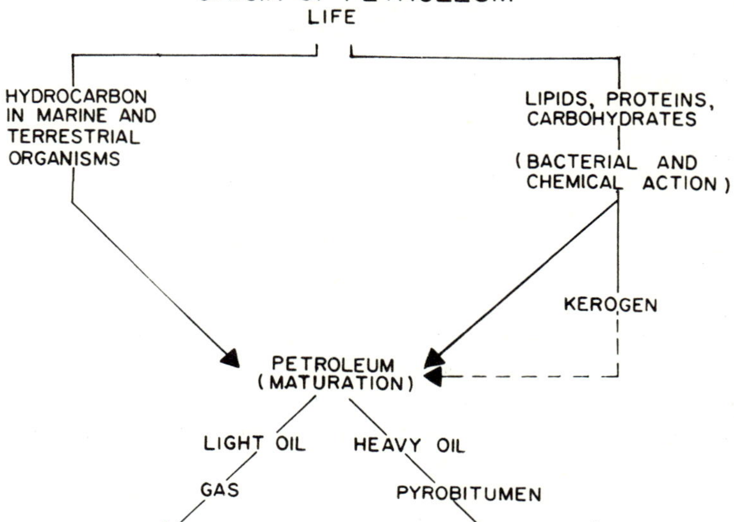

Fig. 1-1. Schematic diagram of the origin and maturation of petroleum.

TABLE 1-II

Distribution of hydrocarbons and organic matter in nonreservoir rocks (After Hunt [1])

Rock type	Hydrocarbons (ppm)	Organic matter (wt%)
Shales		
Wilcox, La.	180	1.0
Frontier, Wyo.	300	1.5
Springer, Okla.	400	1.7
Monterey, Calif.	500	2.2
Woodford, Okla.	3000	5.4
Limestones and dolomites		
Mission Canyon Limestone, Mont.	67	0.11
Ireton Limestone, Alta.	106	0.28
Madison Dolomite, Mont.	243	0.13
Charles Limestone, Mont.	271	0.32
Zechstein Dolomite, Denmark	310	0.47
Banff Limestone, N.D.	530	0.47
Calcareous shales		
Niobrara, Wyo.	1100	3.6
Antrim, Mich.	2400	6.7
Duvernay, Alta.	3300	7.9
Nordegg, Alta.	3800	12.6

TABLE 1-III

Representative chemical composition of natural gas, oil, asphalt, and kerogen

Material	Elemental composition (wt%)				
	C	H	S	N	O
Natural gas	76	23	0.2	0.2	0.3
Oil	84	13	2	0.5	0.5
Asphalt	83	10	4	1	2
Kerogen	79	6	5	2	8

of subunits similar to asphaltic bitumens. Kerogen is the major organic component in source rocks and shales in general.

Biomolecules such as amino acids, sugars, lipids, and phenols are the starting material from which the kerogen originates. Kerogen constitutes the

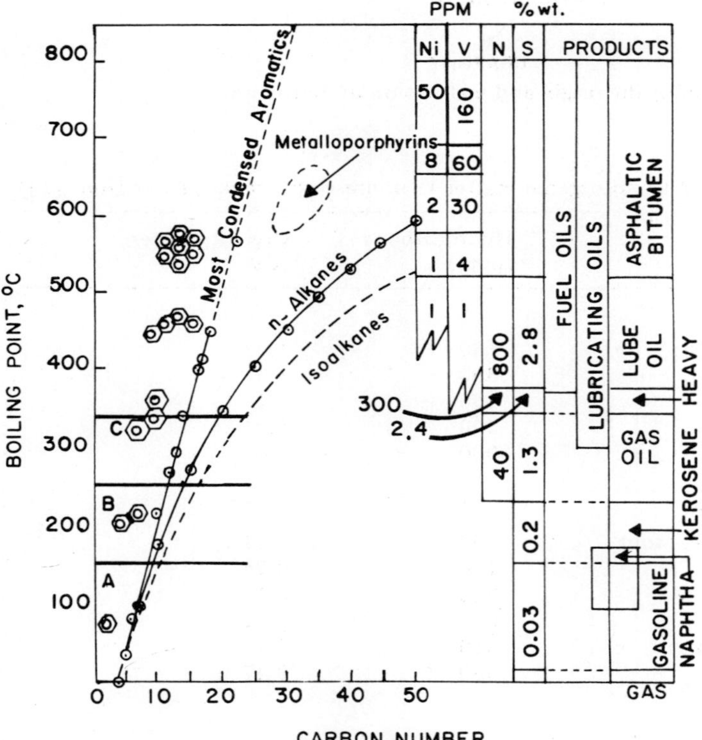

Fig. 1-2. Relationship between boiling point and carbon number for pure hydrocarbons (left), and composition of various petroleum fractions (right). (After Erskine and Whitehead [5].)

bulk of organic carbon in the sedimentary rocks of the earth's crust and is present in a highly dispersed form.

During diagenesis, new condensates or polymers are synthesized from the breakdown products of former biochemical macromolecules. Degradation gives rise to randomization of the original well-defined structural order of living matter.

The total amount of organic carbon in living matter of the earth constitutes $2 \cdot 10^{11}$ tons, which approaches that of the annual bioproduction. Of this amount, $0.3 \cdot 10^{11}$ tons are attributed to primary production of phytoplankton. The oceans contain $2 \cdot 10^{12}$ tons of C_{org}, present both in living and dead matter, either in dissolved or suspended state [8]. The amount of C_{org} in the stratosphere is around $1.5 \cdot 10^{16}$ tons, of which $1 \cdot 10^{16}$ tons belong to the continental and the rest to the oceanic sectors. The organic carbon in all coals is around $30 \cdot 10^{12}$ tons, i.e., about 0.3% of the total C_{org} in the stratosphere, whereas oil shales contain $10 \cdot 10^{12}$ tons of C_{org}. Soils contain 10^{12} tons of C_{org}. In sediments, there are $200 \cdot 10^{12}$ tons of C_{org} in autochthonous (indigenous) and $n \cdot 10^{14}$ tons in parautochthonous and allochthonous microoil. As far as continental commercial petroleum deposits are concerned, there are $2.5 \cdot 10^{12}$ tons of C_{org} in oil and $2 \cdot 10^{12}$ tons in natural gas. The petroleum deposits in the oceanic sector are estimated at $1.5 \cdot 10^{12}$ tons C_{org} for oil and $1 \cdot 10^{12}$ tons for natural gas. One should also mention that the amount of dissolved organic substances in the sea constitutes only a few milligrams per liter [9]. According to Hunt [10] and Gehman [11], the amounts of total organic matter in shales, carbonates, and sandstones average 2.1%, 0.29%, and 0.05%, respectively.

In terms of carbon distribution on earth, the amount of kerogen present in sediments (shales and sandstones) is, according to Skirrow [12], at least 1000 times greater than the biomass of living organisms (Table 1-IV).

As shown in Table 1-V, the ratio of hydrocarbons to organic matter is much lower in ancient shales than in carbonates. The content of organic matter in Recent carbonate sediments, however, is very similar to that of Recent clays. Possibly, the clays contain primarily lignitic and humic matter

TABLE 1-IV

Carbon distribution on earth (After Skirrow [12])

Source	C_{org} (g/cm^2 of earth surface)
Carbonates	2340
Shales and sandstones	633
Seawater ($HCO_3^- + CO_3^{2-}$)	7.5
Coal and petroleum	1.1
Living matter and dissolved organic carbon	0.6
Atmosphere	0.1

TABLE 1-V

Distribution of hydrocarbons and associated organic matter in Recent and ancient sediments (After Hunt [1])

Sediments	Hydrocarbons (ppm)	Organic matter (wt%)
Clays (Recent) *2	50	1.5
Clays (ancient) *2	300	2.0
Carbonates (Recent) *1	40	1.7
Carbonates (ancient) *2	340	0.2

*1 Gulf of Batabano, Cuba.
*2 Average of samples from several areas.

which survives the period of compaction and diagenesis, whereas carbonates primarily contain proteinaceous organic matter which hydrolyzes during recrystallization (E.T. Degens, personal communication, 1967).

During the early stages of diagenesis, the various organic compounds interact between themselves, with metal ions, and with mineral surfaces. The reactions proceed in the direction of increasing thermodynamic stability:

$$\text{Biomolecules—(early diagenesis)} \xrightarrow{\text{fast}} \text{metastable products—}$$
$$\text{(late diagenesis)} \xrightarrow{\text{slow}} \text{asphalts, C, } CO_2, H_2O, CH_4, \text{ etc.}$$

The thermodynamic stability of organic matter, however, is never reached [13]. It appears that the transformation of biomolecules is kinetically inhibited by the formation of metastable associations during early diagenesis.

In 1934, Treibs [14] identified first structures from living things in crude oils, i.e., porphyrin derivatives of hemin and chlorophyll, which are the animal-blood and green-plant pigments, respectively. The presence in crude oils of optically active substances, which have never been formed except by living things, is good evidence that life processes were involved. A typical composition of crude oil is presented in Table 1-VI.

The ratios of odd- to even-numbered *n*-paraffins in sediments and crude oils are presented in Table 1-VII. Based on these results, Bray and Evans [16] concluded that a sediment can be considered as a source rock for petroleum if enough hydrocarbons were generated to reduce the odd/even ratio to 1.2, which is about the maximum value observed in crude oils.

Carbohydrates, proteins and lipids are quantitatively more important in the plant and animal kingdoms, whereas the phenolic heteropolycondensates, hydrocarbons, and asphalts are more important in sedimentary rocks. E.T. Degens (personal communication, 1974) proposed a classification scheme which includes the major biochemical compounds: (1) amino acids and related material; (2) carbohydrates and their derivatives; (3) lipids, iso-

TABLE 1-VI

Composition of a typical crude oil (After Hunt [15])

Components	Wt%	Components	Wt%
Fraction (molecular size)		*Molecular type*	
Gasoline (C_4—C_{10})	31	Paraffins	30
Kerosene (C_{11}—C_{12})	10	Naphthenes	49
Gas oil (C_{13}—C_{20})	15	Aromatics	15
Lubricating oil (C_{21}—C_{40})	20	Asphaltics	6
Residuum (>C_{40})	24		100
	100		

TABLE 1-VII

Ratio of odd- to even-numbered *n*-paraffins in sediments and crude oils (After Hunt [15])

Source	Ratio of odd- to even-numbered *n*-paraffins
Recent sediments	2.5—5.5
Ancient sediments	0.9—2.4
Crude oils	0.9—1.2

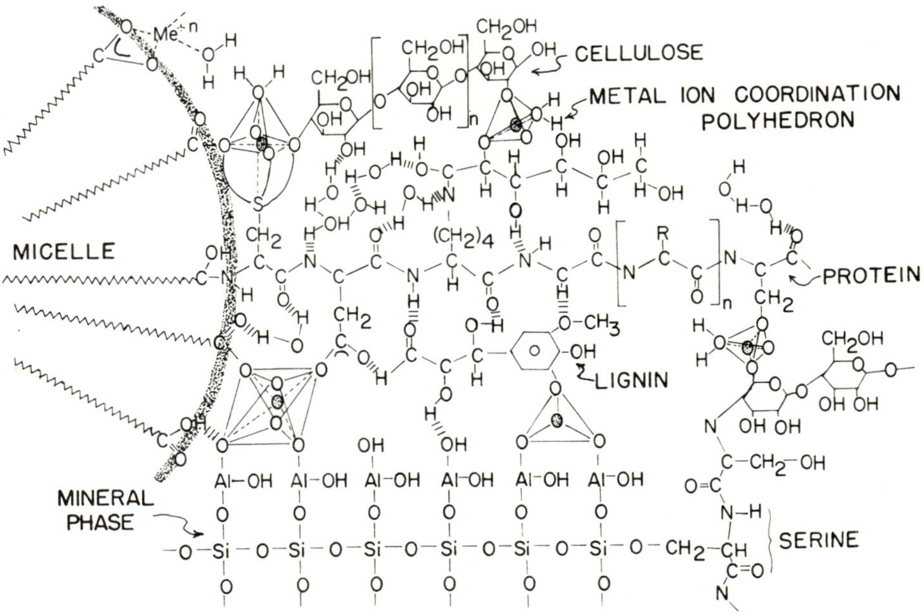

Fig. 1-3. Hypothetical schematic of the carbohydrate- and protein-containing residue of sediment. (After E.T. Degens, personal communication, 1975.)

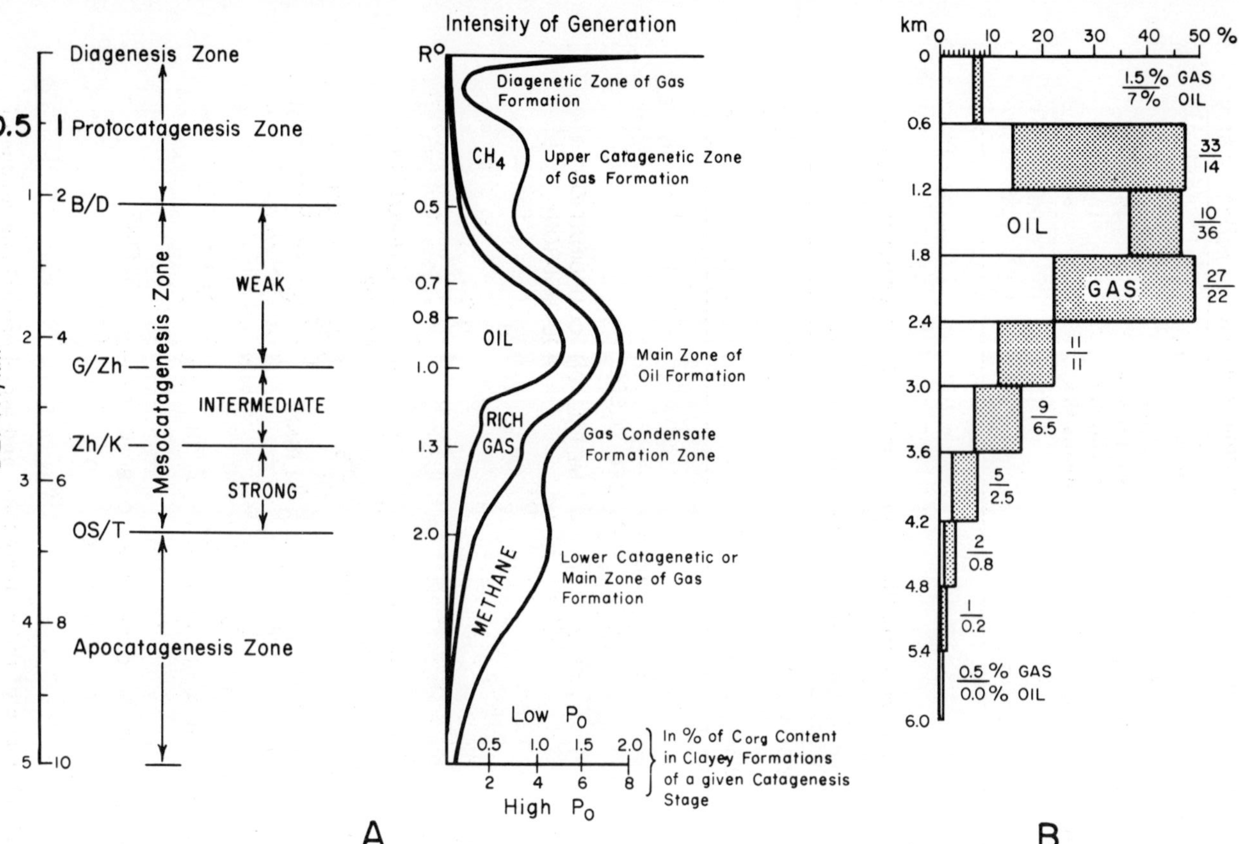

Fig. 1-4. A. General schematic zonation diagram of generation of methane-rich gas (C_2—C_4) and oil from organic matter during lithogenesis. (After A.M. Akramkhodzhaev, N.B. Vassoevich, A.E. Kontorovich, Yu.I. Korchagina, N.V. Lopatin, V.D. Nalivkin, S.G. Neruchev, E.A. Rogozina, V.P. Stroganov, K.P. Chernikov, B. Durand, J. Espitalié, B. Tissot, and other investigations; in: Vassoevich, p. 133 [18].) *B, D, G, Zh, K, OS, T* = coal ranks, used as indicators of catagenetic ("metamorphic") stage of organic matter as developed for coalification stages in Donbass, U.S.S.R.; R^o = reflectance of vitrinite in oil; and P_0 = source rock potential, which at first approximation is determined by H/C atomic ratio. Coal ranks (approximate equivalence of Soviet and American classifications): *D* = sub-bituminous to high-volatile bituminous C; *G* = high-volatile bituminous C to high-volatile bituminous B; *Zh* = high-volatile bituminous A; *K* = medium-volatile bituminous; *OS* = medium-volatile bituminous to low-volatile bituminous; *T* = low-volatile bituminous to semi-anthracite. B. Distribution of oil and gas reserves with depth in giant and average petroleum deposits as of 1/1/1972.

prenoids, and steroids; (4) phenols, quinones, and related substances; (5) heterocyclic compounds; (6) hydrocarbons; and (7) asphalts and related substances.

Degens [17] presented the structural composition and general geochemistry of organic constituents. Organic matter in sedimentary rocks either (1) survives diagenesis (organic molecules which are chemically similar or identical to living matter), or (2) is altered by diagenesis (e.g., branched hydrocarbons, which do not resemble biochemical building blocks).

There is a definite lack of knowledge about the factors which affect the transformation of biomolecules during early diagenesis. One should concentrate on studying the proteins plus carbohydrates, because they constitute about 40—80% of the organic matter of most organisms. Various interactions are possible among the organic molecules and metal ions (and minerals), as presented in Fig. 1-3, which shows a hypothetical schematic of the carbohydrate- and protein-containing residue in sediments.

The relationship between the depths of oil and gas generation and depths of occurrence of the largest petroleum deposits is presented in Fig. 1-4. Intensity of hydrocarbon generation during the various stages of lithogenesis is shown in this figure. This approach presents a possibility of quantitatively estimating the volume of hydrocarbons at these various lithogenetic stages. In preparing this figure, two factors were considered: (1) differences in geothermal gradient and, in general, in thermal history of deposits (double depth scale); and (2) difference in magnitude of oil and natural-gas source-rock potentials (two almost extreme scales of hydrocarbon generation).

On using Fig. 1-4, determination of the quantity of generated-hydrocarbon gases and/or oil up to a certain catagenetic stage represents only a rough approximation. In the first place, it is necessary to take into consideration facies-genetic type of original organic matter. For example, in the case of its "humic" nature, it is necessary to use upper generation scale and exercise great care in using generation scales of other hydrocarbons besides methane. Secondly, the following rule should be used in measuring horizontal segments, which enables determination of the quantity of generated hydrocarbons: one must use vertical intervals of 0.5 km for the left scale and 1-km intervals for the scale to the right. In this case the total sum obtained should be reduced 2—2.5 times. If the vertical intervals used are 0.25 and 0.5 km, respectively, then the values must be reduced 4—5 times. It should be remembered, however, that the above-described scheme does not take into consideration large losses of hydrocarbons that accompany their generation.

Tar sands

By definition, a tar sand is a sedimentary rock (consolidated or unconsolidated) that contains bitumen (solid or semisolid hydrocarbons) or other heavy petroleum that, in natural state, cannot be recovered by conventional petroleum-recovery methods. This condition usually applies to oils having a gravity less than 12° API.

In a paper on the source of oil in tar sands of the Athabasca River, Alberta, Canada, Link [19] stated that these sands of Lower Cretaceous age probably derived their bituminous content from the underlying coral-reef reservoirs of Upper Devonian age. The latter come into contact and near-contact with the overlying sands at the unconformity between the Lower Cretaceous and the Devonian. The escape of oil and gas through fissures and fractures from the Devonian reservoirs into the Cretaceous tar sands occurred during and after their deposition. Athabasca tar sands are treated in detail in Chapter 5.

The quartzose tar sands are of fluviatile and lacustrine origins with the highest contents of tar being present in the clean, well-sorted fluviatile sands. The maximum tarry-oil content is 18—20% by weight of saturated sand [47]. Up to 90% of pore space is occupied by oil, and sands appear to be water wet. The Lower Cretaceous McMurray Formation consists of 10—13-m thick coarse to very coarse river deposits, overlain by 30—45-m thick deposits of sand and silt. The lower section commonly contains coarse gritstone and fine conglomerate composed of poorly rounded pebbles [47].

In Chapter 4, the term "tar sand" is discussed and the distribution of both major and minor deposits of the world are shown. Details of the stratigraphy, lithology, facies, and structure of twenty major tar accumulations in Canada, Venezuela, the Malagasy Republic, the U.S.A., Albania, Trinidad, Rumania, and the U.S.S.R. are given together with reserve figures where known. The relationships of tar-oil impregnation to the geological setting and presumed origin of the oil are discussed. A brief resume of other world occurrences and the present state of commercial development of tar sands are also presented in Chapter 4.

It is estimated that more than 300,000 million bbl (57,700 million m^3) of oil are present in Athabasca oil sands in the area north of Fort McMurray. With additional similar accumulations in other areas of northern Alberta, the total potential resources of oil in place are probably around 500,000—600,000 million bbl. According to Conybeare [47], however, the economic limit may prove to be less than 200,000 million bbl (31,800 million m^3).

In an excellent paper entitled "Recovery of Oil from Athabasca Oil Sands and from Heavy-Oil Deposits of Northern Alberta by In-Situ Methods", Mungen and Nicholls [48] pointed out that in addition to the vast reserves of bitumen in the oil sands of northern Alberta in the Athabasca, Wabasca and Peace River areas, there are also large reserves of heavy oil in areas like Cold Lake and Marten Hills. Although the chemical properties of heavy oil and bitumen are quite similar, their viscosities are different. Bitumen is essentially immobile at usual formation temperatures, whereas the heavy oils have some mobility. The properties of Athabasca oil sands and their bitumens are presented in Table 1-VIII.

In-situ processes require the use of hydrocarbon solvents or the applica-

TABLE 1-VIII

Properties of Athabasca oil sands and their bitumens (After Mungen and Nicholls [48])

Sands	
Grain size	—200 mesh
Porosity	30—40%
Permeability of:	
most tar zones	200—300 mD
clean sand from tar zone	several darcies
silt zones	a few millidarcies
Saturation	0—90% bitumen (remainder water; with little or no gas saturation)
Bitumen content	0—18 wt% basis
Bitumens	
Specific gravity	1.08
Viscosity at: 200° F	1000 cP
50° F	2,000,000—5,000,000 cP
Hydrogen/carbon ratio	1.44

tion of heat in order to provide means of reducing the viscosity of the bitumen. The application of heat can be achieved either directly by the injection of steam or indirectly by air injection for the combustion of part of the in-situ bitumen. Many experimental programs are being carried out now both in the laboratory and in the field on the in-situ recovery techniques.

Acknowledgment

Partial support from A.G.A. BR-18-12, PRF 627AC-2, NSF GI-35683, AER-74-23797, and ERDA, E(29-2)-3619, E(49-18)-2031 and E(29-2)3758 is acknowledged.

AN OVERVIEW OF THE POTENTIAL OF TAR SANDS

TODD M. DOSCHER

The feasibility of producing a synthetic crude from a particular tar-sand accumulation is an exponential function of the size of the accumulation. Further, when the technical operations for recovering the tar or bitumen are considered, it becomes apparent that most of the tar-sand accumulations described in the literature will not be economically exploited in the foreseeable future. The only two accumulations that do have sufficient size and for which the technology is on a relatively advanced basis are those in Canada and Venezuela.

The significant effect of size occurs because of the large capital investment required for both extracting and upgrading the tar. This problem is exacerbated by distance from supply and marketing centers. A surface accumulation along the California coast could efficiently yield 10,000—15,000 bbl/day of an overhead fraction (by coking or other visbreaking operations), which could be further upgraded in refineries in the environs of Los Angeles or San Francisco. Taking into account extraction and refining efficiencies, a resource of some 200 million bbl would be required to support such an operation for twenty years. If the distance to refinery centers were significantly greater than a few hundred miles, or if daily production were significantly greater, then a grass-roots refinery installation would probably be required. The minimum resource base would now jump to approximately 1 billion bbl to economically justify the required capital investment. If the tar-sand accumulation was in the subsurface so that surface mining was out of the question, then the minimum size of the resource would have to be some 50% larger than the values quoted because of the overall lower recovery efficiency of saleable product realized in in-situ operations.

Over and above the question of the size of the accumulation is the availability of technology for either surface extraction or in-situ production of tar-sand accumulations. Well-consolidated tar-sand-bearing rocks can probably not be economically mined because of the economic burdens of intensive ripping and blasting required prior to excavation. Further, inasmuch as some variation of a hot-water process is probably the only feasible scheme for surface extraction, the rock would have to be comminuted to grain-size particles to facilitate the flotation of the tar. Finally, even a 100% tar-saturated rock having a porosity of 20% would contain less than 10% tar by weight. Thus, degree of consolidation, porosity, and tar saturation become limiting factors on the economic feasibility of recovering a syncrude from surface accumulations of tar sands.

For those accumulations which meet the specifications already suggested, a remaining hurdle is the nature and content of the "fines" that are associated with such accumulations. If the fines are comprised of clay minerals, the conventional hot-water separation techniques do not reject these materials quantitatively. A significant fraction will be carried over and the residue fractions will contain a significant percentage of these minerals. As a result, the residues will not be suitable fuels for power- or steam-generation facilities and their economic value will be nil. The malicious effect of the clays will also be apparent in the quality of the rejected sand in that their inclusion will represent an economic loss of mined material and possibly interfere with the environmental safety of disposal plans.

Successful research and development efforts on improvements in the efficiency of the hot-water separation process would reap great rewards. It is tacitly assumed in this discussion that optimal mining and upgrading schemes can be developed based on existing technology and accumulated empirical knowledge of such operations.

No full-scale in-situ operation for recovering bitumens from tar sands has yet been implemented, although more than a decade earlier the Shell Oil Company of Canada, Ltd. requested permission to undertake development of such a scheme to produce 100,000 bbl/day of a syncrude. In the intervening years, the required capital investment for such a project had multiplied severalfold, whereas the anticipated selling price of the syncrude has not risen significantly above originally anticipated prices. These considerations, together with the still present risks in such a venture and the increased restrictions on product disposition, have probably prevented any renewed interest in commercial-scale demonstration projects.

The technical problems encountered in recovering bitumen from tar sands by in-situ operations are profound. All such processes conceived up-to-date involve the use of heat to reduce the viscosity of the bitumen. Although steam-injection operations have had a rather wide degree of success when used in reservoirs containing viscous oils (e.g., 12°API, and greater), and in-situ combustion operations have had more limited success, the application of these techniques to tar-sand accumulations involves a quantum jump in technology. Because of the lack of any significant depletion, injection of steam or air into tar-sand accumulations requires high pressures, fracturing, or injection into intervals containing a high water saturation. In the absence of synergistic influences, high temperatures must be reached in the reservoir to fluidize the tar. The high sulfur contents of most bitumens and sulfates of associated brines result in (1) some hydrogen sulfide liberation upon the introduction of high-pressure steam, and (2) significant production of sulfur oxides during in-situ combustion operations. In-situ combustion production operations must be achieved at very high gas/oil ratios, and steam production operations entail overcoming the effects of flashing steam because of high production temperatures. All of these factors increase the technical problems, reduce displacement and thermal efficiencies, and decrease the overall economic feasibility of applying in-situ recovery schemes to the recovery of bitumen from tar sands. Successful exploitation must await the development of improved and novel in-situ processes for the tar sands.

Finally, it must be borne in mind that the reserves of recoverable syncrude cannot be reliably estimated until sound geological studies are merged with a sound concept of the mechanics by which an in-situ process will function. Until this is achieved, reserves cannot be estimated, because it is not known what degree of continuity of deposits, what minimum or maximum thickness, and what minimum tar saturation, porosity, and permeability are required for successful economic exploitation. Thus, pending the completion of successful research and development efforts it is necessary to be very wary of reports of the potential of in-situ recovery schemes, and surface-mining schemes, on increasing the world reserves of useable petroleum. The future needs of the world for petroleum liquids and the indicated resources of some of the tar-sand accumulations require that such research and development proceed with the full support of all involved organizations and governments.

References

1 J.M. Hunt, "Distribution of Hydrocarbons in Sedimentary Rocks", *Geochim. Cosmochim. Acta, 22,* 37—49 (1961).

2 J.M. Hunt and G.W. Jamieson, "Oil and Organic Matter in Source Rocks of Petroleum", *Bull. Am. Assoc. Pet. Geol., 40,* 477—488 (1956).

3 G.T. Philippi, "Identification of Oil Source Beds by Chemical Means", *Int. Geol. Congr., 20th, Mexico, Rep.,* pp. 25—38 (1956).

4 T.F. Yen and G.V. Chilingarian, *Oil Shale,* Elsevier, Amsterdam, 289 pp. (1976).

5 R.L. Erskine and E.V. Whitehead, "Composition and Analysis of Marine Pollutants", *Iran. J. Sci. Technol., 3(4),* 221—243 (1975).

6 T.F. Yen and S.R. Sprang, "ESR g-Values of Bituminous Materials", *Am. Chem. Soc., Div. Pet. Chem., Prepr., 15(3),* A65—A76 (1970).

7 T.F. Yen, "Structure of Petroleum Asphaltene and Its Significance", *Energy Sources, 1(4),* 447—463 (1974).

8 N.B. Vassoevich, A.A. Trofimuk, A.A. Geodekyan, N.A. Eremenko and V.Ya. Trotsyuk (Editors), *Combustible Rocks (Problems of Geology and Geochemistry of Naphtides and Bituminous Rocks),* Rep. Sov. Geol., Int. Geol. Congr., 25th, Nauka, Moscow, 144 pp. (1976).

9 E.D. Goldberg, "Marine Geochemistry", *Annu. Rev. Phys. Chem., 12,* 29—48 (1961).

10 J.M. Hunt, *Some Observations on Organic Matter in Sediments,* Paper presented at the Oil Scientific Session, "*25 Years Hungarian Oil*", Oct. 8—13, Budapest (1962).

11 H.M. Gehman Jr., "Organic Matter in Limestones", *Geochim. Cosmochim. Acta, 26,* 885—897 (1962).

12 G. Skirrow, "The Dissolved Gases — Carbon Dioxide", in: J.P. Riley and G. Skirrow (Editors), *Chemical Oceanography*, Academic Press, New York, N.Y., Ch. 7 (1965).

13 M. Blumer, "The Geochemical Significance of Fossil Porphyrins", in: "Symposium on the Technology and Chemistry of Oil Shale", Brazilian Academy of Sciences, Caritiba, Brazil, Dec. 1971, *An. Acad. Cienc. Braz., 46(1)*, 77—81 (1974).

14 A. Treibs, "Chlorophyll- und Häminderivate in bituminösen Gesteinen, Erdölen, Edwachsen und Asphalten, Ein Beitrag zur Entstehung des Erdöls", *Ann. Chem. Liebigs, 510:* 42—62 (1934).

15 J.M. Hunt, "The Origin of Petroleum in Carbonate Rocks", in: G.V. Chilingar, R. Fairbridge and H.J. Bissell (Editors), *Developments in Sedimentology, 9B, Carbonate Rocks*, Elsevier, Amsterdam, pp. 225—251 (1967).

16 E.E. Bray and E.D. Evans, "Distribution of *n*-Paraffins as a Clue to the Recognition of Source Beds", *Geochim. Cosmochim. Acta, 22*, 2—15 (1961).

17 E.T. Degens, *Geochemistry of Sediments*, Prentice Hall, Englewood Cliffs, N.J., 352 pp. (1965).

18 N.B. Vassoevich, "Contribution of Scientists of the Academy of Sciences and Development of Petroleum Geology (250-Year Jubilee of Academy)", *Izv. Akad. Nauk S.S.S.R., Ser. Geol., No. 5*, 123—134 (1974).

19 T.A. Link, "Source of Oil in 'Tar Sands' of Athabasca River, Alberta, Canada", *Bull. Am. Assoc. Pet. Geol., 35(4)*, 854—864 (1951).

20 A.M. Akramkhodzhaev and N.B. Vassoevich, "Present-day status of problem of terrigenous type petroleum source rocks", in: *Status and Problems of Soviet Lithology, III*, Nauka, Moscow (1970).

21 G.A. Amosov (Editor), *Questions of Oil and Gas Migration*, V.N.I.G.R.I., Leningrad, Publ. No. 370, 169 pp. (1975).

22 E.A. Bars and S.S. Kogan, *Organic Matter in Underground Waters in Petroliferous Areas (Analysis and Interpretation)*, Nedra, Moscow, 91 pp. (1965).

23 A.I. Bogomolov and L.I. Khotyntseva (Editors), *Handbook on Analysis of Oils*, V.N.I.G.R.I., V.N.I.G.N.I., and Ministry of Geology of the U.S.S.R., Nedra, Leningrad, 299 pp. (1966).

24 F.L. Chukhrov, "Towards question of isotopic fractionation of sulfur during lithogenesis", *Litol. Polezn. Iskop., No. 2* (1970).

25 N.A. Eremenko (Editor), *Geology of Petroleum (Handbook), I, Principles of Geology of Petroleum*, Gostoptekhizdat, Moscow, 592 pp. (1960).

26 A.S. Gadzhi-Kasumov and A.A. Kartsev, *Professional Gas—Oil Geochemistry*, Nedra, Moscow, 128 pp. (1975).

27 G.A. Gladysheva, V.P. Kozlov and L.V. Tokarev, *Experiment of Studying Geochemistry of Organic Matter of Coal-Bearing Deposits of Lower Carbon of Permian Along-Kam' Area in Relation to Petroleum Genesis*, G.O.S.I.N.T.I., Moscow, 60 pp. (1959).

28 L.A. Gulyaeva, *Microelements in Caustobioliths and Sedimentary Rocks*, Akad. Nauk S.S.S.R., Nauka, Moscow, 128 pp. (1965).

29 A.Ya. Krems, B.Ya. Vasserman and N.D. Matveivskaya, *Conditions of Formation and Regularities in Distribution of Oil and Gas Deposits*, Nedra, Moscow, 332 pp. (1974).

30 A.V. Kudel'skiy and K.I. Lukashev, *Formation and Migration of Oil*, Akad. Vysheyshaya Shkola, Minsk, 134 pp. (1974).

31 S.P. Maksimov (Editor), *Questions of Theory and Practice in Geology*, V.N.I.G.N.I., Moscow, 248 pp. (1970).

32 S.P. Maksimov (Editor), *Primary and Secondary Migration of Oil and Gas*, Ministry of Geol. U.S.S.R., Moscow, V.N.I.G.N.I., Trans., No. 178, 330 pp. (1975).

33 S.M. Manskaya and T.V. Drozdova, *Geochemistry of Organic Matter*, Nauka, Moscow (1964).

34 M.F. Mirchink et al. (Editors), *Origin of Oil and Gas and Formation of Their Deposits*

(Symposium of All-Union Meeting on Genesis of Oil and Gas, 1968), Nedra, Moscow, 631 pp. (1972).
35 I.S. Sarkisyan, *Investigation of Dispersed Bitumens With the Aid of Ultraviolet Microscopy*, Nauka, Moscow, 78 pp. (1970).
36 T.L. Simakova and Z.A. Kolesnik, *Bacteria of Formation Waters, Oils, and Rocks of Petroleum Deposits of U.S.S.R.*, Gostoptekhizdat, Leningrad, Publ. No. 199, 88 pp. (1962).
37 V.A. Sokolov, *Processes of Formation and Migration of Oil and Gas*, Nedra, Moscow, 276 pp. (1965).
38 G.I. Teodorovich (Editor), *Mineralogy and Facies of Bituminous Formations of Several Areas in U.S.S.R.*, Akad. Nauk S.S.S.R., Moscow, 245 pp. (1962).
39 G.I. Teodorovich (Editor), *General Principles of Formation of Bituminous Formations Based on Example of Volga—Urals Province*, Nauka, Moscow, 202 pp. (1965).
40 P.P. Timofeev and L.I. Bogolyubova, "Secondary transformations of organic matter in different facies conditions", *Litol. Polezn. Iskop., (Lithol. Miner. Deposits) No. 5* (1966).
41 V.A. Uspenskiy et al. (Editors), *Handbook on Analysis of Bitumens and Dispersed Organic Matter in Rocks.* V.N.I.G.R.I., V.N.I.G.N.I., and Ministry of Geology of U.S.S.R., Nedra, Leningrad, 315 pp. (1966).
42 N.B. Vassoevich, A.A. Trofimuk, N.A. Eremenko, N.V. Lopatin and S.P. Maksimov, *Combustibles (Problems of Geology and Geochemistry of Naphtides)*, Rep. Sov. Geol., Int. Geol. Congr., 24th, Nauka, Moscow, 119 pp. (1972).
43 N.B. Vassoevich, "Origin of Petroleum", *Vest. Mosk. Univ., No. 5*, 3—23 (1975).
44 V.C. Vyshemirskiy, A.E. Kontorovich and A.A. Trofimuk (Editors), *Migration of Dispersed Bitumoids*, Akad. Nauk S.S.S.R., Siberian Branch, Nauka, Novosibirsk, 149, 167 pp. (1971).
45 L.G. Weeks, "Habitat of Oil and Some Factors that Control It", in: L.G. Weeks (Editor), *Habitat of Oil*, Am. Assoc. Pet. Geol., pp. 58—59 (1958).
46 T.F. Yen, "Chemical Aspects of Metals in Native Petroleum", in: T.F. Yen (Editor), *The Role of Trace Metals in Petroleum*, Ann Arbor Science, Ann Arbor, Mich., pp. 1—30 (1975).
47 C.E.B. Conybeare, *Geomorphology of Oil and Gas Fields in Sandstone Bodies, Developments in Petroleum Science, 4*, Elsevier, Amsterdam, 341 pp. (1976).
48 R. Mungen and J.H. Nicholls, "Recovery of Oil from Athabasca Oil Sands and from Heavy Oil Deposits of Northern Alberta by In-Situ Methods", *Proc. 9th World Pet. Congr.*, Panel Discussion, *22(2)*, 11 pp. (1975).

Chapter 2

GEOCHEMISTRY AND ORIGIN OF NATURAL HEAVY-OIL DEPOSITS

S.R. SILVERMAN

Introduction

The heavy-oil deposits encountered at relatively shallow burial depths represent a very significant proportion of the total known petroleum resources of the world. Geographic locations and dimensions of such deposits are presented and discussed in Chapter 4 of this volume. These deposits, which have also been called "tar", "pitch", "asphalt", "native asphalt", "bitumen", and "inspissated" deposits, are also commonly referred to as "oil sands". Elaborate terminologies and classifications of these heavy-oil deposits abound [1], but such schemes are descriptive rather than genetic in the sense that they promote arbitrary delineations based on specific chemical and physical properties of the organic components and, to some extent, on the nature of the inorganic mineral components. Their principal shortcoming is their inability to identify the physical, chemical, or biological processes responsible for forming the heavy-oil deposits. Consequently, such classifications are of little value in establishing relations between heavy-oil deposits and precursor substances.

It can be shown, however, that these heavy oils are not unique products generated under unusual circumstances from special sources, but are normal products of petroleum evolution that can readily be explained by established concepts of petroleum formation and alteration.

Current concepts of petroleum formation and alteration

The petroleum-formation process is generally accepted as the thermal transformation of complex soluble (bitumen) and insoluble (kerogen) organic substances dispersed in source sediments [13,16,17]. Like most typical thermal-transformation reactions, the petroleum-generation rate increases with increasing temperature and, consequently, with increasing burial depths of the generative sediments. This scheme has been illustrated diagrammatically by Tissot et al. [16]. Their diagram, which includes data on bitumen compositions in sedimentary sequences from various localities, is reproduced here as Fig. 2-1. It shows the relative amounts of hydrocarbons generated as a function of burial depth or subsurface temperature and also the nature of the hydrocarbons formed at three levels of maturation. The upper row of composition plots (*n*-alkanes, cycloalkanes, and aromatics)

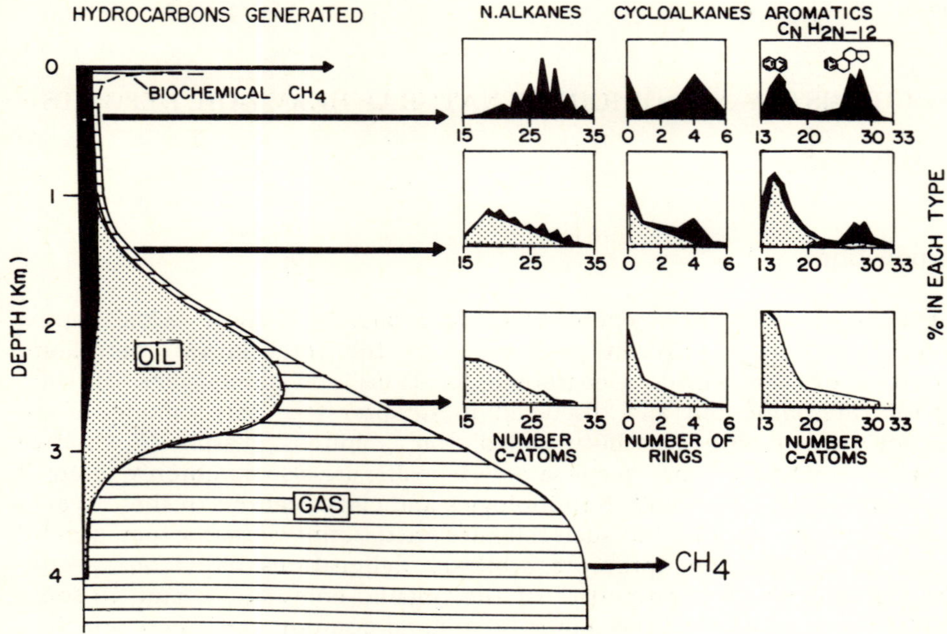

Fig. 2-1. General scheme of hydrocarbon generation. Depth scale represented is based on examples of Mesozoic and Paleozoic source rocks and may vary depending on nature of original organic matter, burial history, and geothermal gradient. (After Tissot et al. [16].)

indicates the composition of hydrocarbons shortly after deposition, but before the onset of oil generation. The second row depicts the hydrocarbon makeup at an early stage of oil generation. The bottom-row compositions represent the products formed during the principal phase of oil formation. The illustrated sequence of compositional changes, implying a progressive transformation of more-complex molecules of each hydrocarbon class to form simpler, lower-molecular-weight hydrocarbons, is the generally accepted direction of maturation or thermal diagenesis of sedimentary organic matter.

Nonhydrocarbon organic constituents also experience compositional changes during burial diagenesis. The highly complex nonhydrocarbon components such as asphaltenes are believed to be the products of polymerization and disproportionation reactions, which proceed simultaneously with the thermal-decomposition reactions responsible for forming the light hydrocarbons [4,15]. In contrast to the hydrocarbon-diagenesis pattern, asphaltenes become more complex with increasing maturity. This is confirmed by the observed increase in molecular weight and aromaticity [20] and the increase in C_R/C_T * ratio [4] of asphaltenes with increasing geologic age or burial depth.

In addition to the maturation changes identified above, petroleums also undergo nonthermal transformations subsequent to their accumulation in reservoirs. The latter changes are usually restricted to shallow oil deposits located near surface outcrops or near faults, which provide access to surface waters containing dissolved oxygen. Jones and Smith [10] noted, in their study of the chemical compositions of 310 crude oils from the Permian Basin in Texas and New Mexico, that a significant group of these oils were depleted in paraffinic constituents and gasoline contents and had lower API gravities relative to other oils from the same province. These peculiar oils were similar in two other respects, i.e., they were associated with relatively fresh waters and occurred in shallow reservoirs. Jones and Smith suggested that these compositional abnormalities could be due to water leaching and bacterial destruction of specific hydrocarbon components of the oils. Winters and Williams [19] noted similar abnormalities among oils from the Powder River Basin in the vicinity of the Montana—Wyoming border. Using gas—liquid chromatographic analysis of the saturate fractions of the normal and abnormal oils, they established that a major difference between the two types was the *n*-paraffin content. The normal crudes contained the typical complement of *n*-paraffins, whereas these components were partly or completely absent in the abnormal oils. Waters coproduced with the *n*-paraffin-depleted oils were cultured and found to contain aerobic microorganisms which were subsequently shown to be capable of consuming *n*-paraffins.

Subsequent studies [3] of a suite of six Williston Basin oils produced from a fresh-water invaded zone of the Mission Canyon Formation in Saskatchewan further attested to the effects of bacterial action and water washing on oil composition. Pertinent analytical data for the least altered and the most altered oils from this sequence are shown in Table 2-I.

The overall change in composition and character of the two end members of this sequence illustrates the extent of alteration that can be expected by water washing and biodegradation. The gravity (°API) of the whole crude decreased by 22.4 units, paraffin content of the C_{15+} saturate fraction experienced a 6.7-fold decrease, and the asphaltene content of the C_{15+} oil fraction increased by a factor of three. Another significant change to be expected from water-washing—biodegradation alteration is a large reduction in gasoline content. The gasoline-range hydrocarbon content of the High Prairie crude oil is given as 0.82%. Unfortunately, the Bailey et al. [3] publication does not contain the measured gasoline content of the least degraded (Browning Field) oil; however, from other compositional parameters revealed for this oil, its gasoline content can be assumed to be about 30%.

---

* C_R = Residual Carbon (after pyrolysis at 1000°C) and C_T =Total Organic Carbon.

TABLE 2-I

Analytical data for two Mission Canyon (5th Zone) crude oils, Saskatchewan, Canada (After Bailey et al. [3])

	Least altered oil	Most altered oil
Field	Browning	High Prairie
Producing depth (ft)	4164—4166	3886—3901
API gravity	37.6	15.2
% Sulfur (whole crude)	1.13	2.99
% Paraffins in C_{15+} saturate fraction	46.5	6.9
% Saturates in C_{15+} oil fraction	47.1	19.1
% Aromatics in C_{15+} oil fraction	37.5	43.3
% Asphaltenes in C_{15+} oil fraction	5.4	16.2

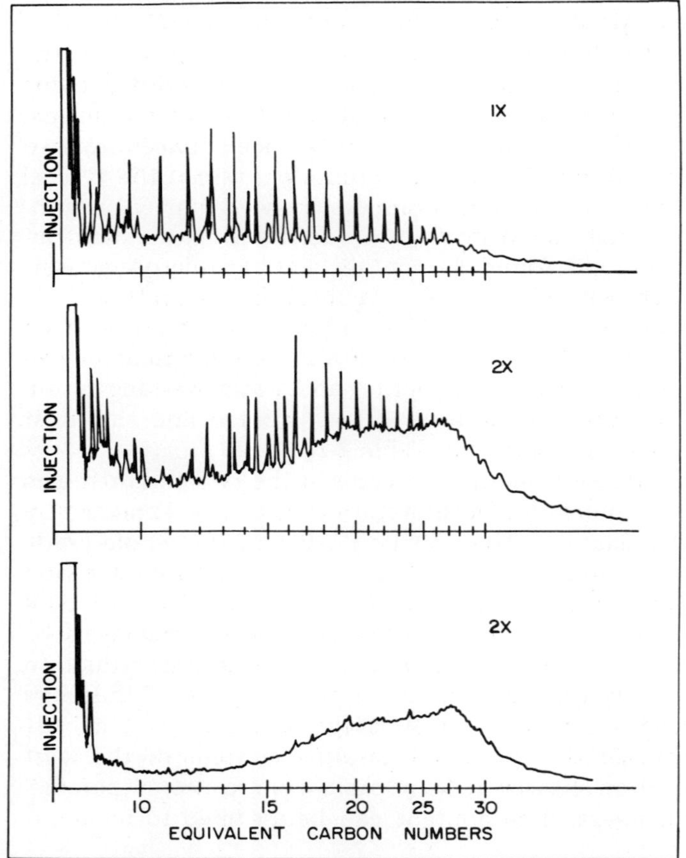

Fig. 2-2. Gas—liquid chromatograms of unaltered crude oil (top) and the same oil after 6 hours (center) and 26 hours (bottom) incubation with a mixed population of micro-organisms. Note change in attenuation in top chromatogram. (After Mechalas et al. [11].)

Additional evidence from the experimental-biodegradation studies on crude oils [2,11] generally substantiate the direction and magnitude of changes observed in the field occurrences discussed above. An example of the extent of *n*-paraffin depletion realized in the experimental biodegradation of a California petroleum [11] is shown in Fig. 2-2. The figure illustrates the change in *n*-paraffin content after 6 and 26 hours of accelerated incubation (high level of aeration was maintained by use of compressed air) of the oil with a culture of mixed microorganisms, which were previously groomed for petroleum utilization by selective culturing on petroleum-based media. The upper chromatogram in this figure is that of the unaltered petroleum; its full complement of pronounced *n*-paraffin peaks is a typical characteristic of unaltered mature petroleums. The middle chromatogram illustrates the partial attrition of *n*-paraffins accompanied by the relative increase in the unresolved nonparaffinic components as indicated by the increased amplitude and skewed nature of the base envelope. In the bottom chromatogram, the elimination of *n*-paraffins is virtually complete. Among the compositional changes attributable to biodegradation, which include the types of changes reported in Table 2-I, the preferential loss of *n*-paraffins is probably the most diagnostic. Chromatograms of the type shown at the bottom of Fig. 2-2, as well as those indicating various intermediate stages of *n*-paraffin depletion, have been obtained for sufficient numbers of natural petroleums to indicate that the biodegradation process has been active on a vast scale.

Diagnostic characteristics of heavy oils

As indicated in the preceding section, petroleum in its subsurface environment is susceptible to two principal alteration mechanisms. One, identified as the maturation process, involves thermal transformations. The other, known as the degradation process, involves chemical changes brought about by bacterial activity and water washing. Maturation results in an increase in saturate-hydrocarbon and gasoline contents, and a decrease in nonhydrocarbon (NSO * compounds) content. Biodegradation produces an opposite effect: saturate-hydrocarbon (especially *n*-paraffins) and gasoline contents are diminished, whereas the NSO content (especially asphaltenes) increases.

Inasmuch as heavy-oil deposits are typically low in saturate and gasoline-range hydrocarbons and high in NSO and asphaltene contents, they possess the qualifications both of very immature oils and of highly degraded oils. In other words, based on the general characteristics just cited, it is not possible to establish whether the deposits are products of low-level thermal maturation or degradation. Two specific properties of typical heavy-oil deposits, viz., viscosity and *n*-paraffin content, indicate that these oils are degraded

* NSO = nitrogen-, sulfur-, and oxygen-containing organic compounds.

residues of relatively mature petroleums and are not the products of an early or immature stage of oil generation.

A prominent characteristic of heavy oils is their very high viscosity or immobility, which prevents them from being recovered by the conventional production methods used to produce the more fluid (medium to high API gravity) oils encountered at greater depths. Petroleum viscosity, however, is a function of its composition. This is indicated by the data presented in Table 2-II, which represents measurements on distillation fractions of about 130 crude oils from worldwide localities. These petroleums were selected as typical examples of mature petroleums in the 29—36° API-gravity range. The average composition of this group of oils, in terms of the arbitrary distillation-range designations shown in Table 2-II, is approximately 30% gasoline, 8% kerosene, 18% gas oil, 26% lube oil, and 18% residuum. Viscosities of this whole petroleum group range from 1.5 to 7 cSt (centistokes) at 130° F.

Viscosities of typical heavy oils are of the order of 10^3—10^4 cSt at 130° F. At the significantly lower temperatures prevailing at the shallow depths that characterize many of these deposits, viscosities are in the vicinity of 10^5—10^6 cSt. Fig. 2-3, used by Winestock [18] to identify the major problems associated with Athabasca heavy-oil recovery, is reproduced here to show the viscosity—temperature relations for the Athabasca and Cold Lake heavy oils of Alberta, Canada (also see Chapter 8). The illustrated viscosity—temperature plot includes, for comparative purposes, the viscosity of a 34° API-gravity petroleum (Redwater crude). These viscosity characteristics can be used in reconstructing the geologic history of heavy-oil deposits.

These deposits generally occur in high-permeability, high-porosity sand reservoirs. Sandstones of this nature are not associated with the quantities and types of indigenous organic matter that are regarded as necessary ingredients of petroleum source rocks. Consequently, the heavy oils must have been generated elsewhere and subsequently migrated to their present locations. The petroleum-generation scheme shown in Fig. 2-1 and the viscosities of the petroleum fractions presented in Table 2-II suggest that significant quantities of gasoline- and kerosene-range hydrocarbons must

TABLE 2-II

Viscosities and molecular dimensions of petroleum distillation fractions

Fraction	Distillation range (° F)	Molecular dimensions (carbon number)	Viscosity (cSt at 130° F)
Gasoline	<400	C_4—C_{10}	<0.6
Kerosene	400— 465	C_{11}—C_{12}	1.2—1.6
Gas oil	465— 650	C_{13}—C_{20}	1.9—3.8
Lube oil	650—1000	C_{21}—C_{40}	13 —140
Residuum	>1000	>C_{40}	8500—600,000

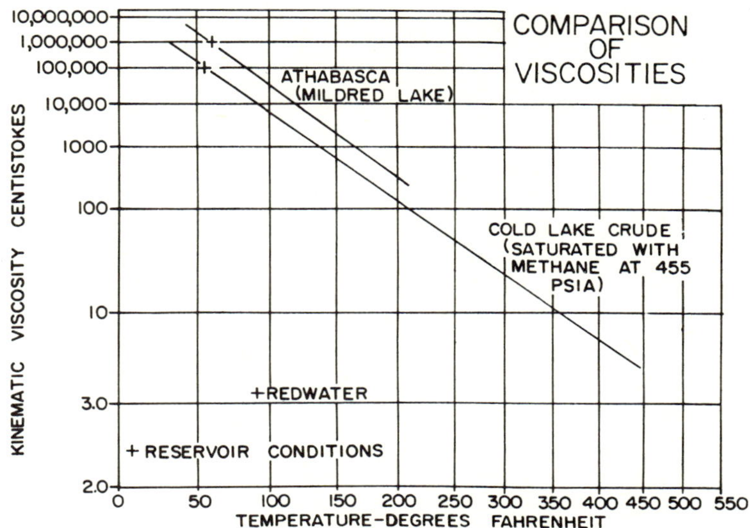

Fig. 2-3. Viscosity—temperature plot of Athabasca and Cold Lake heavy oils, Alberta, Canada. Viscosity of 34°API-gravity Redwater crude oil is shown for comparison. (After Winestock [18].)

have been generated in order to offset the relatively high viscosities of the C_{20+} components, if the products were once sufficiently mobile to migrate into their present reservoirs. The heavy oils presently saturating the shallow sand reservoirs definitely do not contain adequate concentrations of these fractions to provide the degree of mobility required for long-distance migration. In fact, their present viscosities, which exceed 10^3 cSt, preclude the possibility of extensive migration under most conceivable subsurface conditions. Inasmuch as the bitumen could not have been derived in situ from organic matter deposited with the sand, the properties of the bitumen must have been appreciably altered subsequent to its formation and emplacement.

The second specific characteristic of heavy-oil deposits that implies a degradative origin is the absence or near-absence of *n*-paraffins in their saturate fractions. Abnormally low *n*-paraffin contents, approaching the degree of depletion illustrated by the bottom chromatogram of the experimentally biodegraded oils in Fig. 2-2, have been demonstrated for a number of extensive heavy-oil deposits. This condition has been reported for the Cherokee Group heavy oils in southeastern Kansas [7], for the Lower Cretaceous heavy-oil deposits of Alberta, Canada [6,9,12], and for the Cretaceous tar deposits in the Aquitaine Basin of southwestern France [4,5].

The *n*-paraffin depletion is a definitive criterion for affirming the role of biodegradation in petroleum alteration. It cannot be misconstrued as an indication of immaturity because, although the total saturate-hydrocarbon content of immature petroleums is low, n-paraffins constitute a major portion of the saturate fraction. The latter situation has been confirmed by analyses of Recent sediment extracts and other shallow-sediment extracts in which the n-paraffins are not depleted in relation to the total saturate-hydrocarbon content (note, for example, the n-alkane composition plot representing the earliest generation stage in Fig. 2-1).

Other heavy-oil deposits, and these should be more appropriately identified as asphaltic deposits, can be formed by nondegradative processes. An example is the so-called "deasphalting" process [8], in which the simultaneous generation of light paraffins and asphaltenes by normal thermal-maturation mechanisms creates a condition of instability, because these substances cannot coexist in a single liquid phase. This incompatibility is resolved by the separation of the asphaltic components from the liquid oil phase [15]. Asphaltic deposits resulting from this separation process contain only minor amounts of saturate hydrocarbons, which are probably entrained by the precipitating asphaltic components. Inasmuch as high-molecular-weight n-paraffins are prominent constituents of the entrained saturates, these asphaltic deposits can readily be distinguished from biodegraded oils. Another difference between the two is that the asphaltic deposits are usually restricted to deeper zones in the sedimentary column.

Asphaltic deposits are not as extensive as the biodegraded-oil deposits. One reason for this is that the former are derived from a relatively small fraction of the oil, whereas biodegradation attacks and gradually alters the entire body of a pooled petroleum. The importance of biodegradation and its role in forming the vast heavy-oil deposits throughout the world has only been fully appreciated during the past half decade. In fact, there is now reason to believe that primary oils are predominantly paraffinic in nature and that all medium-gravity naphthenic petroleums, as well as the heavy oils which have been the main subject of this chapter, are formed by biodegradation of the primary oils [14].

References

1 H. Abraham, *Asphalts and Allied Substances, I,* Van Nostrand, Princeton, N.J., 6th ed., pp. 52—63 (1960).

2 N.J.L. Bailey, A.M. Jobson and M.A. Rogers, "Bacterial Degradation of Crude Oil: Comparison of Field and Experimental Data", *Chem. Geol., 11,* 203—221 (1973).

3 N.J.L. Bailey, H.R. Krouse, C.R. Evans and M.A. Rogers, "Alteration of Crude Oil by Waters and Bacteria — Evidence from Geochemical and Isotope Studies", *Bull. Am. Assoc. Pet. Geol., 57 (7),* 1276—1290 (1973).

4 J. Connan, K. Le Tran and B. van der Weide, "Alteration of Petroleum in Reservoirs", *Proc. 9th World Pet. Congr., 2,* 171—178 (1975).

5 J. Connan and B.M. van der Weide, "Diagenetic Alteration of Natural Asphalts", in: L.V. Hills (Editor), *Oil Sands — Fuel of the Future, Can. Soc. Pet. Geol. Mem., 3,* 134—147 (1974).

6 G. Deroo, B. Tissot, R.G. McCrossan and F. Der, "Geochemistry of the Heavy Oils of Alberta", in: L.V. Hills (Editor), *Oil Sands — Fuel of the Future, Can. Soc. Pet. Geol. Mem., 3,* 148—167 (1974).

7 W.J. Ebanks Jr. and G.W. James, "Heavy-Crude Oil Bearing Sandstones in the Cherokee Group (Desmonesian) in Southeastern Kansas", in: L.V. Hills (Editor), *Oils Sands — Fuel of the Future, Can. Soc. Pet. Geol. Mem., 3,* 19—34 (1974).

8 C.E. Evans, M.A. Rogers and N.J.L. Bailey, "Evolution and Alteration of Petroleum in Western Canada", *Chem. Geol., 8 (3),* 147—170 (1971).

9 D. Jardine, "Cretaceous Oil Sands of Western Canada", in: L.V. Hills (Editor), *Oil Sands — Fuel of the Future, Can. Soc. Pet. Geol. Mem., 3,* 50—67 (1974).

10 T.S. Jones and H.M. Smith, "Relationships of Oil Composition and Stratigraphy in the Permian Basin of West Texas and New Mexico", in: *Fluids in Subsurface Environments — A Symposium, Am. Assoc. Pet. Geol. Mem., 4:* 101—224 (1965).

11 B.J. Mechalas, T.J. Meyers and R.L. Kolpack, "Microbial Decomposition Patterns Using Crude Oil", in: D.G. Ahearn and P. Myers (Editors), *The Microbial Degradation of Oil Pollutants, La. State Univ. Publ., LSU-SG-73-01,* 67—79 (1973).

12 D.S. Montgomery, D.M. Clugston, A.E. George, G.T. Smiley and H. Sawatzky, "Investigation of Oils in the Western Canada Tar Belt", in: L.V. Hills (Editor), *Oil Sands — Fuel of the Future, Can. Soc. Pet. Geol. Mem., 3,* 168—183 (1974).

13 G.T. Philippi, "On the Depth, Time and Mechanism of Petroleum Generation", *Geochim. Cosmochim. Acta, 29,* 1021—1049 (1965).

14 G.T. Philippi, "On the Depth, Time and Mechanism of Origin of the Heavy to Medium Gravity Naphthenic Crude Oils", *Geochim. Cosmochim. Acta, 41 (1),* 33—52 (1977).

15 S.R. Silverman, "Influence of Petroleum Origin and Transformation on Its Distribution and Redistribution in Sedimentary Rocks", *Proc. 8th World Pet. Congr., 2,* 47—54 (1971).

16 B. Tissot, B. Durand, J. Espitalié and A. Combaz, "Influence of Nature and Diagenesis of Organic Matter in Formation of Petroleum", *Bull. Am. Assoc. Pet. Geol., 58 (3),* 499—506 (1974).

17 N.B. Vassoevich, Yu.I. Korchagina, N.V. Lopatin and V.V. Chernyshev, "Principal Phase of Oil Formation", *Vestn. Mosk. Univ., No. 6,* 3—27 (1969) (English Tansl., *Int. Geol. Rev., 12 (11),* 1276—1296 (1970)).

18 A.G. Winestock, "Developing a Steam Recovery Technology", in: L.V. Hills (Editor), *Oil Sands — Fuel of the Future, Can. Soc. Pet. Geol. Mem., 3,* 190—198 (1974).

19 J.C. Winters and J.A. Williams, "Microbial Alteration of Crude Oil in the Reservoir", in: *Symposium on Petroleum Transformation in Geologic Environments, Am. Chem. Soc., Div. Pet. Chem., Prepr., 14, (4),* E22—E31 (1969).

20 T.F. Yen and S.R. Silverman, "Geochemical Significance of the Changes in the Chemical Structures of the Complex Heterocyclic Components of Petroleum", *Am. Chem. Soc., Div. Pet. Chem., Prepr., 14 (3),* E32—E41 (1969).

Chapter 3

THERMAL EVOLUTION OF NATURAL ASPHALTS

J. CONNAN and B.M. VAN DER WEIDE

Introduction

Petroleum-like substances occur in various forms ranging from light, slightly colored, free-flowing liquids to heavy, dark, viscous semisolids and, finally, black solids. Generally speaking, the liquid products are producible, whereas the solids and semisolids are not; therefore, they are commonly distinguished as crude oils and asphalts, respectively.

The definition of asphalts, expressing simply their physical aspect, is unsatisfactory and essentially ambiguous. Such descriptions as (1) "viscous liquids or solid, low-melting bitumens", or (2) "fusible, solid bitumen easily soluble in carbon disulfide", do not give credit to the fact that the physical properties of asphalts may result from basically different causes. Actually, the asphaltic character of certain accumulations may express their pronounced immaturity as well as their alteration owing to such environmental factors as water washing, oxidation, or sulfurization [3,9]. Inasmuch as both classes of heavy substances cover a broad spectrum of chemical compositions and occurrences, their identification is not evident a priori.

Several authors have noticed recently that asphalt-like fluids and solids may be severely depleted, or even totally devoid of *n*-alkanes [3,4,5]. This feature is thought to be caused by bacterial degradation, a phenomenon which conceivably takes place if reservoir rocks are invaded by meteoric-type waters. Bacteria transported by these waters select their food among the various oil components. Saturated hydrocarbons, in particular *n*-alkanes, have been shown to be preferably assimilated by these microorganisms [6,7]. Inasmuch as the required oxygen may be, in part, derived from sulfate ions present in the invading water, microbial alteration of oils frequently entails a more-or-less marked sulfurization. The simultaneous removal of its light components and the production of hydrogen-poor (and possibly sulfur-enriched) metabolic waste products, which may even be insoluble [6,8,9], would by all means make the oil heavier and more viscous. Thus, altered crude oils will be called *natural asphalts* throughout this chapter.

Asphaltic deposits are fairly common on earth. They occur as impregnations of porous rocks or even of uncompacted sediments. They are frequently encountered as surface shows and seeps, which may assume such impressive forms as the Athabasca tars, the pitch lakes of Trinidad or Bermudez, Venezuela [10], or the remarkably straight asphalt veins, several miles long, found in the Uinta Basin, Utah, U.S.A. [1], and in southeastern Turkey [11]. In oil-bearing sedimentary basins, asphalt accumulations are mostly located along the border, a fact which, incidently, argues in favor of the alteration being brought about by infiltration of meteoric water.

In spite of their abundance and potential energy reserve, asphalts have frequently been considered economically unattractive, owing to the technological difficulties and high cost involved in their production and handling. Yet, from a theoretical point of view, as well as for practical reasons, they represent a challenging object for geochemical studies.

Although asphalts are frequently encountered near the surface, this is by no means a general rule. Asphaltic accumulations have been found at considerable depths [12,13], an observation which sets a puzzling problem as to their origin, because both invasion by meteoric water and bacterial proliferation seem unlikely in those circumstances. Assuming that in the geologic past, biodegradation has been a widespread phenomenon, the resulting asphalts may, if conditions were favorable, have undergone a renewed deep burial. The presence of asphalt in a borehole in the Aquitaine Basin (southwestern France), between 1900 and 4200 m of depth, demonstrates this possibility. The surprising fact remains, however, that the corresponding diagenesis obviously has failed to obliterate the asphaltic character of the fluids, even if it has measurably affected their gross composition. The diagenetic behavior of bacterially degraded cruds oils, therefore, would seem to differ from that of unaltered, immature oils which, reputedly, undergo a significant quality improvement with increasing depth [14,15,16].

It was felt that in-vitro thermal experiments might provide information as to the diagenetic evolution of altered and unaltered crude oils and thus throw some light on these conflicting observations.

Geochemical properties of natural asphalts

The severe depletion in *n*-alkanes, characteristic of natural asphalts, is considered the most reliable evidence of their biodegradation presently available. It is clearly demonstrated by the gas-chromatographic pattern of their saturated-hydrocarbon fraction, a few examples of which are shown in Fig. 3-1. Apart from that common feature, asphalts cover a wide range of chemical properties and gross compositions [10]. Thus, saturate contents varying from 0.5 to 34% and aromatics/saturates ratios ranging from 36 to 0.5 have been recorded (Table 3-I). Plotting these parameters as an x—y diagram (Fig. 3-1), provides a visual means of comparison between samples [12,17]: any point of the approximately hyperbolic curve, representing a random crude oil or asphalt sample, will move towards the abscissa as a result of bacterial degradation and towards the ordinate by thermal alteration.

Asphalts have widely varying sulfur contents (Table 3-I), which, however,

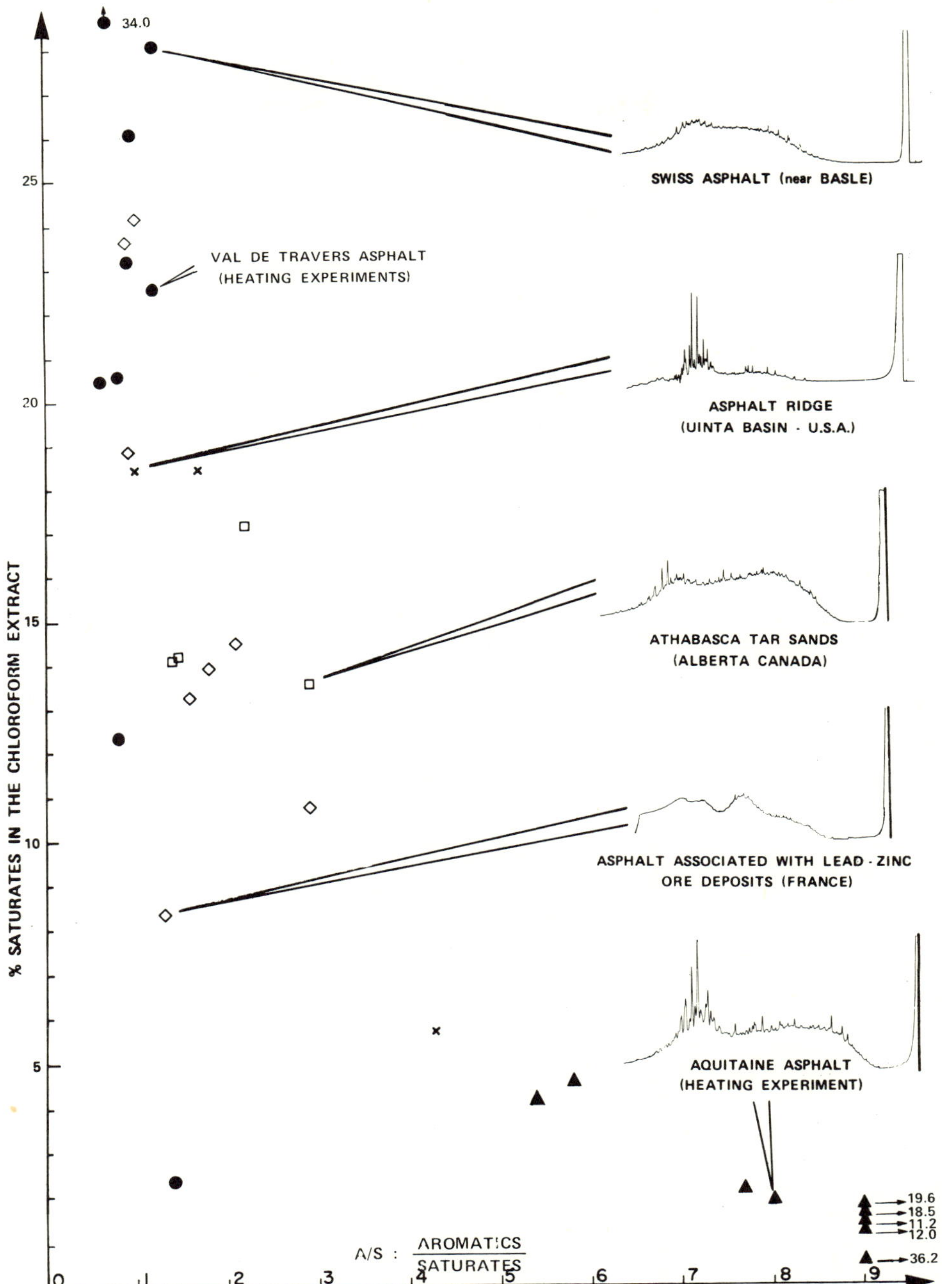

Fig. 3-1. Relationship between content of saturates in the chloroform extract and (aromatics)/(saturates) ratio for various natural asphalts. ● = Swiss asphalts; ▲ = Aquitaine asphalts; □ = Athabasca tar sands; ◇ = asphalts associated with barite or lead—zinc ore deposits; and × = miscellaneous asphalts (Trinidad, U.S.A., S.W. Africa).

TABLE 3-I

Geochemical properties of natural asphalts

REFERENCES OF NATURAL ASPHALTS	GROSS COMPOSITION OF THE CHLOROFORM EXTRACT: SATURATES %	AROMATICS %	RESINS %	ASPHALTENES %	A/S ratio	SULFUR %	$[C_R/C_T]$ A	PERCENTAGE CHLOROFORM EXTRACT IN ROCK
TRINIDAD PITCH LAKE	5,7	24.8	38.5	31.0	4.3			63.9
UNISGOAS ASPHALT VEIN (S.W. AFRICA)	18.5	31.7	32.9	16.9	1.7		0.64	1.5
ASPHALT RIDGE (UINTA BASIN, UTAH, U.S.A.)	18.5	17.8	57.1	6.6	1.0		0.41	15.0
ATHABASCA TAR SANDS (ALBERTA, CANADA)	17.2	38.2	29.3	15.3	2.2	3.54		14.3
	13.6	39.8	40.6	6.0	2.9	5.30		13.6
	14.1	20.3	53.3	12.3	1.4		0.40	11.4
	14.2	20.6	51.5	13.7	1.4		0.40	11.7
ASPHALTS ASSOCIATED WITH LEAD-ZINC DEPOSITS (THANN REGION, N.E. FRANCE)	14.6	30.6	47.9	6.9	2.1		0.57	1.7
	14.0	25.1	52.0	8.9	1.8		0.58	1.3
	10.8	31.8	30.6	26.8	2.9		0.60	
ASPHALTS ASSOCIATED WITH BARYTE DEPOSITS (ST PRIVAT, LODEVE BASIN, FRANCE)	18.9	16.4	17.0	47.8	0.9			6.1
ASPHALTS ASSOCIATED WITH LEAD-ZINC DEPOSITS (LES MALINES MINE, E. BORDER OF THE MASSIF CENTRAL, FRANCE)	13.3	21.1	35.2	30.4	1.6		0.50	97.1
	24.2	23.6	41.6	10.6	1.0		0.60	21.3
	8.4	11.0	31.0	49.6	1.3			11.4
	23.7	21.3	34.0	21.0	0.9		0.46	9.1
ASPHALTS FROM SWITZERLAND: KREUZLIBER TUNNEL	23.2	21.0	32.8	23.0	0.9	0.87	0.58	6.5
LÄGERN SW E. BIRMENSTORF	26.1	24.7	30.3	18.9	0.9	0.62	0.59	5.4
ISTEINER KLOTZ (near BASLE)	28.1	33.7	35.3	2.9	1.2	1.50		
VAL DE TRAVERS MINE	20.5	12.4	53.6	13.5	0.6	0.77	0.51	16.3
VAL DE TRAVERS MINE	22.6	27.9	42.3	7.2	1.2	1.1	0.56	11.0
GOLDINGER TOBEL	2.4	3.4	88.1	6.0	1.4	2.07		0.2
KÖLLINGEN AG	34.0	22.4	35.0	8.5	0.7	0.85	0.55	2.1
ST AUBIN	20.6	16.9	56.5	6.0	0.8	1.66	0.55	
AARETRÄNKI NEAR FULENBACH	12.4	10.0	75.2	2.4	0.8	1.95		

REFERENCES OF NATURAL ASPHALT			GROSS COMPOSITION OF THE CHLOROFORM EXTRACT					SULFUR %	$[C_R / C_T]_A$	PERCENTAGE CHLOROFORM EXTRACT IN ROCK
GEOGRAPHICAL LOCATION	BOREHOLE NAME	DEPTH (m)	SATURATES %	AROMATICS %	RESINS %	ASPHALTENES %	A/S RATIO			
AQUITAINE BASIN (S.W. FRANCE)	Cg.101	2682 2693	1.9	16.2	48.7	33.2	8.0	7.9	0.47	94.2
	Air.1	1910	4.1	22.2	42.7	31.0	5.4		0.45	
		1920	4.6	26.7	40.3	28.4	5.8		0.45	
	Lee.1	1680.2	1.8	35.3	43.5	19.4	19.6	10.6	0.41	3.7
		1926.4	1.7	31.4	39.8	27.1	18.5	9.8	0.44	5.5
		1930.8	0.5	18.1	34.6	46.8	36.2	10.2	0.45	8.9
		2046 2050	1.4	15.8	40.4	42.4	11.2	10.0	0.44	21.0
		2545 2548	2.1	16.1	44.6	37.2	7.7	9.1	0.45	
		2584	1.2	14.4	43.8	40.6	12.0	9.0	0.43	

$[C_R / C_T]_A$: carbon ratio of asphaltenes

do not necessarily indicate a more-or-less intense alteration; they may reflect the sulfur content of the original crude oil as well.

One may expect asphalt properties to be related to the maturity level of the crude oil at the beginning of biodegradation. For high polymers involving a tridimensional network, such as kerogen, that level may be defined by the so-called C_R/C_T ratio [18]. It seems that a similar description might be obtained for crude oils, asphalts and rock extracts through the C_R/C_T ratio of their high-molecular-weight (i.e., asphaltene) fraction [19]. Its value varies from about 0.10 to 0.85, as documented by a large number of measurements carried out on petroleum and related substances from the five continents [19]. Inasmuch as for natural asphalts the values recorded are in the 0.40–0.65 range (Table 3-I), these asphalts presumably originate from immature to moderately mature oils.

Concept of artificial maturation

Organic diagenesis expresses the inescapable trend of sedimentary organic matter towards its chemical equilibrium. It involves large numbers of chemical reactions, the rates of which are temperature dependent. The exceedingly slow process of diagenetic evolution at normal temperatures will be, therefore, considerably speeded up at higher temperatures, a statement which has been frequently confirmed by the analysis of organic matter with increasing depth [20,21,22]. Hence, artificial maturation of organic matter, i.e., its exposure to adequate temperatures for increasing lengths of time, may be expected to simulate natural evolution.

Required temperature

The choice of the required, adequate temperature is delicate because it has to reconcile the opposing imperatives of accuracy and speed. One may reasonably assume that at temperatures, which are not extremely high, equilibrium will involve compounds having low standard free enthalpy, such as methane, carbon oxide and dioxide, water, hydrogen sulfide, nitrogen, ammonia, and elemental carbon. Owing to the temperature dependence of reaction rates, however, temperature is liable to exert a strong influence on the pathways leading to equilibrium, i.e., on the intermediate states of the organic system, which are observable through chemical analysis. It is clear that at higher temperatures an observable but increasingly unrealistic simulation will be obtained, whereas at lower temperatures simulation will be more realistic but too slow to be observed. A compromise, therefore, must be reached.

A lower temperature limit was found to be 250° C. Experiments carried out at this temperature did not reproduce known diagenetic phenomena in their entirety (see pp. 37, 39). On the other hand, a temperature of 375° C is obviously too high. Experiments on *n*-alkanes resulted, at these high temperatures, in extensive cracking and formation of abundant amounts of olefins [23]. Inasmuch as olefins are only minor components in diagenetically altered sediments, the latter process seems unrealistic.

At 300° C, both asphalts and crude oils did undergo significant evolution without going as far as to cause immediate cracking of the generated alkanes (see pp. 41, 43). This was checked by experiments on pure *n*-eicosane, 93% of which was recovered unchanged after a six months' exposure to that temperature [24]. Consequently, 300° C has been tentatively chosen as the optimum temperature for conducting simulation experiments.

Choice of samples

The interpretation of in-vitro experiments will be possible only if, to a reasonable extent, they reproduce natural diagenesis. The presence of asphalt in the Aquitaine tar belt, at a depth of between 1900 and 4200 m, enabled comparison with an asphalt sample heated at constant temperature for various lengths of time (Fig. 3-8).

Differences in diagenetic behavior between biodegraded and unaltered crude oils will be the more significant if the samples are of comparable origin. This requirement was once more fulfilled by the Aquitaine Basin sediments which, in addition to asphalts, contain an accumulation of unproducible, heavy, asphalt-like, unaltered crude. The altered and unaltered oils are equally immature as documented by the C_R/C_T ratio of their asphaltene fraction (Table 3-II).

Finally, the in-vitro evolution of Aquitaine asphalt was compared with that of an equally biodegraded asphalt of very different composition sampled in the well-known Val de Travers mine in the Swiss Jura [25]. The latter substance exhibited, in particular, a higher level of maturity as documented by the C_R/C_T ratio of its asphaltenes (Table 3-II).

TABLE 3-II

Characteristics of samples selected for in-vitro experiments

Description and location of samples	Saturates (%)	Aromatics (%)	Resins (%)	Asphaltenes (%)	Sulfur (%)	$[C_R/C_T]_A$ ratio	Maturity level
Unaltered crude oil, Aquitaine Basin, southwestern France	11.2	40.2	35.0	13.4	5.4	0.48	immature
Natural asphalt, Aquitaine Basin	1.9	16.0	48.2	32.9	7.9	0.47	immature
Natural asphalt Val de Travers, Canton of Neuchâtel, Switzerland	22.6	27.9	42.3	7.2	1.1	0.56	moderately mature

Criteria for the interpretation of in-vitro experiments

Evolution, whether natural or artificial, will be reflected in the gradually changing chemical composition of the organic matter. Although, for obvious reasons, detailed organic analysis cannot be achieved, changes will occur in various characteristic and experimentally obtainable properties of the organic matter. This fact enables the geochemist to describe the sedimentary organic matter in terms of its type and degree of diagenetic evolution. The search for new criteria is an object of permanent concern in organic and, particularly, in petroleum geochemistry.

For the in-vitro evolution of asphalts, the following criteria have been considered adequate:

(1) The solubility in chloroform, which will disclose the generation of an insoluble coal-like fraction called *pyrobitumen.*

(2) The gross composition, i.e., the percentages of saturated and aromatic hydrocarbons, resins, and asphaltenes, determined by a standard procedure [12]. From the gross composition, secondary parameters, such as the ratio of aromatics to saturates (A/S), may be computed.

(3) The composition of the saturated-hydrocarbon fraction, including concentrations of *n*-alkanes and branched-cyclic alkanes, as well as the distribution patterns within each of these classes of compounds. As to the *n*-alkane fraction, attention is paid to both the width of its distribution pattern in terms of carbon numbers and the relative amounts of odd- and even-carbon-numbered alkanes. The latter parameter is frequently expressed by the carbon preference index (CPI) which, for the whole *n*-alkane spectrum, defines the ratio of the concentrations of odd- and even-numbered paraffins [26,27]. The odd—even preference (OEP) curve [28], which represents the odd/even ratio for each *n*-alkane individually, clearly provides more detailed information; therefore, it has been used by the writers.

Branched- and cyclic-alkane fractions from different extracts are compared by visual inspection of their gas-chromatographic patterns, which, usually, show various characteristic peaks or groups of peaks. Thus, immature extracts contain considerable amounts of tetra- and pentacyclic naphthenes, ranging from C_{25} to C_{35} and belonging to the sterane and triterpane series (Figs. 3-8 and 3-9) [29]. Amongst these, members of the (17αH, 21βH) hopane series are remarkably abundant [30]:

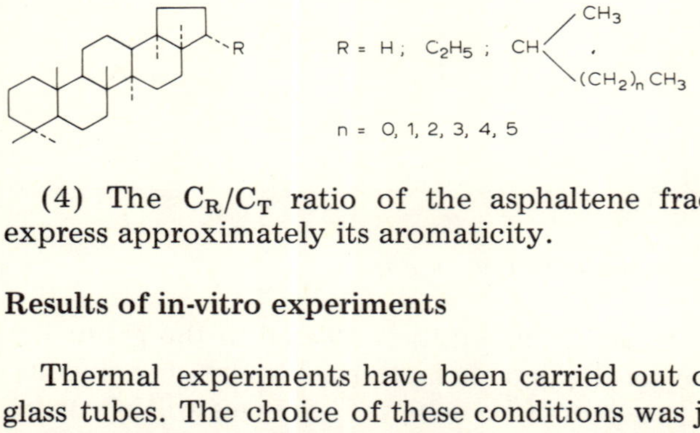

R = H; C_2H_5; $CH\langle^{CH_3}_{(CH_2)_n CH_3}$

n = 0, 1, 2, 3, 4, 5

(4) The C_R/C_T ratio of the asphaltene fraction, which is believed to express approximately its aromaticity.

Results of in-vitro experiments

Thermal experiments have been carried out on samples vacuum-sealed in glass tubes. The choice of these conditions was justified by the experimental facilities they entail, although it is realized that they represent a poor imitation of the natural habitat of oils.

Experiments at 250°C involved the aforementioned Swiss asphalt and the heavy, immature, asphalt-like, but actually unaltered, crude oil from the Aquitaine Basin. In spite of their essentially different origin and composition, both substances responded similarly to the heat treatment. No increase in content of saturates was noticed in either case, even after six months of

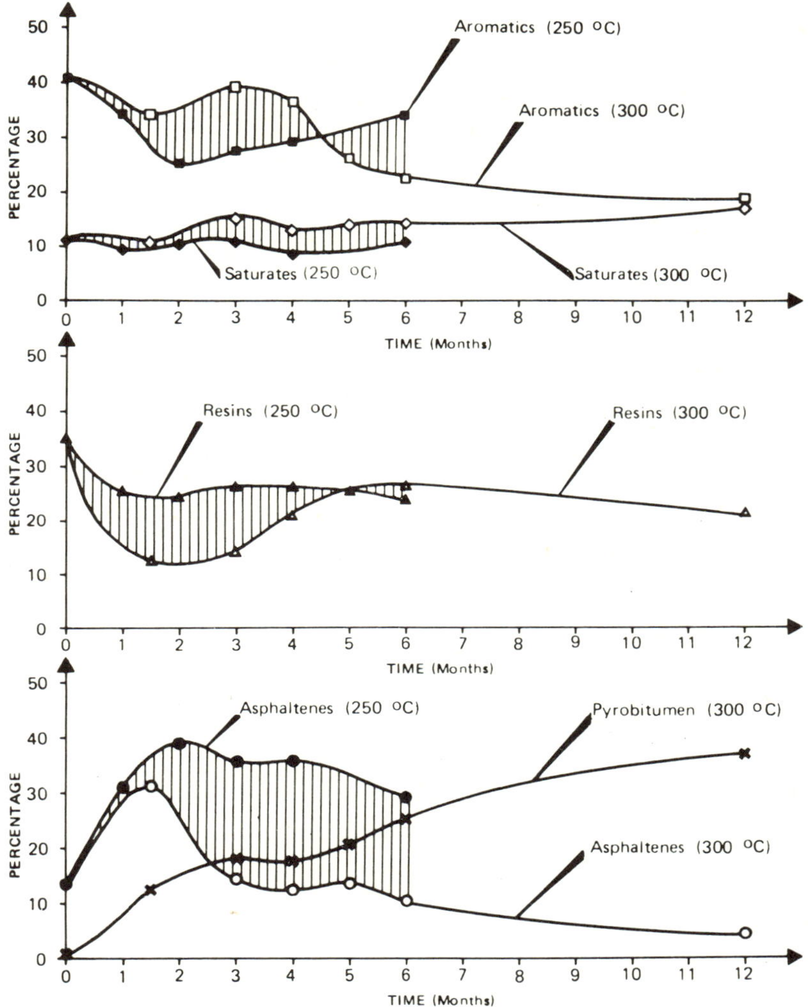

Fig. 3-2. Change with time in gross composition of Aquitaine crude oil by thermal treatment at 250° and 300°C.

heating (Tables 3-III and 3-IV). Moreover, the alkane distribution within the saturate fraction remained virtually unchanged, as documented by the OEP curves (Fig. 3-4a) and the gas chromatograms of the branched-cyclic alkane fractions (Figs. 3-5 and 3-6). It is inferred that, at the given temperature and in the available period of time, generation of alkanes would not take place to any measurable extent.

Observable evolutionary phenomena were limited to structural rearrange-

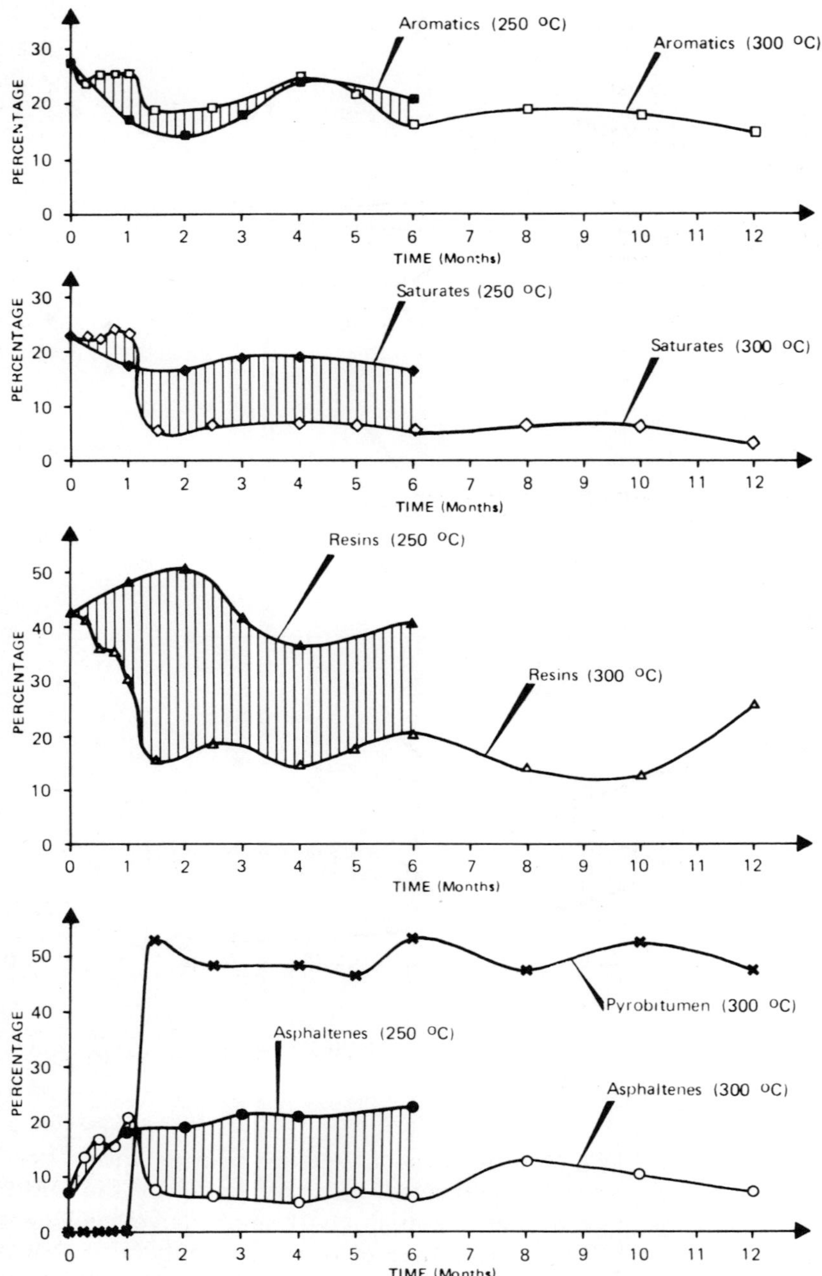

Fig. 3-3. Change with time in gross composition of Val de Travers asphalt by thermal treatment at 250° and 300°C.

TABLE 3-III

Gross composition of Aquitaine crude oil heated at 250°C for various lengths of time

HEATING TIME AT 250 °C	CHLOROFORM EXTRACT				
	SATURATES %	AROMATICS %	RESINS %	ASPHALTENES %	A / S
STARTING MATERIAL	11.2	40.3	35.0	13.4	3.6
1 MONTH	9.6	34.0	25.3	31.1	3.5
2 MONTHS	10.3	25.8	24.9	39.0	2.5
3 MONTHS	10.6	27.4	26.2	35.7	2.6
4 MONTHS	8.6	29.5	26.2	35.7	3.4
6 MONTHS	11.6	34.2	24.8	29.4	2.9

TABLE 3-IV

Gross composition of Val de Travers asphalt heated at 250°C for various lengths of time

HEATING TIME AT 250 °C	CHLOROFORM EXTRACT				
	SATURATES %	AROMATICS %	RESINS %	ASPHALTENES %	A / S
STARTING MATERIAL	22.6	27.9	42.3	7.2	1.2
1 MONTH	17.2	17.2	47.5	18.1	1.0
2 MONTHS	16.1	14.0	50.9	19.0	0.9
3 MONTHS	18.9	17.6	41.7	21.8	0.9
4 MONTHS	18.7	24.1	36.1	21.1	1.3
6 MONTHS	16.7	20.9	40.1	22.3	1.2

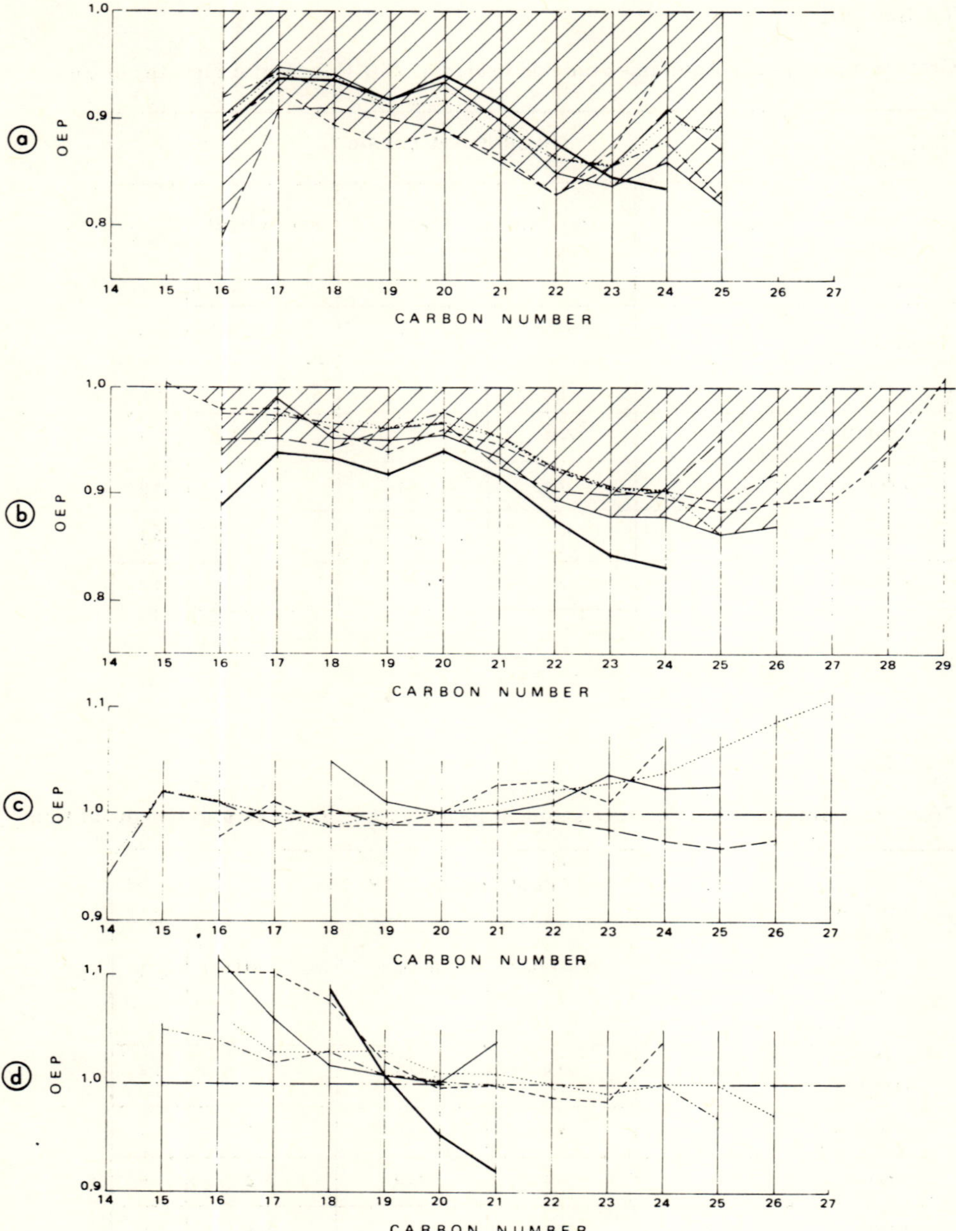

Fig. 3-4. OEP curves for unheated and heated samples.
a. Aquitaine crude oil at 250°C: solid, heavy line = unheated sample; solid, light line = sample heated for 1 month; dashed (short) line = sample heated for 2 months; dashed (long) line = sample heated for 3 months; dotted line = sample heated for 4 months; and dashed-dotted line = sample heated for 6 months.

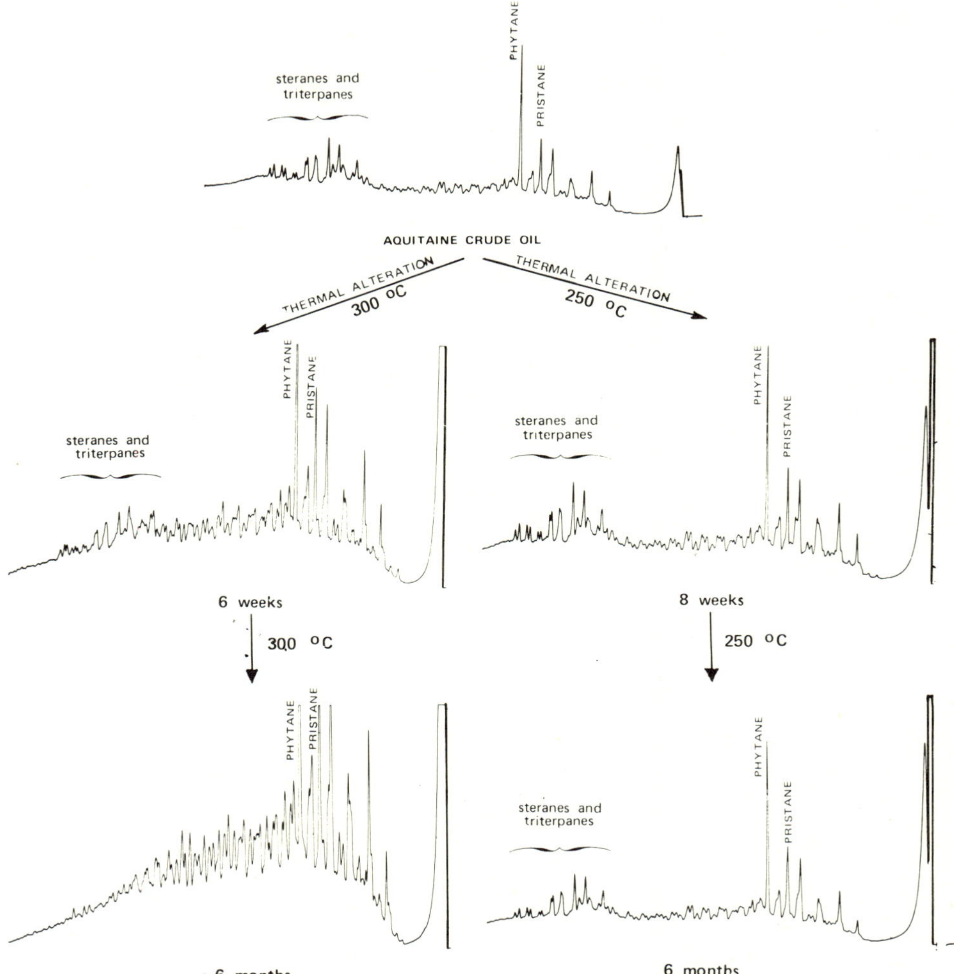

Fig. 3-5. Gas chromatograms of the branched-cyclic alkane fraction before and after heating Aquitaine crude oil at 250° and 300°C.

---

b. Aquitaine crude oil at 300°C: solid heavy line = unheated sample; solid light line = sample heated for 6 weeks; dashed (short) line = sample heated for 3 months; dashed (long) line = sample heated for 4 months; dotted line = sample heated for 5 months; and dashed-dotted line = sample heated for 6 months.

c. Val de Travers asphalt at 300°C: solid light line = sample heated for 1 week; dashed (short) line = sample heated for 4 weeks; dashed (long) line = sample heated for 6 weeks; and dotted line = sample heated for 5 months.

d. Acquitaine asphalt at 300°C: solid heavy line = unheated sample; solid light line = sample heated for 3 weeks; dashed (short) line = sample heated for 6 weeks; dotted line = sample heated for 2 months; and dashed-dotted line = sample heated for 6 months.

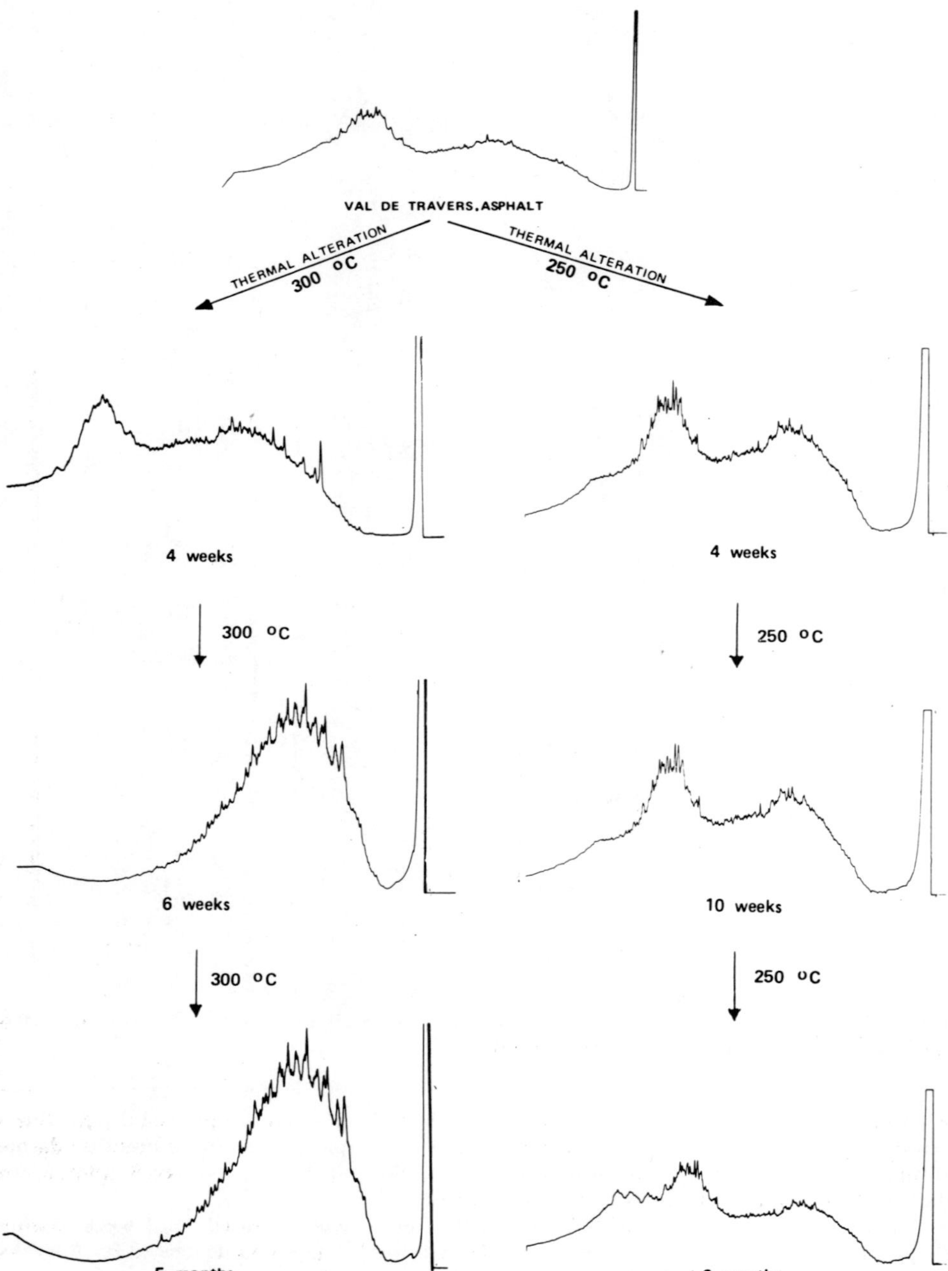

Fig. 3-6. Gas chromatograms of the branched-cyclic alkane fraction before and after heating Val de Travers asphalt at 250° and 300°C.

ments among the more polar components (Tables 3-III and 3-IV). Thus, during the first month, the asphaltene content increased by a factor 2—3, mostly at the expense of resins and aromatics. In contrast, the next five months of heating resulted in rather insignificant fluctuations of the gross composition.

In summary, in-vitro experiments at 250° C seemed unable to duplicate naturally known diagenetic phenomena in their entirety, within a reasonable time. It was assumed that at 300° C reaction rates would be high enough to obtain a better simulation, and that assumption proved correct. Both crude oil and asphalt samples did indeed undergo significant modifications within two months of heating at that temperature. In the following sections, the results of thermal experiments in these conditions, on the three samples selected (see p. 32), are briefly summarized.

Unaltered, immature Aquitaine oil at 300° C (Table 3-V)

During the first six weeks of heat treatment, the thermal alteration of Aquitaine crude oil was not unlike that at 250° C. Once again, the content of saturates remained unchanged, whereas the asphaltene content increased considerably at the expense of the aromatic and, particularly, of the resin fraction. This time, however, a few observations suggest that the evolution

TABLE 3-V

Gross composition of Aquitaine crude oil heated at 300°C for various lengths of time

HEATING TIME AT 300 °C	CHLOROFORM EXTRACT + PYROBITUMEN					A / S
	SATURATES %	AROMATICS %	RESINS %	ASPHALTENES %	PYRO-BITUMEN %	
STARTING MATERIAL	11.2	40.3	35.0	13.4	0	3.6
6 WEEKS	10.2	34.0	12.2	31.1	12.4	3.3
3 MONTHS	15.0	39.0	13.0	14.7	18.2	2.6
4 MONTHS	13.2	36.3	20.5	12.9	17.0	2.7
5 MONTHS	14.0	26.0	25.5	13.9	20.6	1.9
6 MONTHS	14.9	22.4	26.6	10.1	25.9	1.5
12 MONTHS +8 MONTHS (250 °C)	18.4	19.1	21.5	4.8	36.2	1.0

took a different rhythm. Thus, a small part of the organic matter had been converted to coal-like, insoluble pyrobitumen [31,32]. In addition, both *n*-alkane and branched-cyclic alkane contents had been slightly, but significantly, altered (Figs. 3-4b and 3-5). The latter modifications in particular seem to foreshadow an artificial evolution reminiscent of natural diagenesis, as confirmed by the subsequent experimental results.

The period of heating, extending from six weeks to one year did indeed display generation of saturated hydrocarbons. Amongst these, the *n*-alkanes covered an increasingly wide range, whereas the predominance of their even-numbered members was noticeably reduced (Fig. 3-4b). The four- and five-ring naphthenes, including hydrocarbons of the hopane series [30], disappeared gradually with formation of lower-molecular-weight compounds presumably belonging to homologous series (see p. 39).

As to the more polar fractions, the most striking phenomenon consisted in the gradual disappearance of the asphaltenes and the simultaneous precipitation of increasing amounts of pyrobitumen (Fig. 3-2), which after one year accounted for one third of the nongaseous organic matter. Although these observations strongly suggest a relationship between asphaltenes and pyrobitumen, generation of the latter obviously involves the other components as well, as documented by their fluctuating percentages.

Aquitaine asphalt at 300° C (Table 3-VI)

Compared with the unaltered Aquitaine crude, the biodegraded Aquitaine asphalt definitely showed a different diagenetic-alteration picture during the first six-week period. Thus, while the asphaltene content remained practically constant, the contents of other fractions were significantly modified (Fig. 3-7). Especially during the first week of thermal treatment, saturated and aromatic hydrocarbons were generated in relatively high proportions, whereas the resin content decreased correspondingly. Afterwards, evolution progressed more gradually.

Evolution was demonstrated once more by the *n*-alkane spectrum, the original C_{15}—C_{23} range of which extended to C_{14}—C_{26} (Fig. 3-4d). The branched-cyclic fraction displayed the usual disappearance of steranes and triterpanes owing to the generation of lower-molecular-weight compounds (Fig. 3-9).

It should be emphasized that artificial evolution observed during this period compares favorably with natural asphalt diagenesis at a depth of 1900—3800 m (see p. 45). Although the changes in gross composition were not as clear-cut as in the thermal experiments (a fact which might be ascribed, at least in part, to interfering migratory phenomena), the general trend is comparable, as documented by the x—y diagrams (Fig. 3-10). Moreover, the alkane-distribution patterns are surprisingly alike (Fig. 3-8). Connan [33], while studying the artificial thermal evolution of an Australian

TABLE 3-VI

Gross composition of Aquitaine asphalt heated at 300°C for various lengths of time

HEATING TIME AT 300 °C	CHLOROFORM EXTRACT + PYROBITUMEN					A / S
	SATURATES (%)	AROMATICS (%)	RESINS (%)	ASPHALTENES (%)	PYRO-BITUMEN (%)	
STARTING MATERIAL	1.9	16.0	48.2	32.9	1.0	8.4
1 week	4.3	30.7	32.3	31.0	1.6	7.1
2 weeks	4.5	32.2	28.7	32.2	2.3	7.1
3 weeks	5.2	34.2	27.4	33.2	0.1	6.5
4 weeks	6.0	32.4	26.5	34.0	1.0	5.4
6 weeks	7.1	36.6	21.4	33.8	1.1	5.1
2 months	8.3	33.3	12.3	38.6	7.4	4.0
3 months	8.3	30.8	9.5	17.5	33.8	3.7
4 months	8.4	31.2	9.6	17.2	33.6	3.7
5 months	8.3	31.8	10.5	25.2	24.2	3.8
6 months	8.9	32.2	8.7	21.4	28.8	3.6
8 months	3.6	21.1	11.7	8.1	55.5	5.9
10 months	3.5	26.4	14.5	3.1	52.5	7.6
12 months	3.6	27.4	10.7	3.6	54.6	7.6

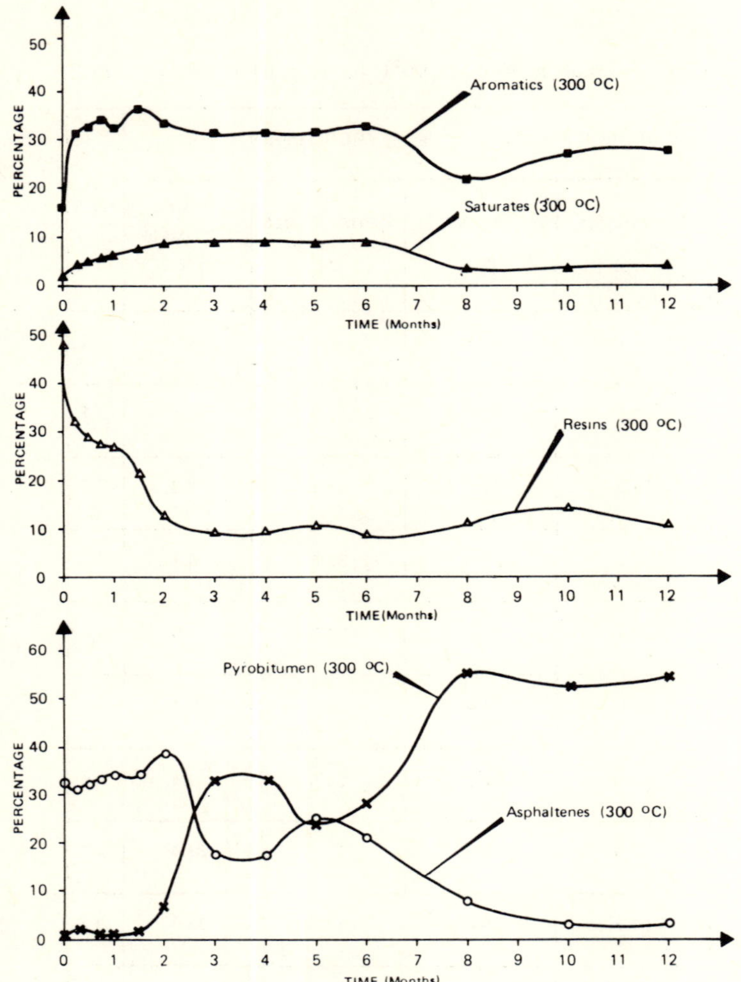

Fig. 3-7. Change with time in gross composition of Aquitaine asphalt by thermal treatment at 300°C.

lignite sample, found a similar striking correlation with the natural diagenesis of terrestrial organic matter observed in boreholes in New Zealand. It would appear that, notwithstanding obvious kinetic and environmental differences, in-vitro experiments do, under the described conditions, rather accurately simulate the maturation of asphaltic impregnations. This supports the idea that those stages of organic diagenesis for which no reference samples are available, may be studied with reasonable confidence through thermal simulation.

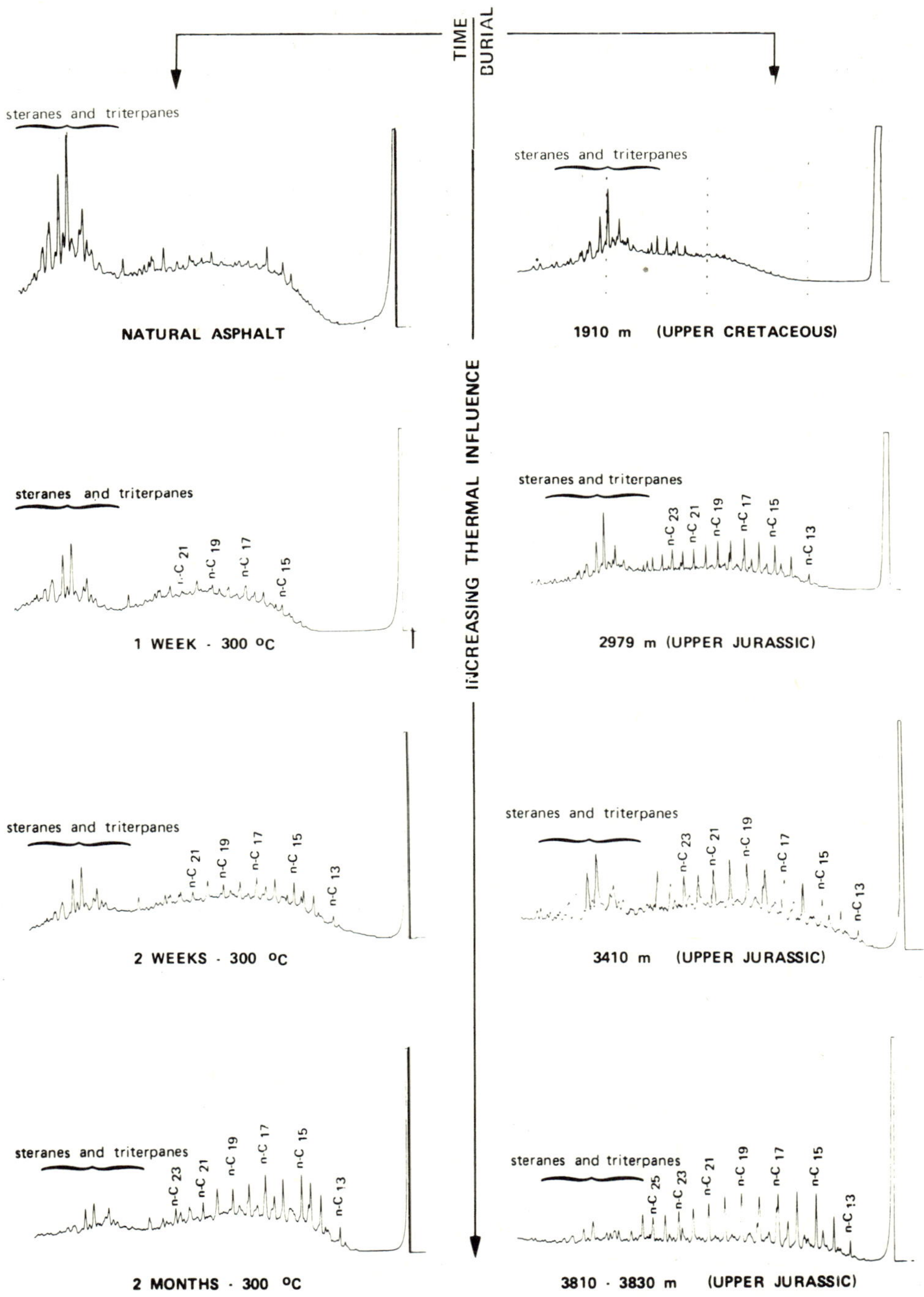

Fig. 3-8. Artificial and natural evolution of the total alkane fraction in Aquitaine asphalt.

Although artificial evolution continued apparently along the same lines during the next two weeks, careful observation discloses major structural changes of the organic matter. Thus, the content of asphaltenes increased slightly, while pyrobitumen began to precipitate. Simultaneously, *n*-alkanes reached their widest distribution (Fig. 3-4d), whereas steranes and triterpanes disappeared altogether (Fig. 3-9).

Generation of pyrobitumen continued until, after eight months, roughly half of the organic matter had been converted. Again, this conversion took place mostly at the expense of the asphaltene fraction (Fig. 3-7).

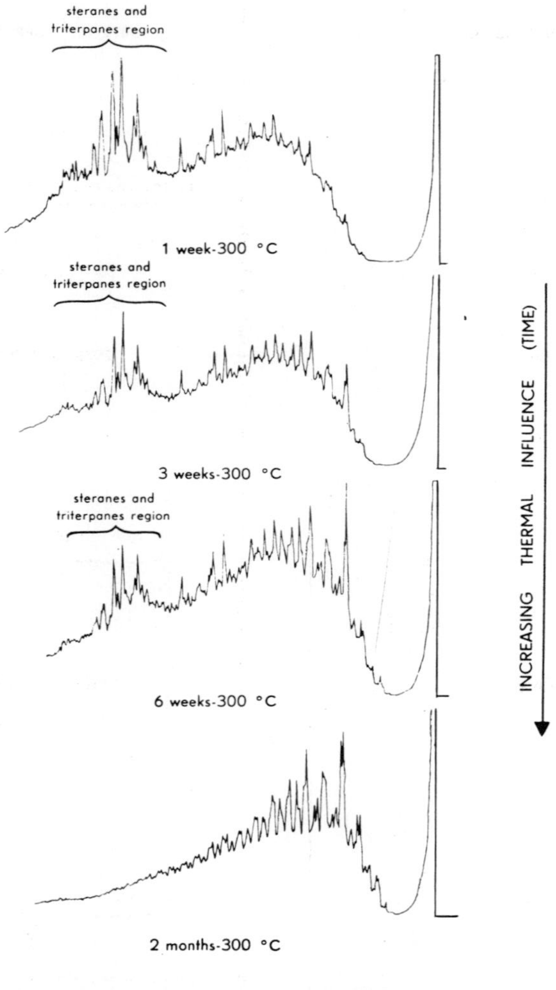

Fig. 3-9. Gas chromatogram of branched-cyclic alkanes from an Aquitaine asphalt sample heated at 300°C.

TABLE 3-VII

Gross composition of Val de Travers asphalt heated at 300°C for various lengths of time

HEATING TIME AT 300 °C	CHLOROFORM EXTRACT + PYROBITUMEN					A / S
	SATURATES %	AROMATICS %	RESINS %	ASPHALTENES %	PYRO-BITUMEN %	
STARTING MATERIAL	22.6	27.9	42.3	7.2	–	1.2
1 week	22.2	23.6	40.5	13.7	–	1.0
2 weeks	22.2	25.1	35.9	16.8	–	1.1
3 weeks	24.0	25.1	35.3	15.6	–	1.0
4 weeks	23.2	25.4	30.5	20.9	–	1.9
6 weeks	5.5	19.0	15.1	7.9	52.4	3.4
10 weeks	6.5	19.9	18.3	6.7	48.5	3.3
4 months	7.0	24.5	14.7	5.5	48.4	3.5
5 months	6.4	22.5	17.3	7.0	46.9	3.5
6 months	3.7	16.3	20.4	6.2	53.3	4.3
8 months	6.6	19.4	13.6	13.0	47.7	2.9
10 months	6.6	18.0	12.5	10.5	52.5	2.8
12 months	3.7	15.1	25.6	7.9	47.8	4.1

The last six months of heat treatment featured the decrease in the content of saturates, presumably through extensive cracking, as documented by the evolution of gas. No major changes in alkane distribution were observed during this period, apart from the usual shift of the distribution maximum of n-alkanes towards lower carbon numbers (Fig. 3-4d).

Val de Travers asphalt at 300° C (Table 3-VII)

Although both the Swiss and the Aquitaine asphalts present the characteristics of biodegraded oils of similar physical aspect, they differ greatly in gross composition (Table 3-II). It is not surprising, therefore, that they responded quite differently to the heat treatment. It is true that in both cases a considerable loss of resins was observed, but contrary to Aquitaine asphalt, in the case of Swiss asphalt this loss corresponded to the formation of asphaltenes. No hydrocarbons were generated (Fig. 3-3), whereas the alkane distribution remained unchanged (Figs. 3-4c and 3-6).

After four to six weeks of heating, the Swiss asphalt exhibited a dramatic evolution. During this very short period, half of the organic matter was converted to insoluble pyrobitumen. Simultaneously, the content of saturates in the extract was severely reduced, and its alkane composition was greatly modified. Thus, the n-alkane content increased by a factor of 20, whereas its spectrum reached its widest range (Fig. 3-4c). High-ring naphthenes suddenly disappeared and gave rise to the familiar picture of lower homologous series.

During the rest of the year, no striking modifications of the gross composition were noticed. The alkane content of the saturate fraction remained virtually constant.

Geochemical implications

As has been pointed out previously, in-vitro experiments do represent a reasonable simulation of natural diagenesis. The analytical data obtained suggest that unaltered and altered crudes display differences as well as similarities in their diagenetic behavior. The three samples studied exhibited generation of chloroform-insoluble pyrobitumen, the extent and rate of which, however, varied with the sample. From the Aquitaine crude oil, pyrobitumen was precipitated fairly slowly and the amounts formed did not exceed 36% of the organic matter after one year at 300° C. In sharp contrast with this, over 50% of the Swiss Val de Travers asphalt converted to pyrobitumen within two months.

In-situ precipitation of coal-like pyrobitumen will reduce reservoir porosity and permeability and, therefore, will hamper production. Although this consequence basically holds true for both altered and unaltered crude oils, the higher rate and abundance of precipitation of pyrobitumens in the case of biodegraded crudes places them in the least favorable position with respect to their producibility.

Another contrast appeared in the evolution of the saturate fraction, which steadily increased in the case of Aquitaine crude oil, whereas in the Swiss asphalt it suddently dropped to very low values. This contrast is clearly visible on the x—y diagram (Fig. 3-10), which features a shift towards the ordinate for the unaltered and towards the abscissa for the biodegraded oil. These shifts, according to the analysis of numerous crudes and rock extracts [12,17]represent an improvement and a deterioration of the oil quality, respectively.

The diagenetic fate of the Aquitaine asphalt turned out to be intermediate between these two extremes: while yielding the same amount of pyrobitumen as the Swiss asphalt, its rate of conversion was definitely lower (Figs. 3-2, 3-3, 3-7). The intermediate behavior is clearly visible on the x—y diagram also: the initial shift towards the ordinate was reversed after six months (Fig. 3-10a). The initial shift, which duplicated that of natural diagenesis (Fig. 3-10b), did not improve oil quality beyond that of unaltered, immature crude oil.

In summary, bacterial degradation occurring at any time during the geologic past, seems to exert an unfavorable influence on crude oils. Even after renewed, deep burial, these altered crudes are unlikely to become producible.

Apart from their obvious practical implications, in-vitro experiments are expected to provide a deeper insight into geochemical phenomena. Putting organic diagenesis on any basis other than purely empirical may, at first sight, seem an impossible undertaking, because numerous reactions involved are too slow to be observed. It should be realized, however, that the evolution of sedimentary organic matter, like that of any isolated system, results from the necessity to decrease its free enthalpy until that function reaches its minimum value. Thus, the overall direction of evolution is defined (see p. 50). Evolution is actually taking place through chemical reactions, each one of which, at any temperature, has its own rate. It would appear, both from common knowledge of chemical reactions and from published data [34,35,36], that, on average, reactions involving functional groups (hydrolysis, esterification, decarboxylation, dehydration and, particularly, condensation) are "easier", i.e., they require lower activation energies than reactions involving rupture of carbon—carbon and carbon—hydrogen bonds. Hence, at all temperatures high enough for both types of reactions to occur, reactions of the first kind will precede the others.

These theoretical considerations are supported by the experimental data. Heating at 250°C did indeed produce mere structural rearrangements among the polar-oil and asphalt components (see pp. 35—40) and, specifically, resulted in the synthesis of high-molecular asphaltenes. Inasmuch as no noteworthy changes in alkane composition were observed, it is inferred that at this

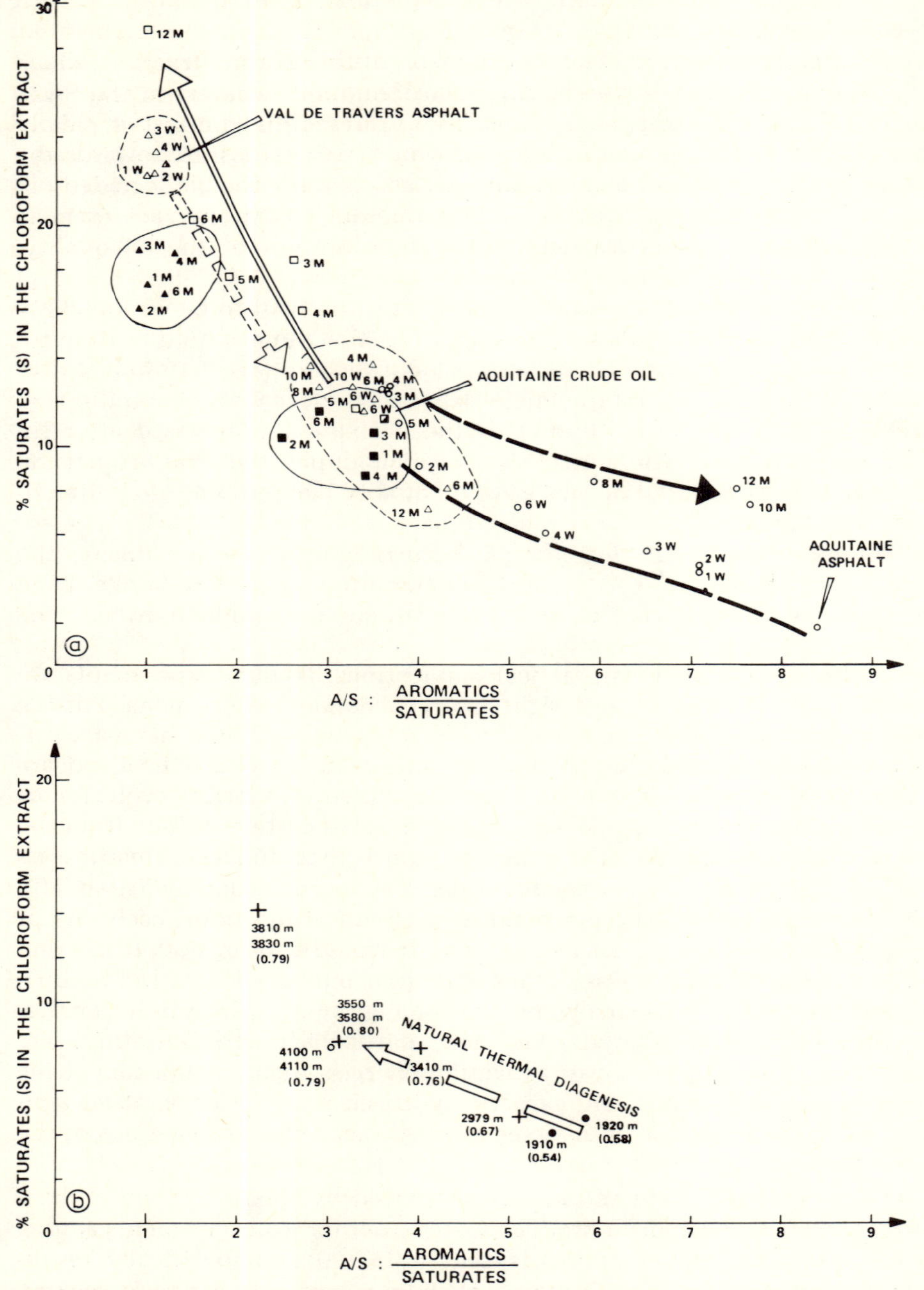

Fig. 3-10. Relationship between content of saturates in chloroform extract and (aromatics)/(saturates) ratio.

temperature cracking reactions were still too slow to be detectable. During the initial four to six weeks at a temperature of 300° C, a similar diagenetic behavior was apparent, although slight changes in the content of saturates did indicate incipient cracking. The latter type of reaction became progressively predominant as evolution proceeded.

Calculation of the chemical equilibrium of organic model systems discloses that, at these experimental temperatures, cracking reactions as such (i.e., involving the production of olefins) can only take place to a limited extent [37]. Secondary reactions are essential to make them proceed quantitatively. Amongst these, hydrogenation by hydrogen donors such as naphthenic and aromatic compounds must eventually prevail [37]. Owing to dehydrogenation, the hydrogen donors can only condense to polyaromatic structures and thus become increasingly insoluble and unfusible. This disproportionation is illustrated by the increasing (1) atomic C/H ratio of kerogen in source rocks [32], (2) content of asphaltenes in rock extracts and crude oils [19], and (3) C_R/C_T ratio of these substances [19]. In the described in-vitro experiments, this mechanism accounts for the appearance of pyrobitumen.

It follows that the successive gross compositions testifying to the diagenetic evolution should depend on the initial composition of the organic matter. If the latter is relatively depleted in hydrogen, i.e., if it is more aromatic, it may be expected to give rise to only minor amounts of C_{15+} alkanes. Disproportionation would, in that case, favor the generation of small alkane molecules, i.e., gas, together with insoluble, coal-like substances. Inversely, although the ultimate fate of hydrogen-rich, predominantly lipid material, will be qualitatively comparable, it would rather convert to non-gaseous hydrocarbons at the price of a slighter coalification.

These ideas, which are supported by the well-known diagenetic differences between higher-plant and algal deposits, allow for the reported in-vitro data as well. In that respect, it should be emphasized that the observed evolution of the three samples studied would be expected to reflect differences in elemental composition, which do not necessarily correlate with differences in gross composition. Aquitaine crude oil is immature and, therefore, relatively rich in hydrogen. Although equally immature, Aquitaine asphalt, is somewhat depleted in this element owing to bacterial degradation. Val de

---

a: ○ = Aquitaine asphalt at 300°C, □ = Aquitaine crude oil at 300°C, △ = Val de Travers asphalt at 300°C, ■ = Aquitaine crude oil at 250°C, and ▲ = Val de Travers asphalt at 250°C.

b: ● = Upper Cretaceous, △ = Lower Cretaceous, + = Upper Jurassic, ○ = Lias, and (0.79) = C_R/C_T ratio of the associated kerogen. Dashed arrow = artificial maturation of the Aquitaine asphalt at 300°C; double solid arrow = artificial maturation of the Aquitaine crude oil at 300°C; dashed (short) arrow = artificial maturation of the Val de Travers asphalt at 300°C; and dashed (long) arrow = natural thermal diagenesis of Aquitaine asphalts.

Travers asphalt is the most dehydrogenated of the three samples, owing to the cumulative effects of maturation and biodegradation (Table 3-II). A pre-existing sequence of increasing dehydrogenation correlates remarkably well with the differences in aromatization disclosed by both the x—y diagrams and by the extent and rate of pyrobitumen precipitation.

More detailed study of diagenetic phenomena supports the postulated mechanism. Organic matter, whether disseminated in rocks or accumulated as crude oil, contains hydrocarbons in both the free and combined state [38,39,40]. The latter will gradually disappear while hydrocarbons are liberated as evolution proceeds. Consequently, more mature oils, although enriched in hydrocarbons, are proportionally depleted in hydrocarbon precursors. Bacterial degradation partly removes free hydrocarbons, specifically n-alkanes and low-molecular-weight branched and cyclic alkanes, from the crude oil. The remaining, nonbiodegraded precursors, mostly contained in the heteroatomic resin and asphaltene fractions, are still capable of liberating additional hydrocarbons in case of renewed diagenesis. The higher the maturity before biodegradation the smaller will be the amount liberated. The improvement of unaltered Aquitaine crude, the slighter and temporary improvement of the Aquitaine asphalt, and the deterioration of the Swiss asphalt might thus be accounted for.

It has been pointed out (see p. 46) that, in addition to the asphaltenes, other oil components are probably involved, directly or indirectly, in pyrobitumen generation. This is remarkably well demonstrated by the Val de Travers asphalt, the sudden conversion of which occurred at the expense of the four fractions, although the asphaltenes and saturates suffered the heaviest losses. The contribution of asphaltenes is logically interpreted in terms of the increasing molecular size of its aromatic building blocks, following dehydrogenation and condensation. As to the saturated hydrocarbons, it is realized that, owing to maturation and biodegradation, they consist principally of high-ring naphthenes (Figs. 3-8 and 3-9). These compounds, contrary to noncyclic alkanes, are capable of both aromatization and cracking, a fact which, incidentally, allows for the rapidly vanishing steranes and triterpanes in all experiments described (Figs. 3-5, 3-6, and 3-9). Naphtho-aromatics formed in this manner will contribute to the aromatic and resin fraction and, eventually, through condensation reactions will give rise to polycyclic, high-molecular structures, as exemplified by asphaltenes and pyrobitumen. The small gas molecules resulting from the simultaneous cracking will possibly promote insolubilization of the pyrobitumen [3].

In summary, the thermal evolution of asphalts and crude oils seems to be governed by essentially the same principle and to display the same features as those created by the diagenetic evolution of any sedimentary organic matter. The generation of saturated hydrocarbons depends on the availability of precursors controlled by initial maturity and biodegradation. The gradual formation of insoluble pyrobitumen, resulting from disproportionation, corresponds to the increasing coalification of kerogen as documented, in both cases, by its increasing reflective power [41,42,11]. As far as their diagenetic behavior is concerned, therefore, biodegraded asphalts relate to unaltered crudes in the same way as reputedly unfavorable, lignitic, detrital or oxidized organic matter relates to favorable algal and planktonic deposits.

Conclusions

In-vitro experiments, conducted at the carefully selected optimum temperature of 300° C, are believed to provide an acceptable simulation of the diagenesis of natural asphalts and crude oils. The selected experimental conditions do not necessarily apply to any organic matter. Experiments on rock samples containing kerogen may, in view of possible catalysis by inorganic components, require different temperatures. Compared to kerogen, such soluble organic substances as crude oils and asphalts, present the advantage of revealing the possible generation of an insoluble, high-molecular pyrobitumen fraction.

It would appear that bacterial degradation, occurring at any time in the geologic past, irreversibly exerts an unfavorable influence on crude oils collected in reservoirs. These altered oils are unlikely, even in case of renewed deep burial, to be reconverted to producible crudes.

The observed diagenetic behavior of unaltered and altered crude oils is consistent with the general concept of organic diagenesis involving a permanent process of disproportionation that gradually supplants functional reactions. The diagenetic differences between unaltered and altered oils are thus entirely comparable to those existing between favorable (sapropelic) and unfavorable (lignitic, detrital) organic deposits.

References

1 G.K. Bell and J.M. Hunt, "Native Bitumens Associated with Oil Shales", in: I.A. Breger (Editor), *Organic Geochemistry*, Pergamon, Oxford, Ch. 8, pp. 333—366 (1963).

2 H. Abraham, *Asphalts and Allied Substances*, *I*, Van Nostrand, Princeton, N.J., 6th ed., pp. 56—57 (1960).

3 C.R. Evans, M.A. Rogers and N.J.L. Bailey, "Evolution and Alteration of Petroleum in Western Canada", *Chem. Geol.*, *8*, 147—170 (1971).

4 J.A. Williams and J.C. Winters, *Symp. Petroleum Transformation in Geologic Environments*, 158th Natl. Meet. Am. Chem. Soc., New York, N.Y. (1969).

5 N.J.L. Bailey, H.R. Krouse, C.R. Evans and M.A. Rogers, "Alteration of Crude Oil by Water and Bacteria — Evidence From Geochemical and Isotope Studies", *Bull. Am. Assoc. Pet. Geol.*, *57*, 1276—1290 (1973).

6 A. Jobson, F.D. Cook and D.W.S. Westlake, "Microbial Utilization of Crude Oil", *Appl. Microbiol.*, *23*, 1082—1089 (1972).

7 N.J.L. Bailey, A.M. Jobson and M.A. Rogers, "Bacterial Degradation of Crude Oil: Comparison of Field and Experimental Data", *Chem. Geol.*, *11*, 203—221 (1973).

8 L.R. Brown, W.E. Phillips and J.M. Tennyson, *Effect of Salinity on the Oxidation of*

Hydrocarbons in Estuarine Environments, Report No. 39762, Water Resources Res. Inst., Mississippi State Univ., 37 pp. (1970).
9 O.A. Radchenko, A.S. Chernysheva and O.P. Bolotskaya, *Trans. All Union Sci. Geol. Res. Inst. Pet. Prosp.*, New Ser. No. 5, *Contributions to Geochemistry*, Gostoptekhizdat, Leningrad, pp. 124—154 (1951).
10 G.D. Hobson, "Oil and Gas Accumulations, and Some Allied Deposits", in: B. Nagy and U. Colombo (Editors), *Fundamental Aspects of Petroleum Geochemistry*, Elsevier, Amsterdam, Ch. 1 (1967).
11 R.F. Lebküchner, F. Orhun and M. Wolf, "Asphaltic Substances in Southeastern Turkey", *Bull. Am. Assoc. Pet. Geol., 56,* 1939—1964 (1972).
12 J. Connan, "Laboratory Simulation and Natural Diagenesis", *Bull. Centre Rech. Pau-SNPA, 6,* 198 (1972).
13 U. Colombo and G. Sironi, "Geochemical Analysis of Italian Oils and Asphalts", *Geochim. Cosmochim. Acta, 25,* 24—51 (1961).
14 H.M. Smith, "Effect of Time and Depth of Burial on Naphtha and Gas Oil Content of Crude Oil", *Proc. Okla. Acad. Sci., 47,* 195—205 (1968).
15 K.K. Landes, "Eametamorphism, and Oil and Gas in Time and Space", *Bull. Am. Assoc. Pet. Geol., 51(6)* (Part I), 828—841 (1967).
16 J.G. McNab, P.V. Smith and R.L. Betts, "The Evolution of Petroleum", *Ind. Eng. Chem., 44,* 2556—2563 (1952).
17 J. Connan and B.M. van der Weide, "Diagenetic Alteration of Natural Asphalts", in: L.V. Hills (Editor), *Oil Sands — Fuel of the Future, Can. Soc. Pet. Geol. Mem., 3,* 134—137 (1974).
18 J.A. Gransch and E. Eisma, "Characterization of the Insoluble Organic Matter of Sediments by Pyrolysis", in: G.D. Hobson and G.C. Speers (Editors), *Advances in Organic Geochemistry 1966*, Pergamon, Oxford, pp. 407—426 (1970).
19 J. Connan, 1972, unpublished data.
20 G.T. Philippi, "Depth, Time and Mechanism of Petroleum Generation", *Geochim. Cosmochim. Acta, 29,* 1021—1049 (1965).
21 M. Louis and B.P. Tissot, "Effect of Temperature and Pressure on the Formation of Hydrocarbons in Kerogenous Clays", *Proc. 7th World Petrol. Congr., 2,* 47—60 (1967).
22 N.B. Vassoevich, Yu.J. Korchagina, N.V. Lopatin and U.V. Chernyshev, "Main Phase of Petroleum Formation", *Int. Geol. Rev., 12,* 1276—1296 (1970).
23 W. Henderson, G. Eglinton, P. Simmonds and J.E. Lovelock, "Thermal Alteration as a Contributory Process to the Genesis of Petroleum", *Nature, 219,* 1012—1016 (1968).
24 C. Esnault, Thesis,University of Pau, France, pp. 32—33 (1973).
25 H. Abraham, *Asphalts and Allied Substances, I,* Van Nostrand, Princeton, N.J., 6th ed., pp. 185—187 (1960).
26 E.E. Bray and E.D. Evans, "Occurrences of Hydrocarbons in Shaly Rock-Source Beds", *Bull. Am. Assoc. Pet. Geol., 49,* 248—257 (1965).
27 E.E. Bray and E.D. Evans, "Distribution of *n*-Paraffins as a Clue to Recognition of Source Beds", *Geochim. Cosmochim. Acta, 22,* 2—15 (1961).
28 R.S. Scalan and J.E. Smith, "Improved Measure of the Odd—Even Predominance in Petroleum", *Geochim. Cosmochim. Acta, 34,* 611—620 (1970).
29 I.R. Hills, G.W. Smith and E.V. Whitehead, "Hydrocarbons from Fossil Fuels and Their Relation with Living Organisms", *J. Inst. Pet., 56,* 127—137 (1970).
30 A. Ensminger, A. van Dorsselaer, Ch. Spyckerelle, P. Albrecht and G. Ourisson, *Actes due 6ème Congrès International de Géochimie Organique*, Paris (1973).
31 M. Louis, "Essais sur l'Evolution de Pétrole à Faible Température en Présence de Minéreaux", in: G.D. Hobson and M.C. Louis (Editors), *Advances in Organic Geochemistry 1964*, Pergamon, Oxford, pp. 261—278 (1966).

32 R.D. McIver, "Maturation of Oil, an Important Natural Process", *Annu. Meet. Geol. Soc. Am., Abstr.*, p. 113 (1963).
33 J. Connan, "Diagenèse Naturelle et Diagenèse Artificielle de la Matière Organique à Eléments Végétaux Prédominants", in: B. Tissot and F. Bienner (Editors), *Advances in Organic Geochemistry 1973*, Technip, Paris, pp. 72—95 (1974).
34 B. Tissot, "Initial Data on the Mechanism and Kinetics of Oil Formation in Sediments. Computer Simulation of a Reaction Flow", *Rev. Inst. Fr. Pét.*, *24*, 470—501 (1969).
35 D.H. Welte, "Petroleum Exploration and Organic Geochemistry." *J. Geochem. Explor.*, *1*, 117 (1972).
36 J. Connan, "Time—Temperature Relation in Oil Genesis, *Bull. Am. Assoc. Pet. Geol.*, *58*, 2516—2521 (1974).
37 C. Leibovici and B.M. van der Weide, "Aspects Qualitatifs de la Diagenèse Organique", in: B. Tissot and B. Bienner (Editors), *Advances in Organic Geochemistry 1973*, Technip, Paris, pp. 368—378 (1974).
38 T.F. Yen, J.G. Erdman and S.S. Pollack, "Investigation of the Structure of Petroleum Asphaltenes by X-ray Diffraction", *Anal. Chem.*, *33*, 1587 (1961).
39 J.P. Dickie and T.F. Yen, "Macrostructures of the Asphaltic Fractures by Various Instrumental Methods", *Anal. Chem.*, *33*, 1587 (1961).
40 A.L. Burlingame and B.R. Simoneit, "High Resolution Mass Spectrometry of Green River Formation Kerogen Oxidations", *Nature*, *222*, 741 (1969).
41 B. Alpern, B. Durand, J. Espitalié and B. Tissot, "Localisation, Charactérisation et Classification Pétrographique des Substances Organiques Sédimentaires Fossiles", in: H.R. von Gaertner and H. Wehner (Editors), *Advances in Organic Geochemistry 1971*, Pergamon, Oxford, pp. 1—28 (1972).
42 H. Jacob, "Petrology of Asphaltites and Asphaltic Pyrobitumens", *Erdoel Kohle, Erdgas Petrochem.*, *20(6)*, 393—400 (1967).

Chapter 4

MAJOR TAR-SAND DEPOSITS OF THE WORLD

P.H. PHIZACKERLEY and L.O. SCOTT

Introduction

The descriptive term "tar sand" has been defined under the Alberta Oil and Gas Conservation Act as sand having a "highly viscous crude hydrocarbon material not recoverable in its natural state through a well by ordinary production methods". Strictly speaking, "tar sand" is an inappropriate term, because tar along with pitch is a substance resulting from the destructive distillation of organic matter, and such an origin for tar in tar sands is rarely implied. Also, in practice, the distinction between tar sands and sands containing heavy crude oil is not easily drawn.

The locations of the major and some of the more interesting and well-known minor tar sands of the world are shown in Fig. 4-1. The age, extent, thickness, bitumen saturation, character of oil, and overburden thickness of twenty major world tar-sand deposits are given in Table 4-I. The deposits are

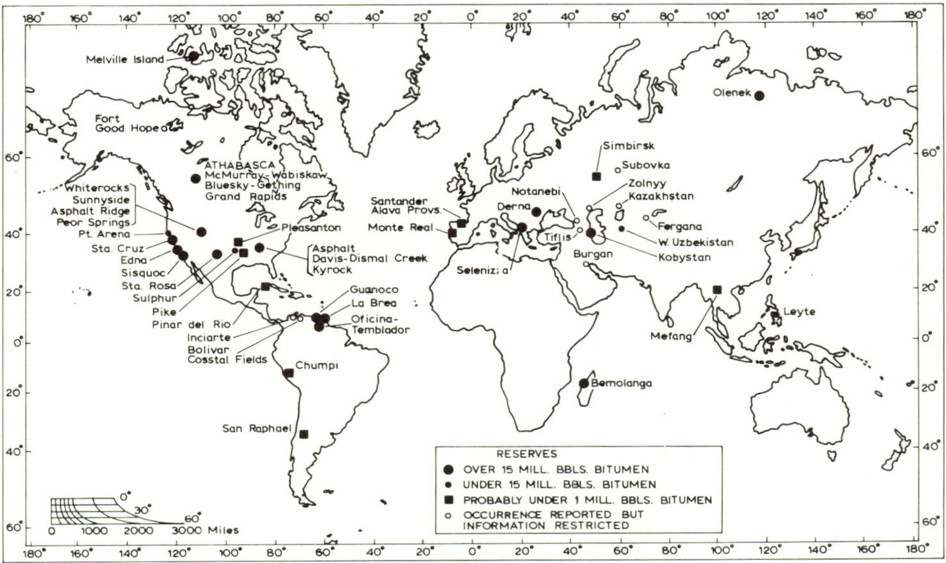

Fig. 4-1. Location map of the major tar sands of the world.

TABLE 4-I

Major tar-sand deposits of the world

Country	Name of deposit	Age of reservoir rock	Areal extent (acres)
Canada	McMurray—Wabiskaw, Alberta		5,750,000
	Bluesky—Gething, Alberta		1,200,000
	Grand Rapids, Alberta		1,100,000
	Total "Athabasca" tar sands	L. Cretaceous	8,000,000
	Melville Island, N.W.T.	Triassic	?
Eastern Venezuela	Oficina—Temblador tar belt	Oligocene	5,750,000
Malagasy	Bemolanga	Triassic	96,000
U.S.A.	Asphalt Ridge, Utah	Oligocene and U. Cretaceous	11,000*
	Sunnyside, Utah	U. Eocene	34,300*
Albania	Selenizza	Mio-Pliocene	5306
U.S.A.	Whiterocks, Utah	Jurassic	1900
	Edna, California	Mio-Pliocene	6595
	Peor Springs, Utah	U. Eocene	1735
Eastern Venezuela	Guanoco	Recent (Alluvial)	1000
Trinidad	La Brea	U. Miocene	126
U.S.A.	Santa Rosa, New Mexico	Triassic	4630+
	Sisquoc, California	U. Pliocene	175
	Asphalt, Kentucky	Pennsylvanian	7000
Rumania	Derna	Pliocene	459
U.S.S.R.	Cheildag, Kazakhstan	M. Miocene	82*
U.S.A.	Davis—Dismal Creek, Kentucky	Pennsylvanian	1900
	Santa Cruz, California	Miocene	1200
	Kyrock, Kentucky	Pennsylvanian	900*

* Statistics asterisked where estimated.

listed in decreasing order of reserves. It should be remembered that statistical information on all the tar-sand areas, except Athabasca, is usually incomplete and some figures represent rough estimates. Reserves at Melville Island are unknown.

The major tar sands are limited to eight countries. The apparently uneven

	Pay thickness		Bitumen sat.	Character of oil		Overburden thickness	In-place reserves	No.
sq.mi.	ft	average	(wt%)	°API at 60°F	(% sulfur)	(ft)	(10^6 bbl)	
9000	0— 300	(175)	—	—	—	0—1900	625,900	
1875	0— 400		—	—	—	700—2600	51,500	
1625	400	(280)	—	—	—	300—1400	33,400	
12,500		(150±)	2—18	10.5	4.5		710,800	1
?	60— 80		?—16	10	0.9—2.2	0—2000*	?	1a
9,000	3— 10			10		0—3000*	200,000	2
150	80— 300	(100)	10		0.7	0— 100	1750	3
—	11— 254	(98)	11	8.6—12	0.5	0—2000	900	4
—	24— 200	(100)						
—	10— 350		9	10 —12	0.5	0— 150	500	5
8	33— 330	(50)*	8—14	4.6—13.2	6.1	shallow	371*	6
3	900—1000		10	12	0.5	nil	250	7
10	0—1200*	(250)	9—16	13	4.2	0— 600*	165	8
3	1— 250	(34)*	9*			shallow	87	9
2	2— 9	(4)	64	8	5.9	nil	62	10
—	0— 270	(135)	54	1 — 2	6.0—8.0	nil	60	11
7	0— 100	(20)	4— 8			0— 40	57	12
—	0— 185	(85)	14—18	4 — 8		15— 70	50	13
11	5— 36	(15)	8—10			6— 30	48	14
—	6— 25		15—22		0.7	shallow	25	15
—		(200)	5—13			shallow	24	16
3	10— 50	(15)	5			15— 30	22	17
2	5— 50	(11)	10—12			0— 100	20	18
—	15— 40	(20)	6— 8			15	18	19

distribution is partly due to the occurrence of natural asphalt in rocks other than sandstone, namely limestone, shale, serpentine, etc., and partly to lack of information. Hence, the major asphalt deposits of France, Germany, Switzerland, Italy, Iran *, Oklahoma and Texas are not included in this review because they occur in limestones.

*Added by the editors.

Reference is made where appropriate to geologically interesting minor deposits; however, only those with reserves of more than 15 million bbl * of bitumen in place have been classed as major accumulations. Reserves from nineteen major tar sands total 915.2 billion bbl, a figure about 10% higher than the most recent estimate [1].

Geology of the major tar sands of the world

The basic geological data on the recorded tar sands of the world are described first in the order given in Table 4-I (with the exception of the Kentucky deposits). A discussion of their origin follows in the subsequent section.

Athabasca deposits, Canada [2,3,4]

The town of McMurray, about 240 miles north-northeast of Edmonton, Alberta, Canada, lies at the eastern margin of the largest tar-sand accumulation in the world (Fig. 4-2). There are actually three major accumulations present as distinct horizons within the Lower Cretaceous deposits (Fig. 4-3). The two smaller deposits are larger than any other known deposit with the possible exception of the Oficina—Temblador tar belt in eastern Venezuela.

According to Conybeare [2], the uppermost 50 ft of the McMurray—Wabiskaw tar sands are Barremian—Aptian in age and overlie 150 ft of more massive sandstones which are Valanginian or older. The Bluesky—Gething deposits are probably of Aptian age and the Grand Rapids tar sands are Albian in age.

The major McMurray—Wabiskaw reservoirs are found towards the base of the formation and are characteristically cross-bedded coarse grits and gritty sandstones, all unconsolidated or cemented by tar. Fine- to medium-grained sandstones and silts occur higher in the sequence. The sandstones are believed to be derived from Precambrian rocks outcropping about 100 miles to the northeast near Lake Athabasca. Both the Bluesky—Gething and Grand Rapids reservoirs are composed of subangular quartz and well-rounded chert grains. Sandstones of both deposits are frequently glauconitic and have a calcareous matrix.

The facies of the lower part of the McMurray—Wabiskaw sands varies from fluviatile to deltaic. Prior to the Hauterivian break in sedimentation, a lake and swamp environment developed probably as a result of subsidence. Later more marine conditions prevailed during deposition of the Wabiskaw Member of the Clearwater Formation. In comparison, the facies of both the Bluesky—Gething and Grand Rapids deposits is generally that of a nearshore

* This figure appears to be too low and some place it at 200 million bbl (editorial comment).

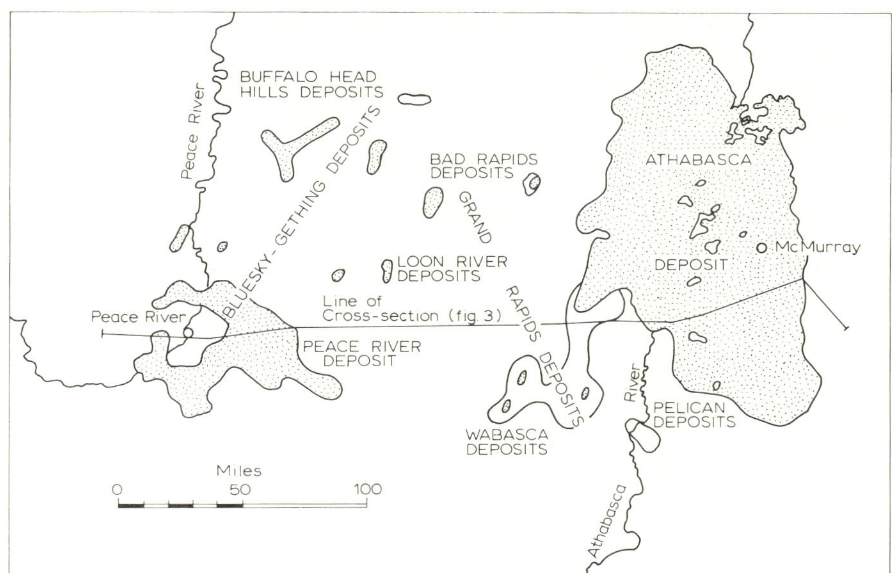

Fig. 4-2. Location of evaluated tar-sand deposits of northern Alberta. (After Alberta Oil and Gas Conservation Board [3].)

environment, although coal deposits occur in the Bluesky—Gething beds.

The McMurray—Wabiskaw and Bluesky—Gething tar sands overlie an irregular topography formed on the surface of westerly dipping Precambrian to Jurassic sediments. This surface is known to control the thickness varia-

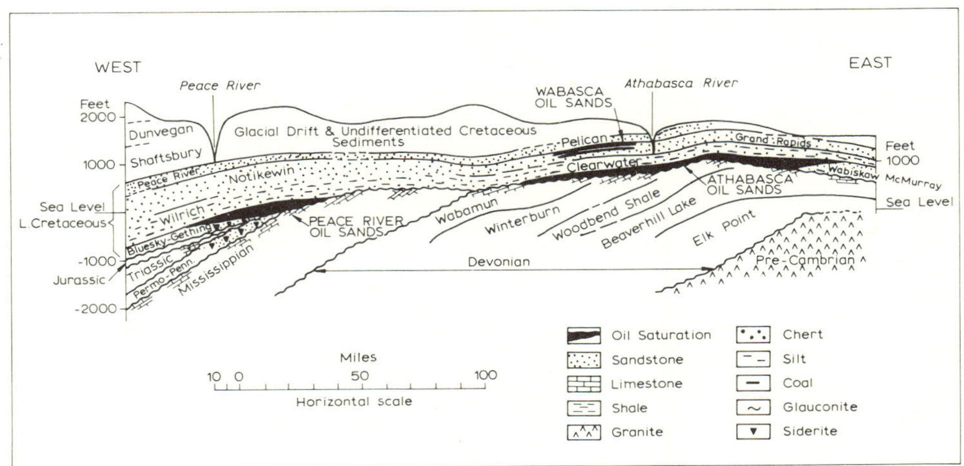

Fig. 4-3. East—west cross-section showing the geological setting of the Athabasca tar sands, Alberta, Canada. (After Alberta and Gas Conservation Board [3].)

tions of the McMurray—Wabiskaw deposit which dips at between 5 and 25 ft/mile to the southwest. These sands are in close proximity to the Devonian—Mississippian unconformity, whereas the Bluesky—Gething sands overlie several unconformities between the Mississippian and Jurassic deposits (Fig. 4-3).

Melville Island, Canada [5,6]

On the north shore of Marie Bay, Melville Island, some 1450 miles north of Edmonton, Alberta (Fig. 4-4), Triassic sandstones of the Bjorne Formation are impregnated with asphalt. Although the sands are not as clean as

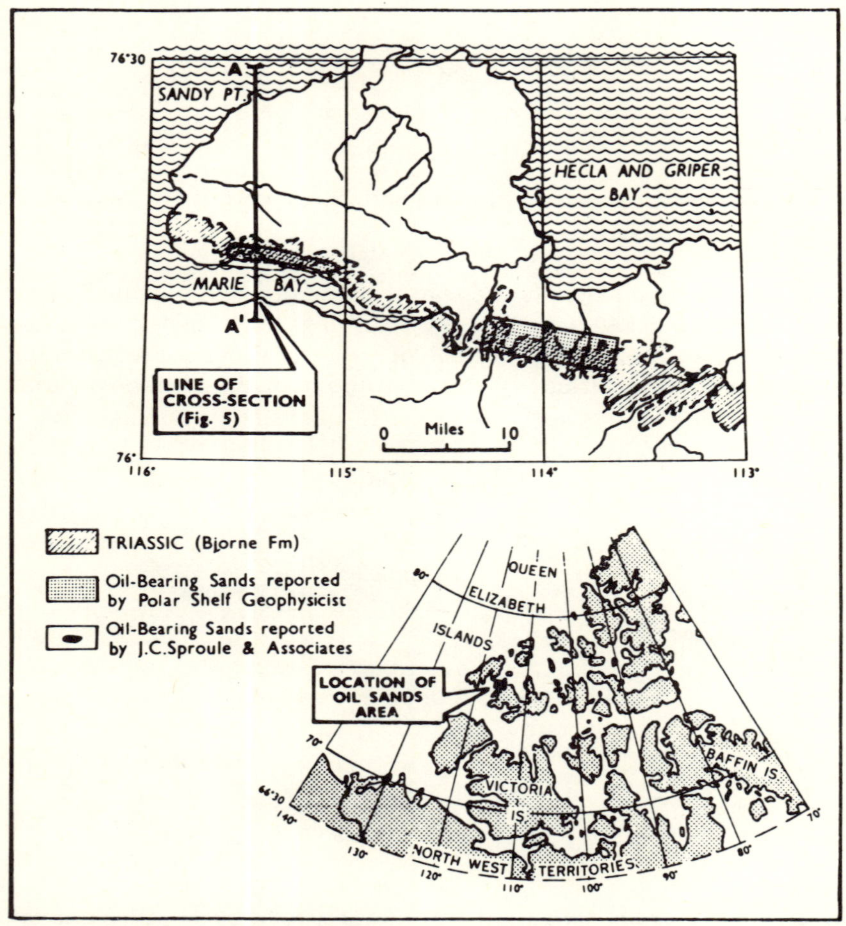

Fig. 4-4. Location of Melville Island tar sands, Canada. (After Sproule and Lloyd [6].)

those of the Athabasca deposits, they represent a similar deltaic facies and are characteristically cross-bedded.

This deposit overlies an unconformity of Triassic age with strongly folded Pennsylvanian—Permian sediments, which outcrop further south along the margin of the Sverdrup Basin (Fig. 4-5). The succeeding Jurassic deposits rest disconformably on the Bjorne Formation. Discovered in 1962, the sands have been seen at intervals along a 60-mile outcrop. It was noted that the richer sands tend to be associated with structurally high areas or are closely related to faulting. If the sands are laterally continuous (a matter still unproved) and allowing for northward impregnation of the Bjorne Formation for about a mile, with only half the average bitumen saturation of the Athabasca tar sands, this deposit could rank third largest in the world.

Oficina—Temblador tar belt, eastern Venezuela [7,8]

Lying north of the Orinoco River (Fig. 4-6), the Oficina—Temblador tar belt in eastern Venezuela, although ill-defined, is thought to contain tar sands occupying as large an area as the McMurray—Wabiskaw deposit. The accumulation is in the southward-thinning beds of the Oligocene Oficina Formation. The sands, where they yield oil further north, are light gray, fine- to coarse-grained and are associated with lignites, silts, and shales representative of a deltaic facies.

Structurally, the Oficina Formation overlaps the Cretaceous Temblador Formation and in the southern part of the tar-sand area rests on the basement of the Guiana shield. The regional dip is 2—4° to the north (Fig. 4-7).

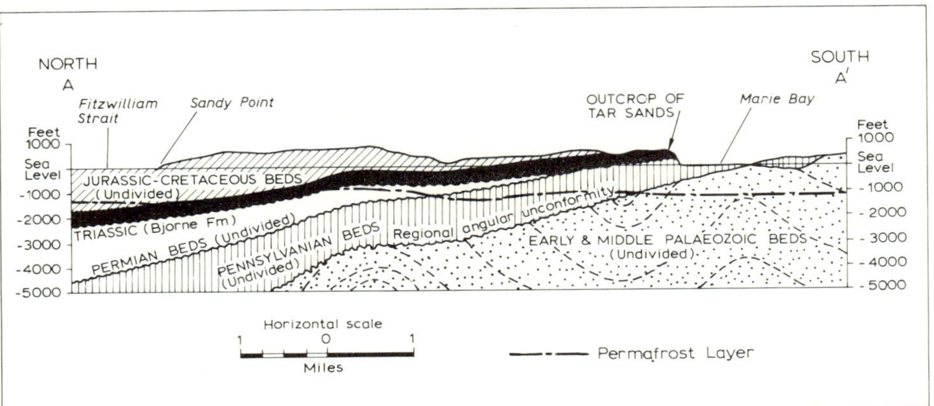

Fig. 4-5. Generalized north—south cross-section showing geological setting of Melville Island tar sands, Canada. (After Sproule and Lloyd [6]; courtesy of *Canadian Oil and Gas Industries.*)

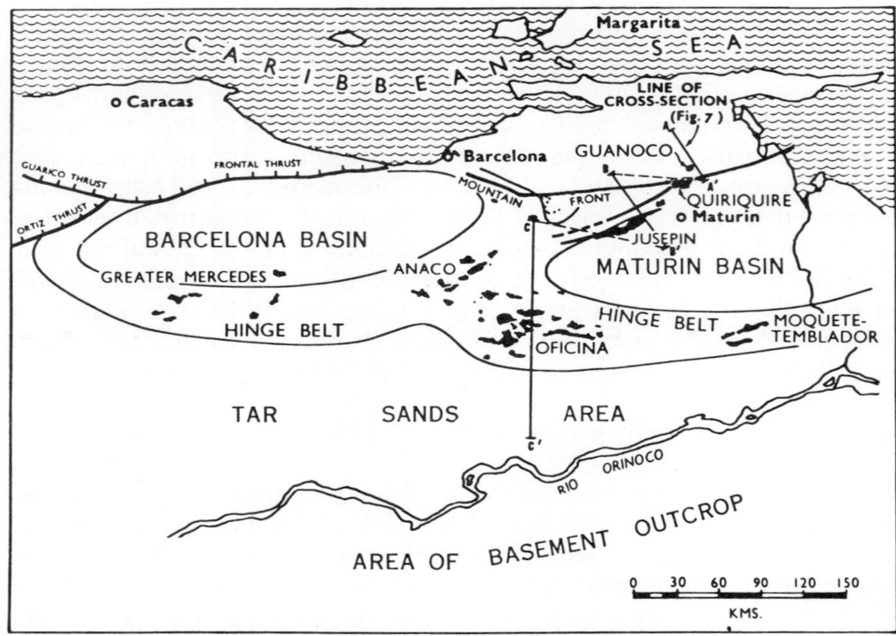

Fig. 4-6. Location of Oficina—Temblador tar sands, eastern Venezuela. (After Dallmus [42]; courtesy of *Tulsa Geol. Soc. Dig.*)

Bemolanga, Malagasy [9]

At Bemolanga, near Morafenobe, western Malagasy, the third largest tar-sand deposit presently known occurs in the upper part of the Lower Isalo Group (Triassic) (Fig. 4-8). The sands are yellow, red, and white, cross-bedded continental sediments of the Karroo System; grits and pebbles are common and the coarser, porous sands are locally succeeded conformably by the Bemolanga clay. The nearest known unconformity in the Bemolanga sequence appears to be at the base of the Aptian deposits which overlie Bathonian limestones. Dykes of Cretaceous or later age cut the tar sands with consequent localized coking of the asphalt (Fig. 4-9).

Structural features are only subtly present in the form of a gentle dome which is, however, considered to be the controlling factor in the location of the accumulation.

Asphalt Ridge, Utah, U.S.A. [10,11,15]

The three major tar-sand deposits occurring in the Uinta Basin, Asphalt Ridge, Sunnyside and Whiterocks in the U.S.A. are especially interesting in that each accumulation has its own characteristics and may be taken as

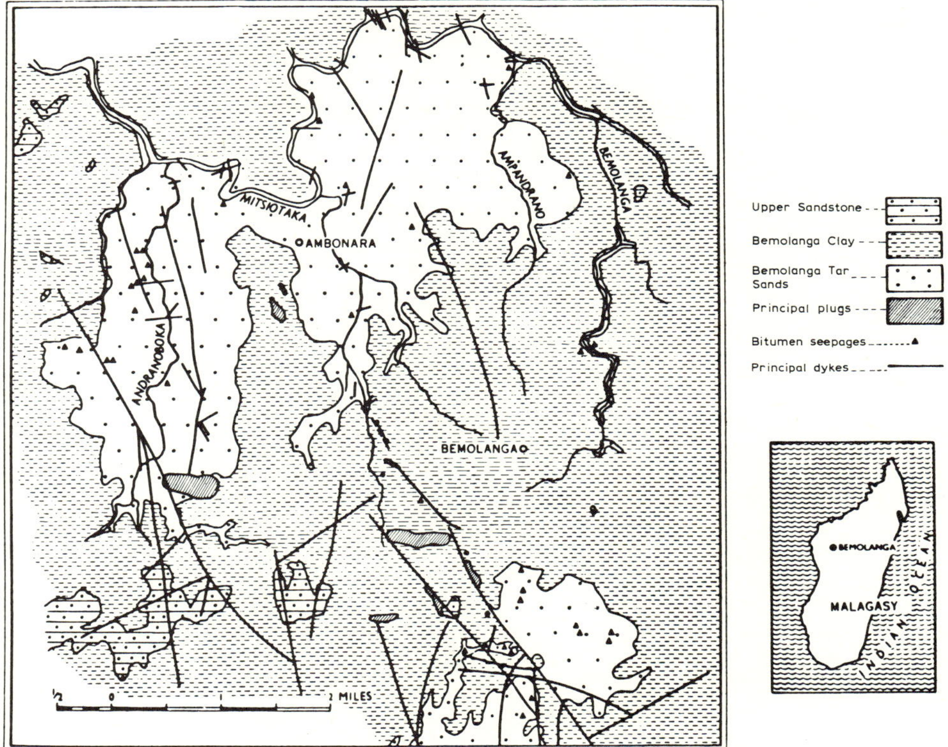

Fig. 4-8. Geological map of Bemolanga tar sands area, Malagasy. (After Kent [9]; courtesy of British Petroleum Company.)

representative of the three types of impregnation discussed later in this chapter.

The largest tar-sand deposit in the U.S.A., according to the latest estimate by Covington [10], occurs at Asphalt Ridge, which is situated on the north flank of the Uinta Basin near Vernal (Fig. 4-10). In these deposits, impregnation occurs in the Upper Cretaceous Mesaverde and the Oligocene Duchesne River formations.

The lower 700 ft of the Mesaverde Formation is represented by two sandstone members, the Asphalt Ridge and Rim Rock, which are separated by about 100 ft of shale. The Rim Rock is a cross-bedded beach deposit. The Asphalt Ridge Member is only locally asphaltic, whereas the succeeding unconformable Duchesne River sandstones and conglomerates contain approximately the same amount of bitumen as the Rim Rock sandstone.

The structure is essentially monoclinal (Fig. 4-11) dipping south to southwest at about 10° in the Cretaceous beds and 6° in the Oligocene deposits.

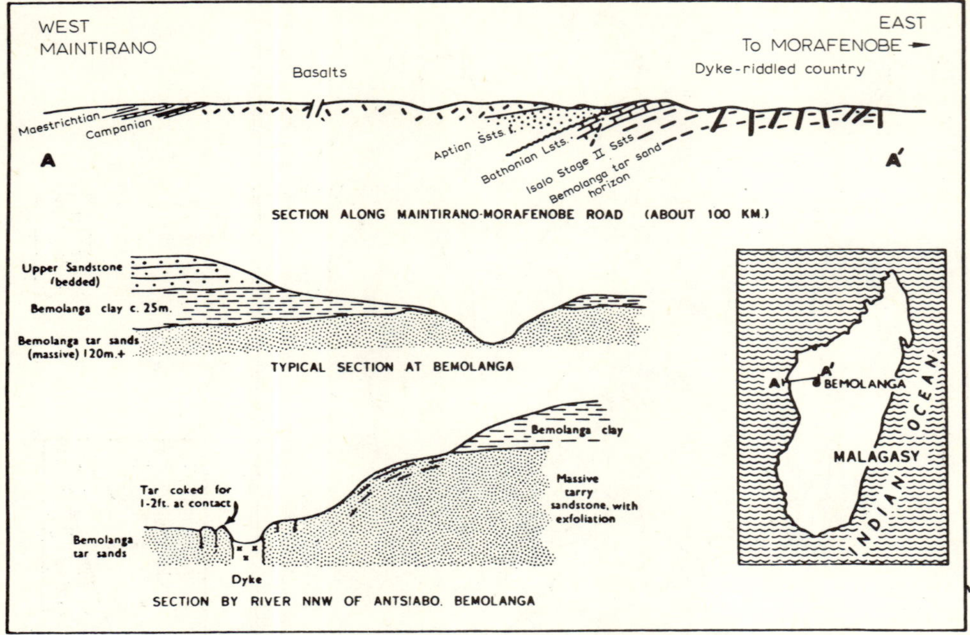

Fig. 4-9. Geological cross-sections in Bemolanga tar sands area, Malagasy. (After Kent [9]; courtesy of British Petroleum Company.)

Sunnyside, Utah, U.S.A. [10,12,15]

About 75 miles southwest of the Asphalt Ridge tar sands (Fig. 4-10), additional deposits are located at Sunnyside, in sandstones of the Wasatch and lower Green River formations of Eocene age. Only the gray—green—purple sands of the Wasatch Formation are economic *, but they are extremely lenticular, occupying broad channels cut in the underlying shales and limestones. These fluviatile deposits are sometimes conglomeratic and are succeeded by lacustrine, more uniform and laterally continuous Green River beds.

The Sunnyside accumulation is in the Uinta Basin on the opposite flank to Asphalt Ridge; dips are between 3 and 10° to the northeast. Faulting and folding are not thought to affect impregnation, which occurs over an interval of 1200 ft in the upper third of the Wasatch Formation.

Selenizza, Albania [13,14]

The largest tar-sand deposits in Europe are those at Selenizza, about 14 km northeast of Valona, Albania (Fig. 4-12). This region also contains

* This is questioned by the editors, however.

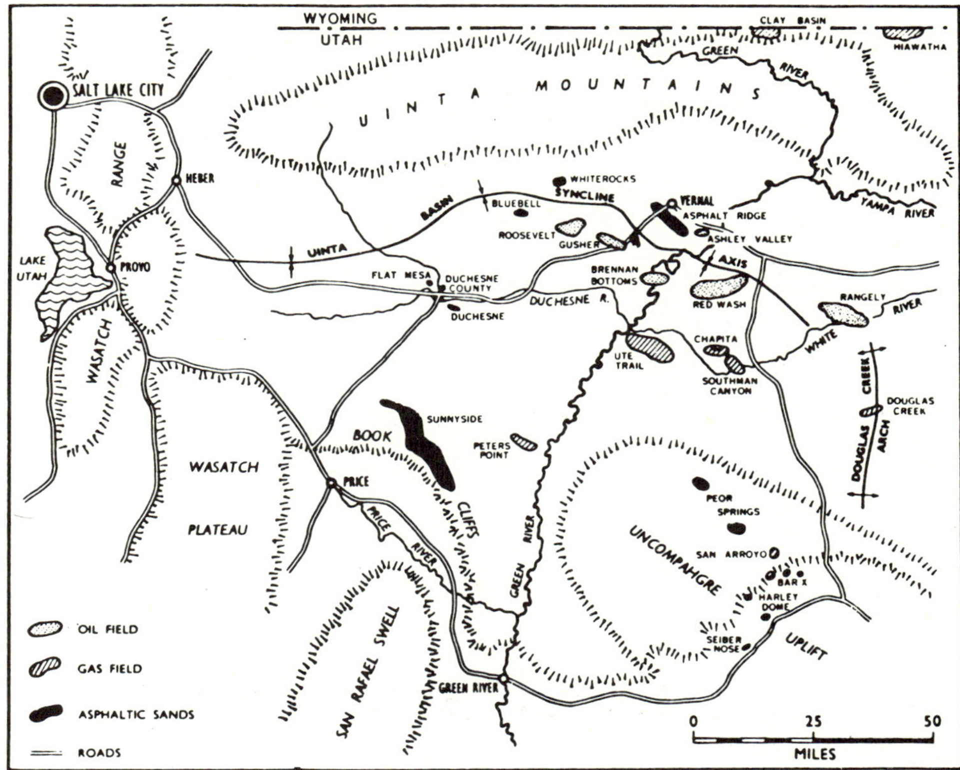

Fig. 4-10. Location of the major tar sands of Uinta Basin, Utah. (After Covington [28]; courtesy of *Bull. Utah Geol. Mineral. Surv.*)

the Patos oil field throughout which extensive asphalt impregnations occur.

The bitumen at Selenizza is present in Middle and Upper Miocene lenticular sands, characterized by a brackish-water fauna. Overlying Pliocene conglomeratic beds are also locally impregnated; these deposits are mostly, but not entirely, marine.

Structure appears to be an important feature, inasmuch as the Selenizza and Patos deposits occupy the crestal portions of a north—south trending anticline. Accumulation is also controlled by transverse faults. Regionally, the Miocene deposits rest unconformably on Eocene limestones (Fig. 4-12).

Whiterocks, Utah, U.S.A. [10,15]

The Whiterocks tar sands, about 20 miles west-northwest of Vernal, occupy a similar regional position in the Uinta Basin to those of Asphalt

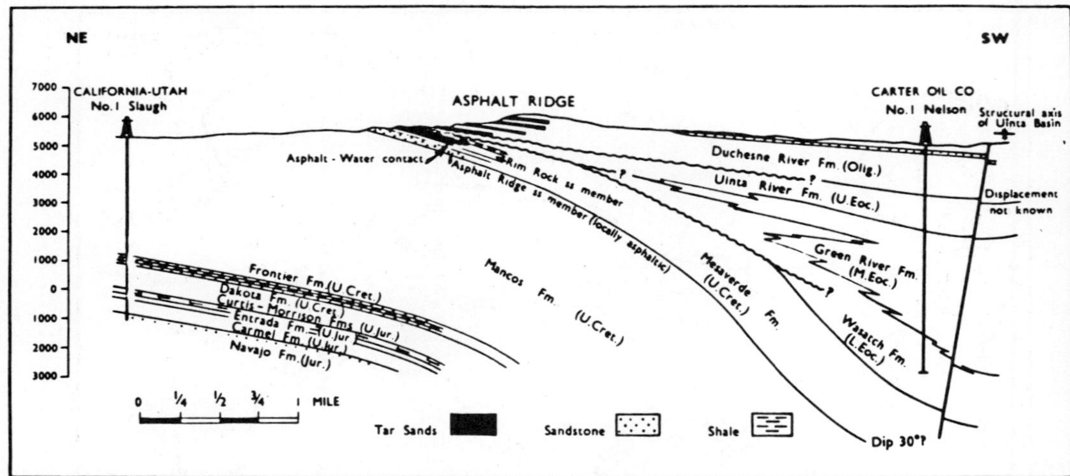

Fig. 4-11. Northeast—southwest diagrammatic geological cross-section in Asphalt Ridge area Utah. (After Covington [10]; courtesy of *J. Pet. Technol.*)

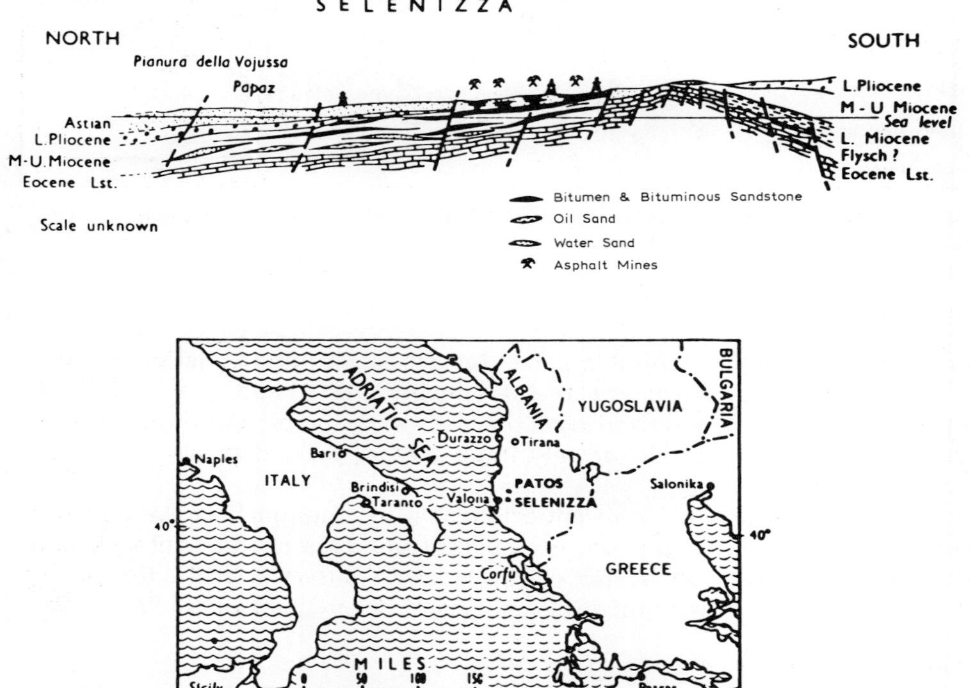

Fig. 4-12. Location map and diagrammatic geological cross-section of the Selenizza area, Albania. (After Ineichen [13]; courtesy of *World Pet.Congr.*, 1951.)

Ridge (Fig. 4-10). They have, however, quite different stratigraphic and structural aspects.

The accumulation is found in the Middle Jurassic Navajo Formation in a clean, fine- to medium-grained sandstone with subrounded clear quartz grains. This formation is about 1000 ft thick and is predominantly an aeolian deposit.

Structure appears important, as the beds are dipping at about 60° to the southeast into the Uinta Basin and faulting is postulated at depth (Fig. 4-13). The degree of impregnation seems to be related to fracturing.

Edna, California, U.S.A. [15,16]

The largest tar-sand deposit in California is situated at Edna and lies 3—8 miles south and southeast of San Luis Obispo in a small structural basin in the Coastal Range district, west of the San Andreas fault (Fig. 4-14).

The reservoir belongs to the Mio-Pliocene Pismo Formation, which is fossiliferous and consists of marine conglomerate, sandstone, diatomaceous sandstone and siliceous shale.

The Pismo Formation unconformably overlies the generally massive, diatomaceous shale of the Monterey Formation (Fig. 4-15) and frequently contains pieces of shale derived from it. The tar sands outcrop in scattered areas on both flanks of a narrow syncline with maximum dips of 40°.

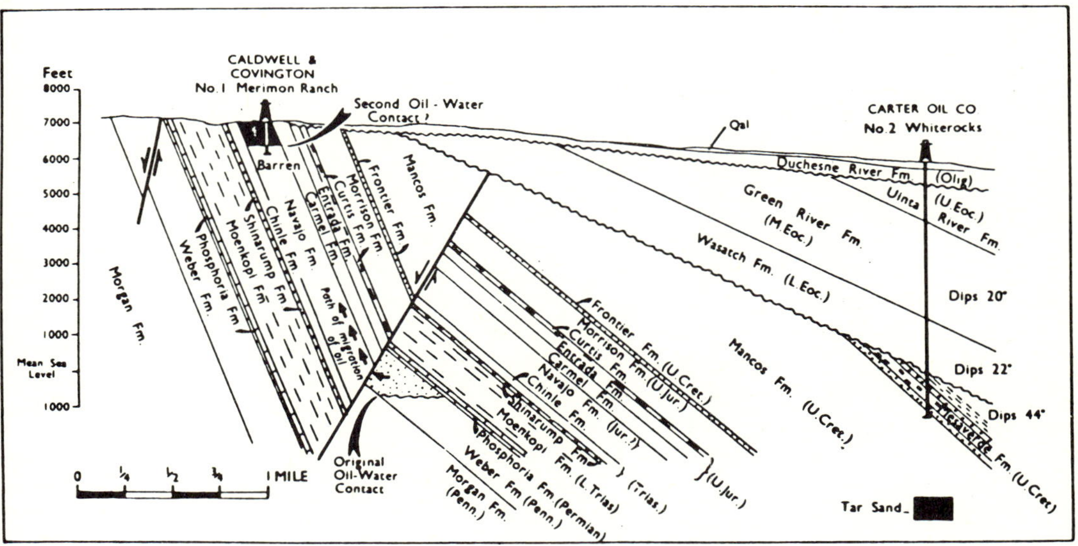

Fig. 4-13. Geological cross-section of Whiterocks area, Utah. (After Covington [10]; courtesy of *J. Pet. Technol.*)

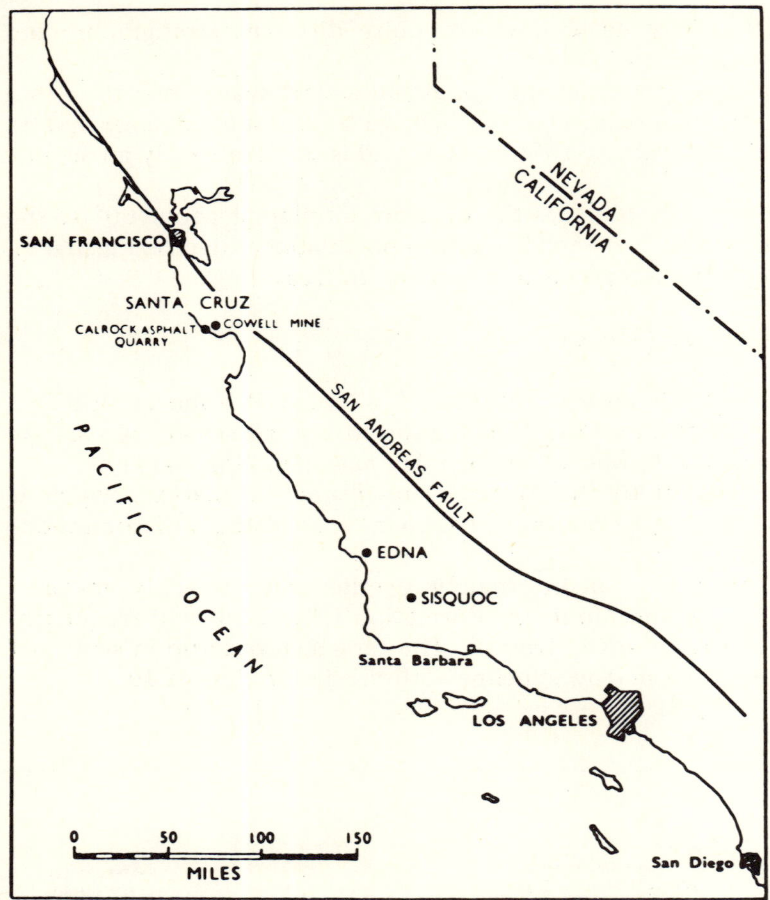

Fig. 4-14. Location of major tar sands in California, U.S.A.

Peor Springs, Utah, U.S.A. [10,15]

The tar sands at Peor Springs, Utah, U.S.A., occur about 60 miles east of those of Sunnyside, which they resemble in most aspects (Fig. 4-10). The impregnation is present in lenticular sandstones of the Eocene Wasatch Formation and there are two main beds from 30 to 85 ft in thickness.

The regional dip off the Uncompahgre uplift is about 4—8° to the northwest. As distinct from Sunnyside, the greatest impregnation occurs on the crests of gentle northwest-plunging anticlinal noses.

Guanoco, eastern Venezuela [17]

The Guanoco asphalt lake is situated 105 miles west of its counterpart in Trinidad (Fig. 4-16). Asphalts are present in Recent alluvial deposits

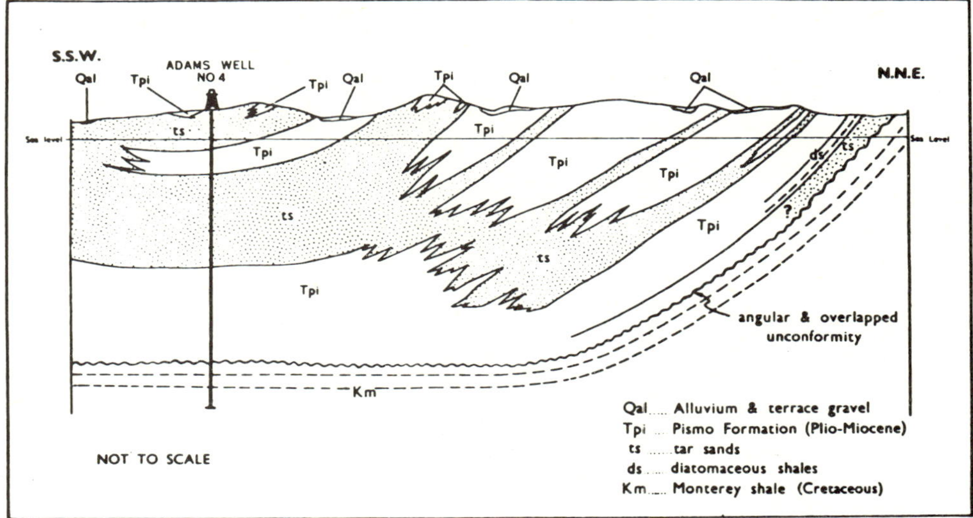

Fig. 4-15. Diagrammatic cross-section through Edna tar sands, California, U.S.A. (After Krueger [40] and Page et al. [41].)

which rest on the Las Piedras Formation of Mio-Pliocene age. This formation dips northwestward towards the Guanoco Lake; it is composed of brackish- to fresh-water sandstones with associated lignites.

The structural picture is simple with the Las Piedras Formation unconformably overlying the marine Upper Cretaceous Guayuta Group, which is composed of highly fractured limestones and argillite (Fig. 4-16).

La Brea, Trinidad, West Indies [18,19]

Situated on the Gulf of Paria, 12 miles west-southwest of San Fernando and 138 ft above sea level (Fig. 4-16), the Trinidad asphalt lake occupies a depression in Miocene Upper Morne l'Enfer sandstones (Figs. 4-17 and 4-18). These rocks (sheet sands) are approximately of the same age as the Las Piedras Formation and have the same lithology and facies.

A complicated structure exists below the lake (Fig. 4-18). The lake is situated above an eroded anticline composed of an argillite core of the Upper Cretaceous Naparima Hill Formation, with remnants of Early Tertiary formations still preserved on the flanks. Between the lake and the crest of the anticline, there are about 4500 ft of reworked Lower Tertiary rocks deposited during the Miocene and later faulted.

Santa Rosa, New Mexico, U.S.A. [15,20]

The Santa Rosa deposit in central New Mexico, U.S.A., occurs in sandstones of the Upper Triassic Santa Rosa Formation. These rocks are a

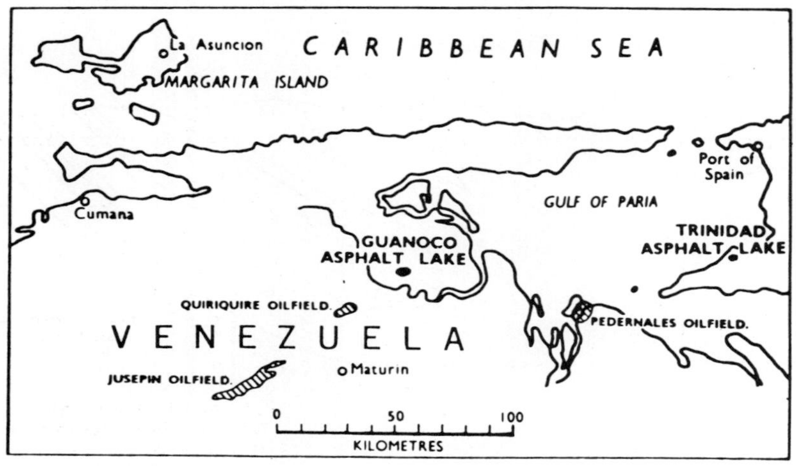

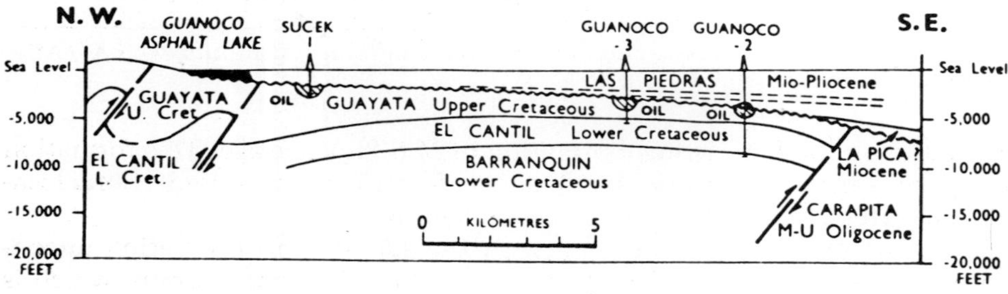

Fig. 4-16. Location map and diagrammatic cross-section of Guanoco area, Venezuela. (After Salvador and Hotz [43]; courtesy of *World Pet. Congr.*, 1963.)

typically thick sequence of red beds consisting of clays, shales, and sandstones. The impregnated beds consist of subangular to subrounded, fine- to medium-grained, micaceous, nonmarine sandstones. The facies is probably fluviatile and was deposited during a moist and temperate climate.

The beds are gently folded into ill-defined north-northeast plunging anticlines. The relationship of the Santa Rosa Formation to the underlying San Andres Limestone of Permian age is disconformable.

Sisquoc, California, U.S.A. [15,16]

The second largest Californian tar-sand deposit occurs in the Upper Pliocene of the Sisquoc region, a 25-square-mile area between the Salmon Hills and the Santa Maria River in Santa Barbara County, California (Fig. 4-14).

Bitumens are present in sandstones of the Pliocene Careaga Formation,

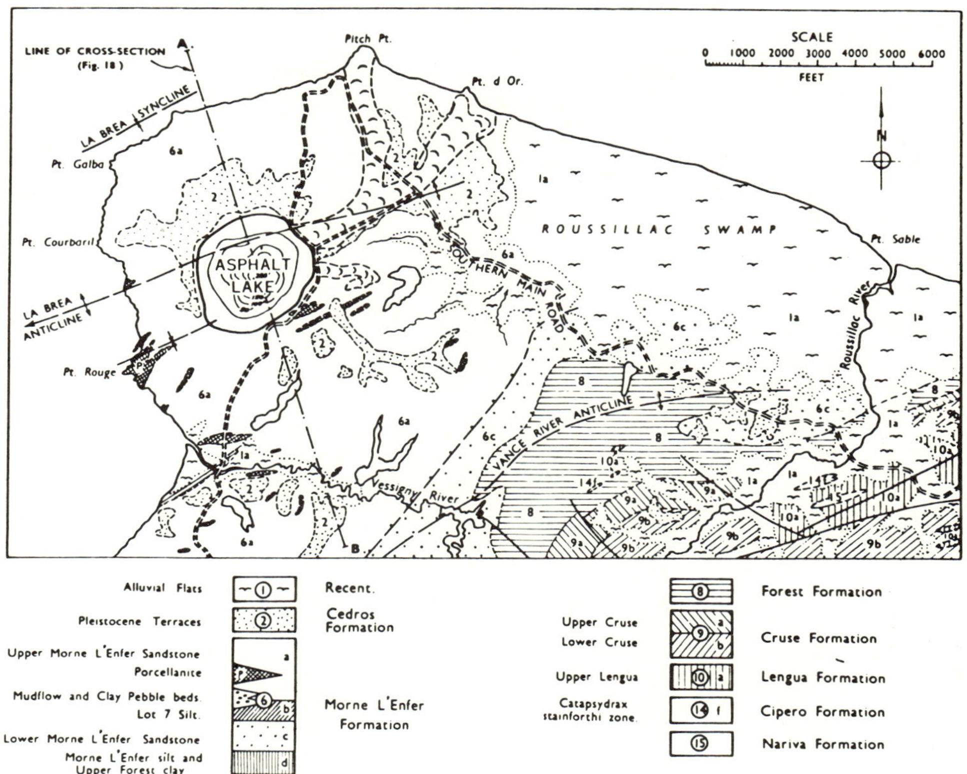

Fig. 4-17. Geological map of the Trinidad asphalt lake area (After Kugler [19]; courtesy of *Pet. Assoc. Trinidad.*)

and there are up to eight individual beds. The quartz grains are subangular to rounded and, generally, medium-grained. There are included pebbles, which were derived from the underlying Monterey Shale.

The relationship with the Monterey Shale is unconformable, the dip divergence ranging from 10 to 15°. The Sisquoc deposits are folded and faulted about an east—west axis without much apparent effect on the degree of impregnation.

Asphalt, Davis—Dismal Creek, and Kyrock, Kentucky, U.S.A. [15,21]

The Asphalt, Davis—Dismal Creek, and Kyrock deposits in Edmonson County, Kentucky, U.S.A. (Fig. 4-19), account for 88 million bbl of the established total Kentucky reserve of 140 million bbl. The tar occurs in sandstones of the Pennsylvanian Pottsville Formation. At Davis—Dismal Creek, sandstones are thin-bedded and shaley, whereas at Kyrock massive conglom-

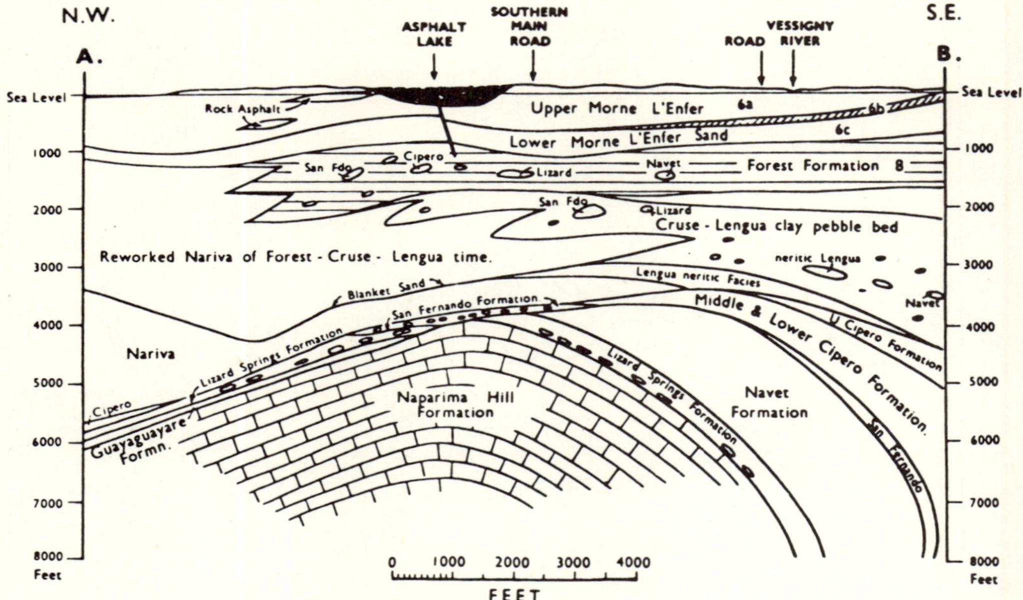

Fig. 4.18. NW—SE geological cross-section through Trinidad asphalt lake area. (After Kugler [19]; courtesy of *Pet. Assoc. Trinidad.*)

erates are present in the basal sandstone. The most southern occurrence at Asphalt, is in the Bee Spring Sandstone. The facies is nonmarine, the Nolin Coal separating the Bee Spring Sandstone from the basal conglomeratic sandstone, which is probably an ancient river-channel deposit.

Dips are between 3 and 4°, and structurally the area lies on the west-dipping flank of the Cincinnati Arch. The section is underlain by the regional Mississippian—Pennsylvanian unconformity.

Derna, Rumania [22]

The Derna deposits are located, along with those of Tataros and others, in a triangular tract east and northeast of Oradia between the Sebes Koros and Berrettyo rivers (Fig. 4-20). The lignitiferous tar sand, occurs in the upper part of the Pliocene Pannonian Formation and the asphalt is characterized by its penetrating odor. The facies of the reservoir rock is nonmarine representing fresh-water deposition during a period of regression.

Structurally, the Tertiary rocks lie in a flat basin floored by crystalline basement. Unconformities are present in the Eocene and Miocene deposits.

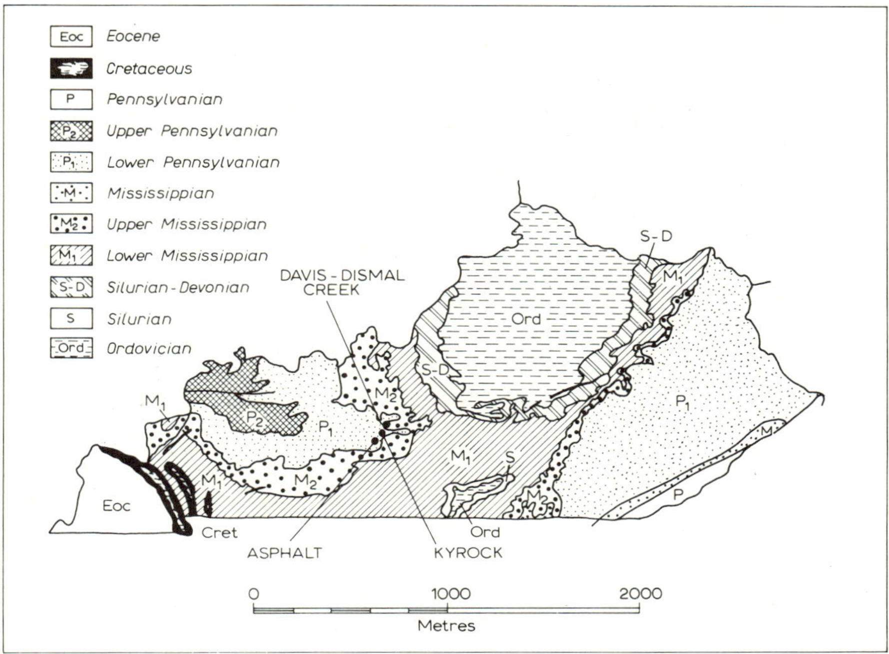

Fig. 4-19. Geological map of Kentucky showing the location of major asphalt deposits. (Courtesy of *U.S. Geol. Surv.*, 1 : 1,500,000 map, 1965.)

Cheildag, Kobystan, U.S.S.R. [23,24,25]

Tar sands occur in the Middle Miocene (*Spirialis* beds of Kabristan) [23] at Cheildag, about 15 miles west-southwest of Baku, Azerbayjan S.S.R. (Fig. 4-1). According to Mirtsching [24], the sands outcrop in the south flank of the Cheildag Anticline. Remaining details are given in Table 4-I. There are other deposits in the area [25].

Santa Cruz, California, U.S.A. [15]

The asphalt in the Calrock quarry and the Cowell mine near Santa Cruz, (Fig. 4-14), 56 miles south of San Francisco, is present in sandstones of the Miocene Monterey and Vaqueros formations, in beds slightly older than the Edna and Sisquoc deposits. The somewhat discontinuous sandstones, which are occasionally gritty, are generally medium-grained, with subangular to rounded quartz. They are slightly indurated and jointed. Peculiar sandstone dykes cut the Monterey Shale.

The rocks are dipping in the southerly direction at between 3 and 7° and

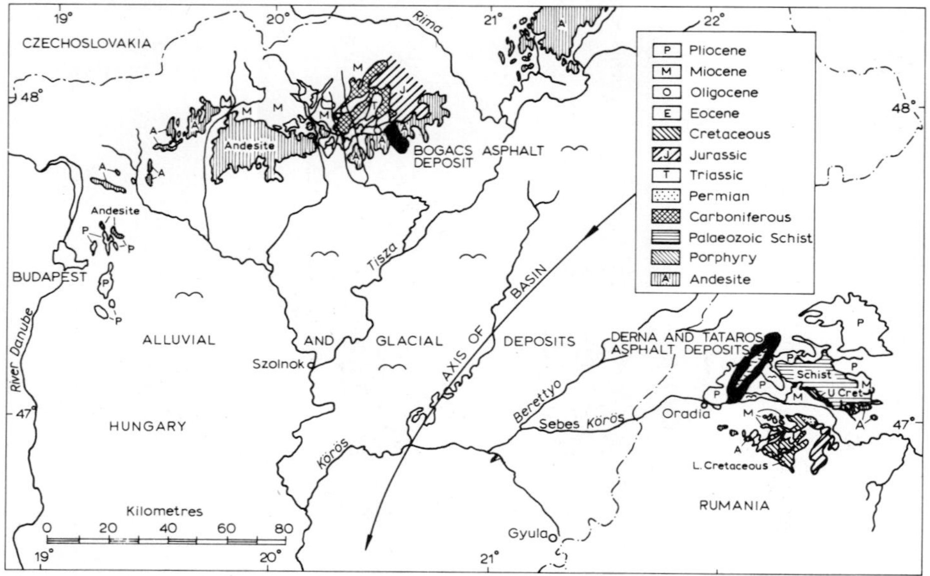

Fig. 4-20. Geological map and location of Derna asphalt deposit, Rumania. (Adapted from 1 : 900,000 Geological Map of Hungary.)

overlie a pre-Cretaceous basement of quartz diorite (Fig. 4-21). Some faulting has been noted.

Geological setting and origin of tar in sands [12]

It is generally considered that the "tar" in tar sands has been derived from a normal oil and there are numerous cases where there are close relationships between tar-sand deposits and oil sands, from which crude oils having different gravities are produced. Some difference of opinion usually occurs in discussing the history of a crude oil up to the time it migrates into a reservoir. The oil may be considered as immature, never "cracked" and, therefore, unlikely to have suffered excessive temperature and overburden pressure; or it may be old, having lost its light ends after sediments were buried to a considerable depth or, possibly, if the oil originated at a considerable depth originally. Some scientists believe that the chemical constituents of the source material, the type of organic matter and environment influence the type of crude oil formed. Many recognize the destructive role of bacteria. In attempts to date the origin of tar-sand oil, evidence from isotope analysis [26] and sulfur content [27] has been utilized. Whereas there are differences of opinion concerning the origin and migration of oil, there is less doubt about the nature of the reservoir rock.

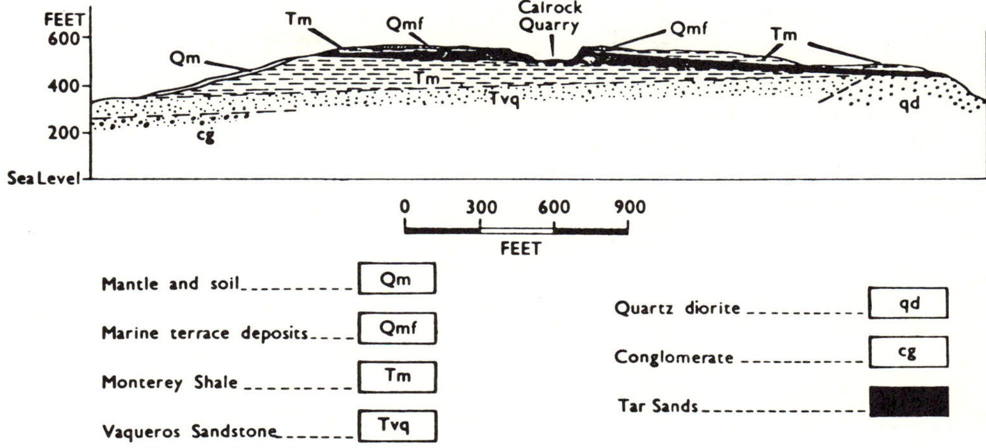

Fig. 4-21. Cross-section through bituminous sandstone deposits at Calrock Quarry near Santa Cruz, California, U.S.A. (After Page et al. [41].)

The geological setting and, in particular, the facies of the reservoir rock of the major tar sands, with few exceptions, are similar. Within the flexible limits of facies terminology, it can be said that most major tar sands are

TABLE 4-II

Mode of entrapment of hydrocarbons in tar sands

Mode of entrapment	Name of deposits, examples
(A) Stratigraphic-trap deposits Structure is of little importance; short distance of migration is assumed.	Sunnyside, Peor Springs, Santa Cruz
Intermediate between (A) and (B)	Athabasca, Edna, Sisquoc, Santa Rosa
(B) Structural—Stratigraphic-trap deposits Folding—faulting and unconformity are equally important.	Oficina—Temblador tar belt, Bemolanga, Asphalt Ridge, Melville Island, Guanoco, and the Kentucky deposits
Intermediate between (B) and (C)	Selenizza, Derna
(C) Structural-trap deposits Structure is very important; long distance migration is assumed; unconformity may be absent.	Whiterock, La Brea

found in rocks that are either nonmarine or not fully marine. They usually occur in estuarine or fresh-water deposits. It has been suggested that bacteria in such sands play an important role in the process of conversion of crude oil into heavy asphalt by polymerization.

Secondly, in the majority of cases, structure in the form of tilting and unconformity seems to control the distribution of tar sands. This would imply some degree of fluidity during migration. This contention also can be supported by the close relationship between a high bitumen content and increase in porosity and permeability, and the frequent presence of an asphalt—water level.

The major tar-sand deposits of the world can be conveniently divided into three groups based on their stratigraphic and structural history. There are some accumulations, however, the mode of entrapment of which falls between the three types presented in Table 4-II.

Stratigraphic-trap deposits

The source of the asphalt in both the Sunnyside and Peor Springs deposits in Utah is considered by Covington [28] to lie in the Eocene Green River Formation. The two accumulations are thought to be stratigraphically controlled, the Green River beds intertonguing with Wasatch Formation deposits. Impregnation at Sunnyside is proportional to the proximity of the sands to the overlying and underlying organically rich shales and limestones. Many sands, highly inspissated with bitumen at the top, grade downwards towards a base that is free of impregnation as if there has been insufficient bitumen to occupy all available pore space.

The Santa Cruz tar sands in California are included in this group, the underlying Monterey Shale being considered as their source. Short distance of migration is assumed for these marine facies. There are no oil fields known to be in close association.

Structural—stratigraphic-trap deposits

There are eight major tar sands in the structural—stratigraphic group, in which entrapment depends about equally on structural and on stratigraphic controls. Of these the largest, but not intensively studied, is the Oficina—Temblador tar belt in eastern Venezuela. It is closely associated with the oil fields of the same name, which have been studied in detail [7]. The geological setting of the oil fields strongly suggests moderate distance of migration within the Oligocene Oficina Formation and that impregnation occurred penecontemporaneously with sedimentation under slight overburden. Towards the rising shield, natural inspissation of the crude oil could be expected. Additionally, any light oil trapped in the fresh-water sands of the Oficina Formation must have been subjected to bacterial action. These conditions appear to be the most favorable for tar accumulation.

The Bemolanga deposits overlying the Malagasy shield occupy a similar structural position to those of the Oficina—Temblador tar belt. No associated oil fields are known to be present and the origin of this tar is not clear. Derivation from the overlying marine Jurassic deposits could be possible, but a source either in the underlying Sakamena Shale or in downdip representatives of the Isalo Group, allowing only moderate distances for migration, is preferred [9].

The tar sands of Asphalt Ridge are perhaps a classic example of the interplay of stratigraphy and structure. The source of the tar is considered to be the Eocene Green River Formation [28]. Migration has occurred updip, probably along unconformities, into the overlying Oligocene Duchesne River Formation and the underlying Rim Rock Member of the Cretaceous Mesaverde Formation (Fig. 4-11). The richest impregnations are associated to some degree with anticlinal noses which are parallel to the Rangely and Ashley Valley oil fields. The closeness of the deposits to the Oligocene—Cretaceous unconformity together with magnitude of porosity and permeability are also important factors. An asphalt—water contact suggests prior mobility of the crude through the reservoir.

Information on the Melville Island tar sand is sparse. According to J.C. Sproule and Associates [6], the seepage probably was derived from the underlying Paleozoic reefs and entered the sands through faults and fractures. Presumably the pre-Triassic unconformity also acted as a conduit.

The Guanoco lake asphalt does not appear to have been derived from the downdip representatives of the Las Piedras Formation or even an older basinward Tertiary facies. The closely associated Guanoco oil field produces heavy oil from shales and fractured argillites of the Upper Cretaceous Guayata Group (Fig. 4-7), and migration from this source along the Tertiary—Cretaceous unconformity is assumed [29].

Finally in this group, the three Kentucky deposits are thought to have received their tar from the Devonian Chattanooga Shale. Normal crude oil is believed to have migrated through the Mississippian deposits into the Pennsylvanian river-channel sands, losing its light ends during pre-Pennsylvanian movements. Proponents of short-distance migration, however, would seek a source in downdip shales above the unconformity.

Structural-trap deposits

Only two major tar-sand accumulations fall into the structural-trap category, i.e., the Whiterocks deposit in Utah and the asphalt lake in La Brea, Trinidad. An in-situ origin for the former (group A) cannot be eliminated. In each case, an argument for vertical migration can be developed.

At Whiterocks, the Eocene Green River shales and limestones probably did not act as source rocks, because there is no evidence of tar impregnation along the sub-Duchesne River unconformity (Fig. 4-13). There is no oil field with which this accumulation can be associated. Covington [28] refers to the Pennsylvanian Weber sandstone as a likely source, which necessitates migration through rather poorly porous and permeable lenses in the Navajo Sandstone host rock. Although there is an asphalt—water contact, the sand grains are coated with tar. There is no variation in saturation with depth, which might indicate in-situ formation.

In Trinidad, the asphalt-lake accumulation is thought to be derived from the argillite of the Upper Cretaceous Naparima Hill Formation, which is of comparable age to the Guayata Group in eastern Venezuela [18]. Regionally in Trinidad, the argillite is known to contain heavy oil. Upon loss of light ends during erosion, after folding in Oligo-Miocene time, the source beds were covered by Miocene clays. Finally, tensional faulting has allowed asphaltic material to migrate and impregnate the overlying Upper Miocene fresh-water sheet sands. There are no known oil fields associated with this deposit.

Athabasca, Edna, Sisquoc and Santa Rosa deposits

Each one of the three Athabasca tar-sand deposits has a slightly different geological setting (Fig. 4-3). The lenticular distribution of the tar in the McMurray—Wabiskaw sands, about which there has been much controversy over problems of origin and migration [2], suggests that stratigraphic entrapment was more important than structural. On the other hand, the Bluesky—Gething tar accumulation appears to have been controlled by structure and pinch-out, whereas the Grand Rapids deposit owes its origin to stratigraphic entrapment alone.

The McMurray—Wabiskaw and Bluesky—Gething tars may be:

(1) Lower Cretaceous in origin, from McMurray—Wabiskaw and Bluesky—Gething equivalents or from the overlying Clearwater Shale, thus necessitating only a short-distance migration.

(2) Pre-Cretaceous in origin, which involves long-distance migration via unconformities.

(3) Formed in-situ from reworking of Cretaceous oil or from humic acids derived from a Lower Cretaceous flora.

Generation during the Lower Cretaceous is currently favored based on the results of various analyses (isotope ratios, sulfur content, and distillation range), in which case these deposits can be regarded as immature oil fields that have never been deeply buried. Migration from the overlying Clearwater Shale into the tar sands is reminiscent of the geological setting of the Asphalt Ridge deposit, but in the latter, downward movement of fluids was presumably enhanced by unconformity. The Clearwater Shale conformably overlies both McMurray—Wabiskaw and Bluesky—Gething tar sands. The origin of the younger Grand Rapids tar is presumed to be either local or a result of migration from the underlying Clearwater Shale.

The tar sands of the Edna district (Figs. 4-14, 4-15) are considered to be derived from the underlying highly organic and petroliferous Monterey Shale and the contemporaneous Pismo Formation [16]. Migration was interrupted by a period of unconformity during which the light ends were lost. This accumulation has a strong stratigraphic aspect and the influence of structure is slight. At Sisquoc, the geological setting is much the same. No oil fields are associated with these deposits.

At Santa Rosa, the source of the tar is thought to be in the Permian San Andres Limestone, from which the oil migrated vertically through sink holes in a karstline surface into the Santa Rosa Formation [15]. Some communication could have been obtained along the pre-Upper Triassic disconformity. No oil field is known to be associated with this tar-sand deposit.

Selenizza (Albania) and Derna (Rumania) deposits

Stratigraphic control of the occurrence of the Selenizza tar sands of Rumania is still important, but the vertical distribution of the tar is also influenced by folding and faulting (Fig. 4-12). The tar accumulation in the Astian deposits occurs in the highest part of the structure, being trapped against the downthrown side of numerous tensional cross faults.

Cretaceous—Eocene limestones are thought by some to be the source rocks for the tar [30]. Post-Eocene folding exposed the reservoirs and migration of the released oil continued throughout the Miocene. Later faulting allowed further impregnation of the Astian sands. The Patos oil field, with its own tar sands, is associated with the Selenizza deposits. Some of the Miocene bitumen may be indigenous [31].

The entrapment of tars at Derna is also partly stratigraphic and partly structural. Probably, the Eocene—Oligocene marls and bituminous limestones acted as source rocks, and migration is thought to have occurred after Miocene folding had produced unconformities [32]. During this period and subsequently after entrapment, the oil was exposed to evaporation, oxidation, and polymerization. Some authorities have favored an in-situ development for this oil in the Mio-Pliocene sands. No oil fields are known to be associated with the Derna deposit.

Other world occurrences of tar sands

There is very little information available to the writers on Soviet tar-sand deposits, of which some are shown on Fig. 4-1. Perhaps the most interesting deposit occurs in the Olenek anticline in the northeast of Siberia, where it is claimed that the extent of asphalt impregnation in the Permian sandstones is of the same order of magnitude (areally and volumetrically) as that of the Athabasca tar sands [33]. Permian asphalt is also reported from sands near

Subovka, U.S.S.R. The mines of the Notanebi deposit occur in Miocene sandstones, which contain 20% asphalt by weight [34]. The Kazakhstan occurrence, near the Shubar—Kuduk oil field, is an asphalt lake in which the bitumen content is estimated to be equal to 95% [35].

The tar sands at Burgan in Kuwait and at the Inciarte and Bolivar Coastal Fields of the Maracaibo Basin are of unknown dimensions. Those at Inciarte have been exploited [22]. In all three, the influence of structure is strong and all are directly or closely associated with large oil fields. The tar sands of the Bolivar Coastal Fields lie unconformably above the oil zones in Miocene beds of brackish-water facies, a lithological environment similar to that of the Oficina—Temblador tar belt.

The small Miocene asphalt deposits in the Philippines on Leyte Island [36] are extreme samples of stratigraphic entrapment in a marine facies and are akin to some of the Californian occurrences. Those of the Mefang Basin in Thailand [37] are in Pliocene lacustrine beds overlying the Triassic deposits unconformably, and their distribution is both stratigraphically and structurally controlled.

Finally, the small accumulation at Chumpi, near Lima, Peru, which occurs in tuffaceous sands, is interesting in that tar is believed to be derived from strongly deformed Cretaceous limestones from which oil was distilled as a result of recent volcanic activity [38].

Stage of commercial development of tar sands

Only four of the major tar-sand deposits discussed, i.e., La Brea, Selenizza, Derna, and Cheildag, are currently being exploited. At one time, the Guanoco deposit operated by the New York and Bermudez Co. was a serious competitor to La Brea. There are no figures available for the Derna and Cheildag operations. Total production for La Brea and export figures for Selenizza tar-sand deposits [39] are as follows (values in tons):

	1961	1962	1963	1964
La Brea, Trinidad	175,326	162,613	169,308	189,246
Selenizza, Albania	27,656	11,220	11,220	not available

The first commercial development of the Athabasca tar sands was started by Great Canadian Oilsands, Ltd. in September 1967. The tar sands were mined by open cast method and the extracted crude oil was pumped to Edmonton at the rate of 45,000 bbl/day. The Oficina—Temblador tar-belt area, the Bemolanga sands, the Utah and some of the Californian deposits have all attracted the interest of oil companies at various times. Pilot schemes, involving steam or hot-water injection or in-situ cracking by fireflooding, have been tried with some technical success, but in the face of competition from conventional crudes their application is as yet uneconomic.

Acknowledgment

The authors thank the Chairman and Directors of The British Petroleum Company, Ltd., London, for permission to publish this chapter.

References

1 United Nations Report by Secretary-General, UN E/4132, *Development of Non-Agricultural Resources* (Jan. 8, 1966).

2 C.E.B. Conybeare, "Origin of Athabasca Oil Sands: A Review", *Bull. Can. Pet. Geol.*, *14*, 145—163 (1966).

3 Alberta Oil and Gas Conservation Board, *A Description and Reserve Estimate of the Oil Sands of Alberta* (1963).

4 M.A. Carrigy (Editor), *The K.A. Clarke Volume — A Collection of Papers on the Athabasca Oil Sands*, Research Council of Alberta, Edmonton, Alta. (1963).

5 J.C. Sproule, *Can. Oil Gas Ind.*, *16(9)*, 1—2 (May, 1963).

6 J.C. Sproule and G.V. Lloyd, in: M.A. Carrigy (Editor), *The K.A. Clarke Volume — A Collection of Papers on the Athabasca Oil Sands*, Research Council of Alberta, Edmonton, Alta., pp. 27—29 (1963).

7 H.H. Renz, H. Alberding, K.F. Dallmus, J.M. Patterson, R.H. Robie, N.E. Weisbord and J. MasVall, "The Eastern Venezuelan Basin", in: L.G. Weeks (Editor), "Habitat of Oil": *A Symposium Am. Assoc. Pet. Geol., Spec. Publ.*, pp. 551—600 (1958).

8 Anonymous, "Shell Eyeing Venezuela's Tar Belt", *Oil Gas J.*, *62(28)*, 64—66 (July 13, 1964).

9 P.E. Kent, British Petroleum Co., Ltd., unpublished report (1954).

10 R.E. Covington, "Some Possible Applications of Thermal Recovery in Utah", *J. Pet. Technol.*, *17*, 1277—1284 (1965).

11 P.T. Walton, "Geology of the Cretaceous of the Uinta Basin, Utah", *Bull. Geol. Soc. Am.*, *55*, 91—130 (1944).

12 C.N. Holmes and B.M. Page, "Geology of the Bituminous Sandstone Deposits near Sunnyside, Carbon County, Utah", *Intermt. Assoc. Pet. Geol., Guideb.*, pp. 171—177 (1956).

13 G. Ineichen, "Petrole Gaz Naturelset Asphaltes du Geosynclinal Adriatique", *Proc. World Pet. Congr.*, *3*, The Hague, Sect. 1, 199 (1951).

14 L. Maddalena and S. Zuber, *Riv. Ital. Pet.*, *52*, 51—75 (1937).

15 Ball Assoc., Ltd. (Comp.), Oklahoma City, Interstate Oil Compact Commission, "Surface and Shallow Oil-Impregnated Rocks and Shallow Oil Fields in the United States", *U.S. Bur. Mines Monogr.*, *12*, 375 pp. (1965).

16 E.W. Adams and W.B. Beatty, *Miner. Inf. Serv. Calif.*, *15(4)* (1962).

17 Creole Petroleum Corp., "Creole Tests its Guanaco Concession", *Oil Gas J.*, *49(33)*, 238—241 (1950).

18 H.H. Suter, *The General and Economic Geology of Trinidad*, B.W.I., H.M.S.O., London, 2nd ed. (1960).

19 H.G. Kugler, *Geological Map and Section of Pitch Lake-Morne l'Enfer Area*, Petrol. Assoc., Trinidad (1961).

20 R.L. Bates, *Bull. N. M. Sch. Mines*, *18*, 307—308 (1942).

21 J.M. Weller, *Kentucky Geol. Surv., Ser. 6, 28,* 199—215 (1927).
22 H. Abraham, *Asphalts and Allied Substances, 1,* Van Nostrand, Princeton, N.J., 6th ed., pp. 194—195 (1960).
23 I.M. Goubkin, "Tectonics of Southeastern Caucasus and its Relation to the Productive Oils Fields", *Bull. Am. Assoc. Pet. Geol., 18,* 603—671 (1934).
24 A. Mirtsching, *Erdol- und Gaslagerstatten der Sowjetunion,* Ferdinand Enke, Stuttgart, 150 pp. (1965).
25 S.G. Salayev, *Dokl. Akad. Nauk Azerb. SSR, 13(2),* 157—161 (1957).
26 D.H. Welte, "Relation Between Petroleum and Source Rock", *Bull. Am. Assoc. Pet. Geol., 49,* 2246—2268 (1965).
27 D.H. Welte, "Relation between Petroleum and Source Rock", *Bull. Am. Assoc. Pet. Geol., 49,* p. 2260 (1965).
28 R.E. Covington, *Bull. Utah Geol. Miner. Surv., 54,* 225—247 (1963).
29 J. Dofour, "General Oil-Geological Problems in Venezuela", *Geol. Rundsch., 45(3),* 759—775 (1957).
30 A. Mayer-Gurr, Unpublished Report (1947).
31 Azienda Italiana Petroli Albania, *Proc. World Pet. Congr., 1,* London, Sect. A, 61, discussion (1933).
32 L. De Loczy, unpublished report (1947).
33 T.N. Kopylova, *Trd. Nauchno-Issled. Inst. Geol. Arkt., 121,* 103—109 (1962); Abstr. by A. Mirtsching, *Erdol Kohle, 17,* p. 475 (1964).
34 H. Abraham, *Asphalts and Allied Substances, 1,* Van Nostrand, Princeton, N.J., 6th ed., p. 208 (1960).
35 Anonymous, *World Pet., 30(3),* p. 110 (1959).
36 D.N. Palacio, "Preliminary Report on the Geology and Röck Asphalt Deposits of Balite, Villaba, Leyette", *11,* 69—110 (1957).
37 W. Whitley, *Oil Gas Int., 5(5),* p. 40 (1965).
38 J.E. Rassmus, "Geological Features of Peru" *Pet. Eng., 30(2),* B3-B31 (April, 1958).
39 *Statist. Summ. Miner. Ind. Colon. Geol. Surv.,* 250, 264 (1959—1964).
40 M.L. Krueger, "Arroyo Grande (Edna) Oil Field (Calif.)", *Calif. Dep. Nat. Resour. Div. Mines Bull., 118,* 450—452 (1943).
41 B.M. Page, M.D. Williams, E.L. Henrickson, C.N. Holmes and W.J. Mapel, "Geology of the Bituminous Sandstone Deposits near Santa Cruz, Santa Cruz County, California", *U.S. Geol. Surv. Oil and Gas Invest. Prelim. Map. No. 27,* scale approx. 1: 3,600 (1945).
42 K.F. Dallmus, "The Geology and Oil Accumulations of the Eastern Venezuelan Basin", *Geol. Soc. Dig.,* Tulsa. Okla., 825—1055 (1963).
43 A. Salvador and E.E. Hotz, "Petroleum Occurrence in the Cretaceous Venezuela", *Proc. 6th World Pet. Congr., 1,* 115—140 (1963).

Chapter 5

ATHABASCA TAR SANDS: OCCURRENCE AND COMMERCIAL PROJECTS

F.K. SPRAGINS

Introduction

The growing energy crisis has focused attention on every known significant source of fossil fuel and has spurred exploratory activity in favorable petroliferous and carboniferous areas on land and beneath the sea throughout the world, including the Athabasca tar-sand deposit of northeastern Alberta, Canada. This deposit, one of major proportions, is now receiving worldwide attention assuring that it will play a leading role in supplying part of the energy requirements of the future.

So far, efforts towards commercial development of the Athabasca tar sands have not been easy and, even though one plant is now operating and another under construction, the future holds little promise of being any easier. As major programs are initiated, new concerns of the politician, the environmentalist and the engineering scientist, coupled with a growing shortage of men and materials, are bound to create major new challenges. In spite of this, however, there is now an even greater determination to bring these sands into successful large-scale production at the earliest possible moment.

Occurrence

The Athabasca tar sands, located in northeastern Alberta, Canada (Fig. 5-1), are the major deposit of oil in a viscous-oil trend stretching across Alberta (Fig. 5-2), from the Alberta—Saskatchewan border in the vicinity of Cold Lake to a point near the town of Peace River. The in-place resources along this trend total almost 900 billion bbl of oil [1]. In today's energy crisis of worldwide proportions any significant accumulation of oil is of interest and cannot be overlooked. In this regard the Athabasca, Cold Lake, and Peace River deposits, all in the central Alberta heavy-oil trend, are receiving detailed attention; however, it is the intention, in this chapter, to deal only in detail with the Athabasca tar sands and the commercial development of this deposit.

As depicted in Fig. 5-3, the Athabasca tar sands are divided into two zones: one where oil deposits are covered by oil-barren surface deposits of from 0 to 150 ft, and one where these overburden deposits range from

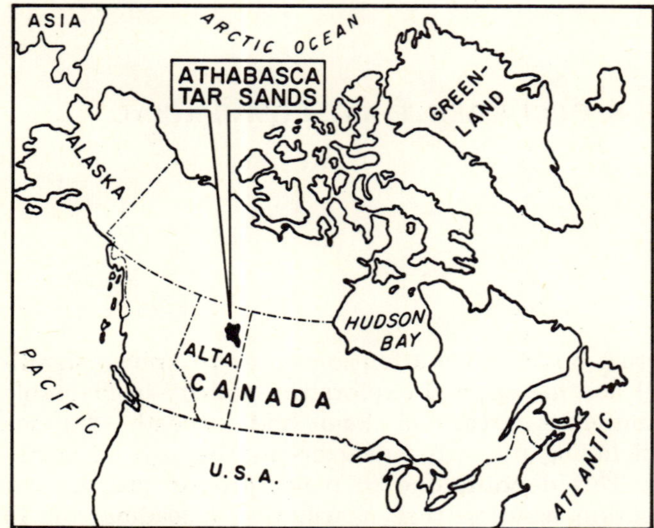

Fig. 5-1. Location of Athabasca tar sands.

Fig. 5-2. Location and extent of major heavy-oil deposits of Alberta, Canada.

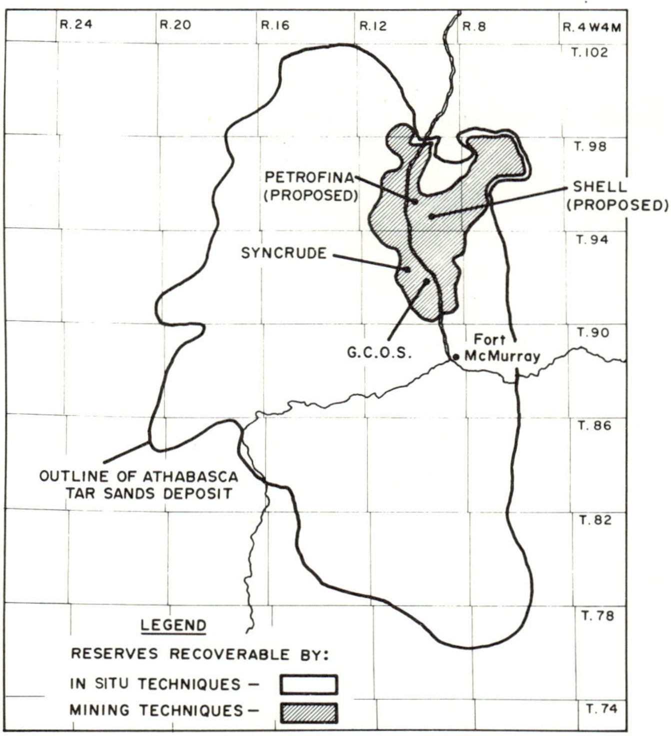

Fig. 5-3. Comparison of mining and in-situ areas of the Athabasca tar sands.

150 ft to more than 2000 ft in thickness. These two zones cover a combined area of approximately 5.75 million acres [2] or about 9000 square miles having approximately 74 and 552 billion bbl of oil in place, respectively. Remaining recoverable synthetic crude oil amounts to about 26.5 billion bbl [1].

A generalized cross-section of the Athabasca tar sands [3], passing roughly across the deposit from southwest to northeast just north of the town of Fort McMurray is illustrated in Fig. 5-4. The oil-bearing Cretaceous sands, overlain immediately above by the Clearwater Formation and underlain by the Devonian unconformity, account for an overwhelming percentage of the oil in place. There are no known occurrences of oil in the Devonian in the Athabasca tar-sand area and only small localized accumulations have been noted in the Clearwater sandstones. The oil-bearing McMurray deposits can be divided into three separate zones: upper, middle, and lower. The upper zone is characterized by fine-grained sandstone impregnated with oil in localized areas. The predominant cementing material is siderite. The middle zone is made up of medium-grained, uniformly oil-impregnated quartz sands, mostly of Cambrian age, and lenticular beds of barren siltstones and shales.

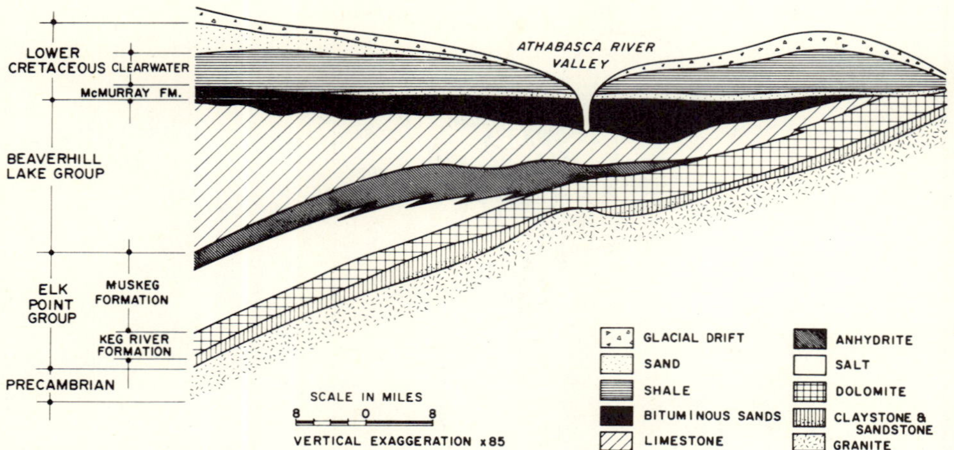

Fig. 5-4. Diagrammatic cross-section of the Athabasca tar sands.

Except in the barren lenses, oil itself is the cementing material in the middle zone. In the third or lowest zone, beds are characterized by coarse-grained to conglomeritic sandstone which is locally richly impregnated with oil. The cementing material in this zone rapidly changes according to local conditions and varies from oil to siderite to marcasite [4]. In all cases a varying amount of formation water occurs along with the oil, varying from 1 to 30%.

The sand grains where the oil occurs are water wet, which is the characteristic that makes the hot-water separation process efficient. Porosities are high, often about 35% [5]; however, fluid conductivities, due to the extremely heavy oil and clays when present, are abnormally low. The sulfur content of Athabasca tar-sand oil is also high — in the range of 4—6% — and contains other complicating components such as significant amounts of nickel—vanadium porphyrins. The array of poor characteristics is not encouraging but this is at least, to some extent, offset by favorable cracking characteristics.

Viscosity of unaltered bitumen (the raw oil) from the Athabasca tar sands is high and has been shown in general to be much greater in the southern part of the area than in the north. Thus, at the Abasand quarry, samples have shown the oil bitumen to have a viscosity of 600,000 P at 50° F, whereas the viscosity of bitumen in the Ells River and Bitumount district is only about 6000—9000 P [6].

The degree of saturation of the oil in the formation is generally measured in weight percent and varies from 0 to as high as about 18% [7] or slightly more. The nature of the tar sands varies from location to location; however, as a specific example taken from Lease 17, one cubic yard of tar sands at 14.3% saturation weighs 1.4130 tons and will produce 0.8784 bbl of oil [8].

Authorities disagree on the saturation at which commercial development can proceed. G.C.O.S. placed the minimum figure at 10% [9], Syncrude at 6% [10] and the Energy Resources Conservation Board uses 5% [11] as its cut-off point for calculating recoverable mineable reserves. There is a great deal of variation both vertically and horizontally in the percent saturation of the oil-bearing zones. The combined oil-bearing zones average about 150 ft in thickness, although in local areas they have been reported to be well over 200 ft.

As examples of local areas considered commercially developable by mining techniques, the G.C.O.S. lease now under production has an average of about 150 ft of tar sand with an average saturation of between 11 and 12%, and an overburden thickness of approximately 53 ft [12]. Reserves are estimated at a quantity to last for at least twenty years at the now allowed production rate of 65,000 bbl/day. The Syncrude lease meets the same general conditions except that it is larger in area and contains enough outlined reserves to last for over 50 years at rated production [13]. The average life of a commercial project site is usually considered to be a minimum of twenty years. Although detailed exploration is still far from being complete, many such sites are known to exist in the mineable area of the Athabasca tar sands.

The parameters of commercial in-situ production sites have not yet been discussed in the literature. In general, however, they should follow about the same conditions in relation to the areal extent, saturation and thickness of beds.

History of commercial development

Fur trader Peter Pond must have been puzzled by the curious black formations outcropping along the banks of the Clearwater River as he paddled his canoe westward in 1778. He obviously liked the area because he set up a trading post at a site which he called "Fort of the Forks" at the confluence of the Athabasca and Clearwater rivers. This site was later to become the town of Fort McMurray [14].

History does not record many of the details of Peter Pond's stay in the area or his reaction to things that were happening in the country around his post. Ten years later in 1788, however, Alexander Mackenzie, one of the greatest explorers of the Canadian northwest, came to relieve Peter Pond. Mackenzie, unlike Pond, kept detailed journals of all that he saw and did. It was from this source that we learned of Canada's first oilmen.

It seems that the Indians then living in the Athabasca area would gather bitumen at the bottom of the tar-sand outcrops, mix the gooey mass with melted resin from spruce trees, and use the resulting material to waterproof the seams in their birch-bark canoes. Mackenzie, being an explorer at heart, was quick to recognize the superior quality of the material used by the

Indians on their canoes. For a sufficient amount of whiskey, he was able to obtain enough of the caulking material for his own canoes. The thing he did not realize at this time was that the material was free in unlimited quantities virtually at the back door of his cabin. In one of his journals, Mackenzie later made reference to the tar-sand outcrops which exist along the banks of the Athabasca and Clearwater rivers for over 100 miles. He described them as "fountains of bitumen".

Probably the first significant technical reference made to the tar sands were those included in reports based on work initially commissioned by the Geological Survey of Canada in 1875 and carried out by the Geological Society of Canada. The first results were reported in 1884 by Dr. Robert Bell and additional results were reported in 1891 by R.G. McConnell. Whether or not it was on the strength of these reports that an exploratory well was drilled in the district by the Geological Survey in 1897 is not known. Needless to say the well was not a "producer", but undoubtedly encountered a thick oil section and, therefore, was at least a "technical success".

Over the years there has been a continuous interest in commercial production from the tar sands. There were those like Count Alfred von Hammerstein [15] who heard of the tar sands on his way to the Klondike and turned his attention to the lure of "black gold". By this time, the deposit had been fairly accurately described and he felt that somewhere within the tar-sand layers there must be pools of oil. Before losing faith in his scheme, von Hammerstein drilled three holes about six miles from Fort McMurray, near Poplar Island. Failure, however, eventually caused him to abandon his short-cut scheme to riches.

Sidney C. Ells [16], a young engineer from the Mines Branch in Ottawa, led a topographic survey party into the Athabasca area. When he and his crew came out of the bush in the fall of 1913 they brought with them a 30-ft scow loaded with nine tons of Athabasca oil samples. The scow was literally dragged up the Athabasca River from Fort McMurray for 240 miles to a point nearest Edmonton — still 100 miles away. The following year, Ells, still intensely interested in the tar sands, returned to the Fort McMurray area and with the aid of 23 teams of horses and sleds brought out 60 tons of tar sand (contained in sacks) over ice and snow in the dead of winter.

In 1915, Ells used some of his valuable material to pave a section of an Edmonton street. This section was still reported to be in good repair and in use in 1950, some 35 years later. Mr. Ells was also responsible for paving some of the sections of road in the Jasper National Park in 1927. These efforts were also successful and reports indicate the sections lasted for many years.

Sidney Ells' efforts related to the tar sands were not confined to building roads. His interest was far deeper. In 1915 he carried out experimental tests at the Mellon Institute in Pittsburg, Pennsylvania, in an effort to determine how the oil might be separated from the sand. From this work he concluded that flotation cells provided the greatest potential for successful separation. Although there have been modifications, this is still the approach being pursued today in the commercial extraction of oil from the Athabasca tar sands.

Another early experimenter in the tar-sand oil recovery was Dr. Karl A. Clark who worked for many years with the Research Council of Alberta and is credited with working out the hot-water extraction process. Dr. Clark's dedication to the tar sands is particularly noteworthy. He had great dreams for the development of this significant Canadian resource and a greater part of his life was spent in solving the technical problems related to its recovery. Perhaps, more than his technical efforts, his greatest personal contribution to Canada was keeping interest in commercial development of the Athabasca tar sands alive until major companies were ready to take up the task. He did so as a professor at the University of Alberta; as a member of the working staff of the Research Council; as a tireless worker in raising funds for tar-sand research projects, both in the laboratory and in the field; and as a publicizer of a cause to which he was fully committed and for which he will, for generations to come, command the respect of those who cross his long and difficult path.

The principle employed by Dr. Clark was a simple one. The tar sand was slurried with hot water (and occasionally chemicals, usually in the form of caustics, according to the nature of the tar sand), introducing this mixture into greater quantities of hot water, removing sand at the bottom and skimming off the oil froth at the top.

Following some initial work that had been undertaken by Alcan Oil Company, it was an easterner from the Maritimes, R.C. Fitzsimmons, who was one of the first to try to put the hot-water process into commercial use. Fitzsimmons acquired the leases at Bitumount, about 50 miles downstream from Fort McMurray, in 1923 and with very limited funds proceeded to build his version of an extraction plant. His product, while not the best, was used in western Canada for several years as a waterproofing material for roofs.

There were a number of name changes and reorganizations but the company which came to be known as Great Canadian Oil Sands and ultimately became a subsidiary of Sun Oil Company probably had its beginning in the Fitzsimmon's undertaking [17].

In 1930, Max W. Ball [18], who was destined to become a Director of the U.S. Department of the Interior, entered the tar-sand picture. Under his guidance Canadian Northern Oil Sands Products, Ltd. was formed. Later, this company became Abasand Oils, Ltd. Over a period of fifteen years and with some government backing the company spent over $ 2 million in its efforts to produce commercial oil.

Abasand [19] was off to a good start in 1936 when a 250-ton/day plant was built on the banks of the Horse River southwest of Fort McMurray. On achieving encouraging results, the plant was increased in size the following year to process 400 tons/day of tar sand and in the summer of 1941 a peak production of 1700 bbl/day of bitumen was achieved. But fire struck late in the fall of that year and the plant burned to the ground. Although the plant was rebuilt, the subsequent operation faltered because of financial difficulties. In 1943, the federal government stepped in and took over the operation, first establishing a 100-ton/day pilot plant and two years later expanding it to 500 tons/day. This plant was also doomed to failure. It burned before ever producing any oil.

In its formative years, Abasand acquired several pieces of choice land including the acreage now being mined by G.C.O.S. Canadian Industrial Gas and Oil purchased Abasand along with its land holdings and when G.C.O.S. started commercial production in 1967 on land in which Canadian Industrial Gas and Oil held an interest, certain royalties started to flow back to the original Abasand stockholders. This was the first revenue Abasand stockholders had received in almost 40 years of effort [20].

In spite of the discovery of major quantities of conventional oil at Leduc in 1947 and the wane of interest in the tar sands, the Research Council of Alberta, under the direction of Dr. Clark carried out major tar-sand tests at Bitumount, from 1948 to 1950, to establish once and for all the commercial feasibility of the hot-water extraction process. Not long after these tests, Royalite Oil Company (later to become the Gulf Oil Canada interest) became interested in the tar sands, took over the Bitumount operation, and eventually attracted Cities Service Company to the venture. (The Cities Service interests in the Athabasca tar sands were first to be administered by Cities Service Athabasca, Inc. and later by Canada-Cities Service, Ltd.) In turn Cities Service attracted Richfield Oil Corporation (later to become Atlantic Richfield, whose interests were to be looked after by Atlantic Richfield Canada Ltd.). Finally the group offered a participating interest to Imperial Oil Limited which was accepted. This consortium was complete by 1959 and constitutes the group now represented by Syncrude Canada Ltd. in the Syncrude Mildred Lake Project. Interest in the project was held on the following basis: Atlantic Richfield Canada Ltd. — 30%; Canada-Cities Service, Ltd. — 30%; Gulf Oil Canada Limited — 10%; and Imperial Oil Limited — 30%. *

Late in the 1950's both Shell Oil Company of Canada and Pan American Petroleum Corporation (now Amoco) initiated in-situ pilot operations in the Athabasca tar sands: Shell near Bitumount about 50 miles north of Fort McMurray and Pan American on the south shore of Gregoire Lake about 35 miles south of Fort McMurray. Both companies enjoyed a reasonable degree of success and both made application (Pan American in the name of

* Arco dropped out, whereas the Province of Alberta, the Province of Ontario, and the Federal Government of Canada joined the project. See addendum.

Muskeg Oil Company) to the Alberta Oil and Gas Conservation Board for permission to build commercial plants; however, both companies withdrew their application when the policies respecting development of the tar sands became discouraging in the mid-1960's.

In its early in-situ work in the Athabasca area Shell followed a chemical—thermal-injection approach in establishing communication within a pattern of injection and production holes. The tar-sand beds were of the order of 200 ft in depth at the point of Shell's experiments, although the saturated zones in the area covered by its commercial application ranged from 500 to 1500 ft below the surface. Shell later carried out extensive experiments in the Peace River heavy-oil deposits where oil-bearing beds are in the order of 2000 ft or more in depth. Though inactive for a period, it is now reported that Shell is reactivating its in-situ field pilot work in the Peace River deposit.

Amoco has concentrated all of its in-situ pilot work in the Gregoire Lake area of the Athabasca tar sands. This company has been more or less continuously active for at least fifteen years, and is still pursuing this approach to heavy-oil production. In its efforts, Amoco has tried many approaches and combinations of approaches including air injection, steam injection and fire flood.

Although not specifically a part of the Athabasca tar-sand development, it is significant to note at this point that Imperial Oil Limited started thermal-stimulation experiments in the Cold Lake deposit in 1964. To date, in the neighborhood of $ 17 million has been spent in this test. Within the past two years a field pilot operation comprised of 23 wells has been producing as much as 1500 bbl/day of crude oil. In 1973, Imperial announced plans to triple the size of the pilot — a step it feels is one step nearer to commercial production.

In the Imperial Cold Lake tests a "huff and puff" approach is employed. Steam at 600° F is forced into the oil zone for about a month at a pressure of 1600 psi (pounds per square inch). The oil is heated, reducing its viscosity so that it can be pumped to the surface. The pumping part of the cycle lasts about three months [21].

In any in-situ approach, whether it be by thermal stimulation or by chemical injection or a combination of the two, a lack of permeability of the formation and the extremely viscous nature of the oil combine to form extremely difficult conditions. Any movement of the oil from a warm zone into a cooler zone only serves to complicate the situation by further blocking movement in the oil horizon. One solution might be to create horizontal fractures between injection and production wells and clean this zone with caustic followed by hot water and steam. This is basically the approach followed by Shell in its in-situ pilot work near Bitumount *.

* See the following patents (editorial comment):
(1) T.M. Doscher and J. Reisberg, applied 11/24/59, issued 2/26/62, *Canadian Patent 639,050.*
(2) T.M. Doscher, applied 1/16/63, issued 6/15/65, *Canadian Patent 711,556.*

Another solution is, of course, the direct application of heat as employed by Imperial in its "huff and puff" aproach, where low permeability slows but does not stop the heating process.

At about the same time that Shell and Amoco were getting underway, an in-situ consortium was formed by Richfield Oil Corporation (later Atlantic Richfield), Cities Service Company (under the name of Cities Service Athabasca, Inc.) and Imperial Oil Limited. This group has carried on tar-sand formation studies, thermal-injection tests and, at one time, devoted a great deal of technical effort towards exploring the feasibility of the detonation of a nuclear device [22] to create an energy source to heat the tar sands to a point where production could be accomplished by conventional means. The proposed nuclear approach, after considerable effort, won acceptance in Alberta but was shelved and never acted upon at the federal level. The technical merits of this test were considered at both the provincial and federal levels. The economic prospects, however, appear to be doubtful due to the lack of a sufficient amount of overburden in the tar-sand area to contain a device of optimum proportions. The proposed test, planned to be located in the Chard area near the southern boundary of the tar-sand deposit, was designed around a 9-kiloton device with some speculation that it would have been possible to go as high as a 20-kiloton device. Actually, on a straight Btu economic calculation, it was indicated that a device of proper size should have been in the megaton range. There is no place in the Athabasca tar sand or in any other Alberta heavy-oil trend deposit with enough overburden to contain such an explosive. To have gone deeper and created a "heat sink" at greater depths would have involved drilling implacement holes in the Precambrian, an extremely difficult task in itself. But in this possible approach, one of the main concerns was the creation of shock waves of earthquake magnitude for a considerable distance in the surrounding area. Radioactive contamination was not considered a major problem in either a shallow or a deep detonation.

In spite of continued laboratory efforts and the activities of inventors and would-be inventors throughout the 1950's, it was not until the 1960's that major sums of money and effort were thrown into the fight to develop significant commercial tar-sand production projects. By 1962, three applications were pending before the Alberta Oil and Gas Conservation Board: one by G.C.O.S. for 31,500 bbl/day of synthetic crude, one by Cities Service Athabasca, Inc. (the operator for the Cities Service—Richfield—Gulf—Imperial consortium before the formation of Syncrude) for 100,000 bbl/day, and one by Shell for 100,000 bbl/day.

In these applications all parties had come to realize that size was an important consideration in meeting the basic requirements for an economically feasible operation. The general attitude of the applicants was that before major production from the tar sands could be established, conventional production in western Canada would begin to level out and additional supplies would be needed to meet growing demand. On the other hand, however, competitors not interested in tar-sand development and government, in a protective attitude for the conventional industry, took an opposing view. As a result, policies were established which imposed serious limitation on tar-sand development. (See Appendix I and II, Policy and Amended Policy of the Alberta Government relating to Development of the Athabasca Tar Sands.)

Of the three pending applications, the one sponsored by G.C.O.S. was heard initially in June, 1960, but called for rehearing in 1962 on technical grounds; however, it was finally approved in 1963 for a throughput of 31,500 bbl/day. Sun Oil Co. gained control of the project and reapplied to the Oil and Gas Conservation Board (later to become the Alberta Energy Resources Conservation Board) for an increase in the allowed throughput from 31,500 to 45,000 bbl/day on the grounds that the lower throughput was uneconomic. This was the same argument that had been used by the four-company consortium and Shell in their respective applications when they argued that the minimum-sized plant should not have a throughput less than 100,000 bbl/day. However, Sun advised the Board that conditions on their lease coupled with their technical approach did warrant a plant at a 45,000-bbl level. This application was approved and the G.C.O.S. plant with controlling interest owned by the Sun Oil Co. was brought on stream in September 1967.

In the meantime the Conservation Board advised the four-company group and Shell that their applications were not being turned down but would be held in abeyance for five years after which time both the four-company group and Shell could reapply. Shell elected to withdraw its interest in tar-sand development, but the four-company group reoriented its activities. A separate company was formed, Syncrude Canada Ltd., to act as operator for the four companies. This company assumed its responsibilities in January 1965. The field pilot operations were terminated and a research and development center was established in Edmonton where a detailed technical reexamination of the tar sands was initiated.

The four-company group, represented by Syncrude, reapproached the Board at the appointed time, with an application to produce 80,000 bbl/day of synthetic crude with initial production to commence in late 1973 or early 1974. This application was heard by the Board in August 1968; however, the Board failed to reach a favorable agreement regarding the application because U.S. markets, which might have been supplied by Athabasca oil, were being eyed by hopeful Alaskan Prudhoe Bay producers. The official tar-sand policy at that time stated, in effect, that synthetic oil from the Athabasca tar sands would supplement and not replace Canadian conventional crude oil. This meant that any future tar-sand producer would have to look for markets not supplied by Canadian conventional oil which, in turn, meant that it was necessary to seek markets outside of North America or, if in the United States, beyond Chicago or beyond the extreme U.S. northwest. Even though the four companies were prepared to proceed in the face of these market restraints, the Board was still concerned that Prudhoe Bay production might preempt intended tar-sand markets which, in turn, would affect markets for conventional Canadian production. In its decision, the Board simply stated that it needed more time to study the effects of the Prudhoe Bay discovery.

Syncrude appealed to the provincial cabinet on the grounds that the Board's plan to study the effects of Prudhoe Bay production on Canadian markets was unrealistic and won the right to have its application reheard. Therefore, in March 1969 Syncrude filed another application which was heard by the Oil and Gas Conservation Board on May 26 and 27, 1969. This time the applicants made every effort to lessen the Board's concern over the early impact of Athabasca production on conventional production. The size of the plant was again set at 80,000 bbl/day of synthetic crude with initial production scheduled for July 1976. This application was recommended by the Board to the Lieutenant Governor in Council for approval on September 30, 1969; however, in the months to follow, increasing labor costs and further resolution of technical problems resulted in a deterioration of the estimated rate of return of the project making it necessary to approach the Board yet another time. In order to get its return on investment back up to an acceptable level, it was necessary to establish the production level at 125,000 bbl/day. The dat for initiation of production was not altered. The application, dated August 1971, was successful with Board recommendation granted as Approval No. 1725, and final approval was confirmed by Order-in-Council No. 244/72 dated February 23, 1972.

The Order-in-Council triggered a concerted effort to complete conceptual engineering and to establish the best possible definitive estimate. During these studies, which were to last well over a year, it became apparent that a shortage of skilled labor would be created in Alberta by construction of the Syncrude plant along with other major construction projects planned for the area. During the 1971 hearing, a construction schedule and completion date had been committed, and with a labor shortage looming, it became increasingly apparent that the project could not be carried out as planned and an application was heard by the Board (renamed by this time The Energy Resources Conservation Board) in April 1973 for an extension of construction time of one year.

Also, during the conceptual engineering stage, it was determined that the use of hydrovisbreaking in the primary stage of upgrading, a yet unproved process in commercial operations, presented a technical risk greater than the applicants were prepared to take. Consequently, at the same time that application was made for construction-time extension, application was made to switch to the use of fluid coking, an up-to-date approach to primary conversion with some commercial background.

The application for extension of construction time and the consideration of technical changes in the plant received Board recommendation for approval on October 3, 1973.

The Order-in-Council, passed by the Alberta Cabinet, stipulated certain conditions under which the Syncrude project could proceed (see Appendix III). Early in the conceptual engineering stage and concurrently with the technical work, negotiations with provincial-government representatives got underway to establish an agreed-upon manner in which the conditions set out in the Order-in-Council could be met. The negotiations eventually resulted in the Letter Agreement dated September 14, 1973 (see Appendix IV). This letter only partially paved the way for commercial construction which originally had been scheduled to begin on September 1, 1973. Limited work did get underway, but with considerable restraint and with the option to back out on the part of the Syncrude group in case developments did not proceed according to plan. In this regard it will be noted in Section 21 of the Letter Agreement that an obligation exists relative to concerns involving pricing and treatment of royalty in relation to federal income tax, both requiring negotiations with federal authorities, and the concern regarding stability of labor which was strictly a provincial matter.

Both the Syncrude and provincial-government representatives became involved in negotiations with federal authorities which resulted in formal statements by the federal Departments of Finance, Internal Revenue, and Energy, Mines and Resources which, in turn, resulted in the amended Letter Agreement between the Syncrude companies and the provincial government dated December 13, 1973 (see Appendix V). The third concern, the one related to stability of labor, had not been resolved and remained a condition of the Agreement. Syncrude participants were prepared to make certain additional commitments on the strength of the Amended Letter, but were not prepared to make an unconditional commitment to complete the project. At the time of the preparation of this document (February 1974), discussions with the Alberta Department of Labor, the unions, and the Alberta Construction Labor Relations Association were continuing in search of an approach which could accomplish the objectives of all concerned. In brief, the government was most anxious that all parties find a solution which would adhere as closely as possible to existing legislation and result in a minimum conflict of interest. The A.C.L.R.A. wanted to circumvent a situation whereby workers on strike in other areas of northern Alberta and on other projects would migrate to the Syncrude project and thus block the possibility of satisfactorily terminating such strikes. Also Syncrude's contractors wanted to bring order to a situation involving as many as fifteen different agreements covering the many crafts involved in the project. Because of major conflicts of interest, it developed that these three groups would not meet at a common bargaining table, leaving the possibility of a completely satisfactory solution very much in doubt.

Although not covered specifically by formal agreement, at least two other concerns are proving of major significance to the Syncrude group: (1) availability of housing and (2) availability of labor.

When field work for the Syncrude project was initiated in December 1973, the town of Fort McMurray was still suffering from the growing pains created by the Great Canadian Oil Sands tar-sand project brought on stream in September 1967. Even though some progress was being made in the construction of apartment blocks, the community was still short by at least 200—300 private residences, not to mention a completely inadequate hotel-space situation. Syncrude expected its contractors to house all construction workers in a camp designed for 3500 people while, on the other hand, construction supervisors and permanent Syncrude staff were expected to find accommodation in Fort McMurray. Disregarding the 200—300 requirements for building lots already existing, Syncrude requirements were expected to reach 400 by the end of the first year of construction and growing to a total of about 1200 by the time plant construction was complete. During the same period, requirements for dwellings for other townspeople was expected to total over 2500.

A number of areas surrounding Fort McMurray was available for development and prior to Syncrude activity in the area, the Alberta Department of Municipal Affairs, through the Alberta Housing Corporation, undertook the task of preparing serviced lots and providing general town planning. An early divergence of opinion developed as to which areas should be developed first; however, it should have been realized that all areas must be developed simultaneously. In spite of early difficulties in initiating the Fort McMurray housing development definite progress began to take shape in early 1974.

To lend some assistance in the development of adequate housing, Syncrude sponsors set up a separate company to be known as Northward Developments Ltd., whose sole responsibility was to do whatever possible to assist in providing more housing in Fort McMurray. Its degree of success is still to be demonstrated.

Additional major projects not related to tar-sand production were announced in Alberta recently, which will serve to aggravate still further an already serious skilled-labor shortage. It would appear that major projects in eastern Canada and throughout the United States will virtually eliminate the possibility of importing skilled labor from those areas. The problem remains, then, for Alberta solution. Aid can possibly be found from three directions: training of new recruits, spacing of projects, and the importation of labor from overseas areas. The planning of training programs is underway. Even though the programs under consideration may be sound in principle, there is no evidence to date that enough will be accomplished in time to relieve the serious shortages looming on the horizon. If the worst situation develops, the degree of overlap between major projects must be lessened; otherwise construction periods will be drawn out, an atmosphere of major labor conflict will be created, and the overall economic return of many projects reduced to a point of unacceptability. At this time Alberta is entering a period of major industrial development including, as well as tar-sand plants, ammonia and fertilizer plants, petrochemical plants, power plants, chemical plants, and pulp and paper plants. To expand too rapidly at this stage could only delay or destroy long-range prospects for industrial advancement of the province.

To make matters more complicated, Shell Canada Limited (and its associate, Shell Explorer Limited) appeared before the Energy Resources Conservation Board with an application to produce 100,000 bbl/day of synthetic crude oil. This application was heard by the Board in July 1973. Seven months later, the decision was still being considered. In this application Shell predicted that the proposed plant could first be brought on stream in January 1980 with full production reached on or before January 1, 1982. In its production scheme Shell intends to use a dragline-mining, hot-water extraction approach.

In view of the extremely high level of construction activity scheduled in Alberta during the middle to late 1970's, it would seem that the final timing of the Shell project must be critically reviewed because of the labor-availability problems.

Still further, Petrofina Canada Ltd. as operator for a group of companies including, besides itself, Pacific Petroleums Ltd., Hudson's Bay Oil and Gas Ltd., Murphy Oil Company Ltd., and Candel Oil Ltd. applied to the Energy Resources Conservation Board in January 1974 for approval to produce 122,500 bbl/day of synthetic crude oil from the Athabasca tar sands. The group intends to use bucket-wheel mining and hot-water extraction to bring about initial production in 1982. Under this plan testing would take place in 1977 with construction initiated in 1978. Interest in the project is shared on the following basis: Petrofina — 35.337%; Pacific — 32.713%; Hudson's Bay — 14.538%; Murphy Oil — 10.487%; and Candel Oil — 6.875%. Petrofina has announced that final decision to proceed is still subject to negotiations with provincial authorities and, of course, the outcome of the anticipated hearing before the Board.

Commercial projects

G.C.O.S. mining project

Based on the new concept that size was important, Great Canadian Oil Sands started construction of its 45,000-bbl synthetic-crude plant in 1963 and placed it on stream in September 1967. A four-stage sequence [23], illustrated diagrammatically in Fig. 5-5, was employed including: mining, material handling, extraction, and upgrading of the heavy oil into a pipelineable, low-sulfur, 38° API synthetic crude. (The final product from tar-sand production is referred to in the literature as "synthetic crude"; however, this

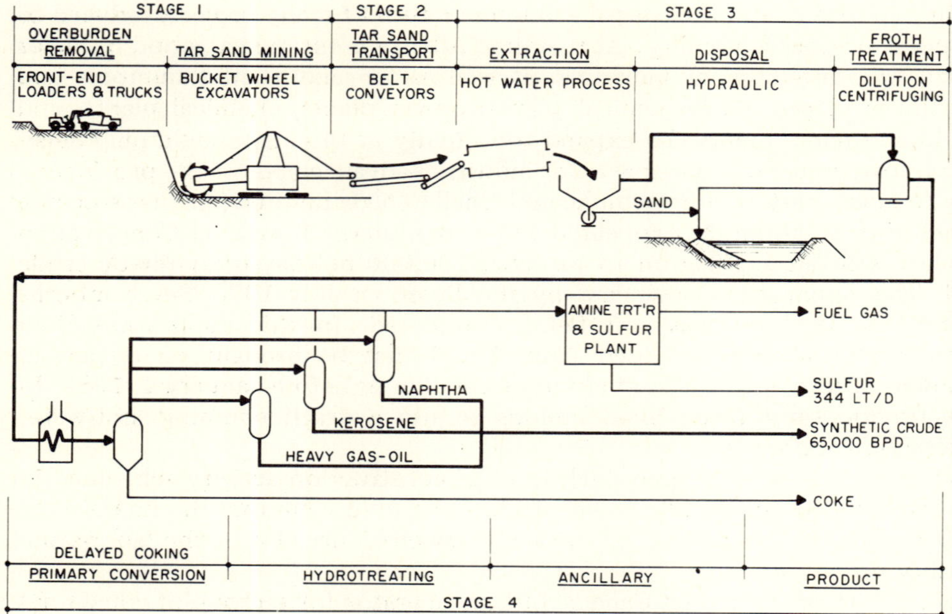

Fig. 5-5. Diagrammatic illustration of the G.C.O.S. plant.

term is a misnomer. It is not synthetic by any meaning of the term but rather, perhaps, a "reconstituted crude".) In the construction of this plant G.C.O.S. and its prime contractor, Canadian Bechtel Limited, were truly pioneers. In order to get to the site, a $ 3.3-million steel bridge had to be constructed, a 20-mile plus access road had to be built, much of the way over virgin muskeg, and before production could be started it was necessary to lay a 266-mile pipeline to Edmonton. The peak construction force employed in building the G.C.O.S. facility reached well over 2000 workmen.

In the G.C.O.S. approach, overburden removal was initially attempted by a large shovel and a fleet of 80-ton trucks. This combination did not prove successful and a switch was made to a fleet of 32-cubic-yard tandem engine scrapers. Although the scrapers were able to remove sufficient quantities of the overburden, plus the advantage of supplemental tar-sand mining when required, the scheme did not meet economic requirements. The overburden-removal scheme was again revamped to the currently employed system. Now, to remove an average of approximately 45 ft of overburden, a fleet of 150-ton trucks are employed, loaded by 15-cubic-yard front-end loaders, which in turn are supplemented by a bucket-wheel excavator when required.

The mining of the approximately 150-ft section of tar-sand ore is accomplished by two giant bucket-wheel excavators of German design, each with a rated capacity of up to 100,000 tons daily, supplemented by a third but smaller bucket-wheel excavator of American design when not required in the removal of overburden. The tar-sand section is divided into two faces, each about 75 ft in height. The bucket wheels operate from the pit floor and from a bench located at the bottom of the upper face.

The material mined by the bucket wheels is fed onto a system of 72-inch wide conveyor belts, mounted on moveable mounts, and transported to the extraction plant at an average velocity of approximately 1000 ft/min. Two main belts, one for each bench, are extended into the pit as the faces progress into the formation. Feeder belts at right angles to the main belts and parallel to the mining faces also follow the faces as they progress, at a distance of 200—300 ft — enough to allow operating room for the excavators and belt loading equipment.

The load from the conveyor belts is dumped into surge bins capable of handling about 10,000 tons. Apron feeders located at the bottom of the bins discharge onto conveyors, which feed four parallel horizontal slurrying drums.

The slurrying drums constitute the beginning of the extraction process. Here hot water and steam are combined with mechanical energy to transform the tar-sand ore into a frothy pulp. As the mass is discharged from the slurrying drums, it is passed over trommel screens where any undigested material is separated. The main body of pulp is then pumped to extraction vessels filled with hot water. Oily froth collects on top and is skimmed off, sand passes out of the bottom, and a middling stream is removed and pumped away to secondary recovery cells where additional oil is recovered. Makeup water is added to keep the cells in balance.

The oily froth is collected from the primary and secondary extraction cells and diluted with gas oil or naphtha and made ready for final cleanup. In order to accomplish this operation, the diluted froth is passed through a centrifuge system where unwanted water and solid matter are removed. The diluted oil from the centrifuge step is then ready for upgrading.

In the upgrading process, the diluted centrifuge product is heated, the diluent recovered, and the remaining oil passed to six delayed-coking drums, each 26 ft in diameter and 95 ft high. The overhead product from these drums is separated into three streams: naphtha, kerosene, and gas oil. Inasmuch as the hydrogenation characteristics of the three streams are different, each is hydrotreated separately so that the final blended products conform to a predetermined set of properties. The blended material constitutes the so called "synthetic crude oil". Although lacking in heavy ends, this crude is readily useable as refinery feedstock. All or a part of the entire stream can be reformed into petrochemical feedstock.

Syncrude project

The Syncrude project, illustrated diagrammatically in Fig. 5-6, has been designed to produce 125,000 bbl/day of synthetic crude oil. Preparation of

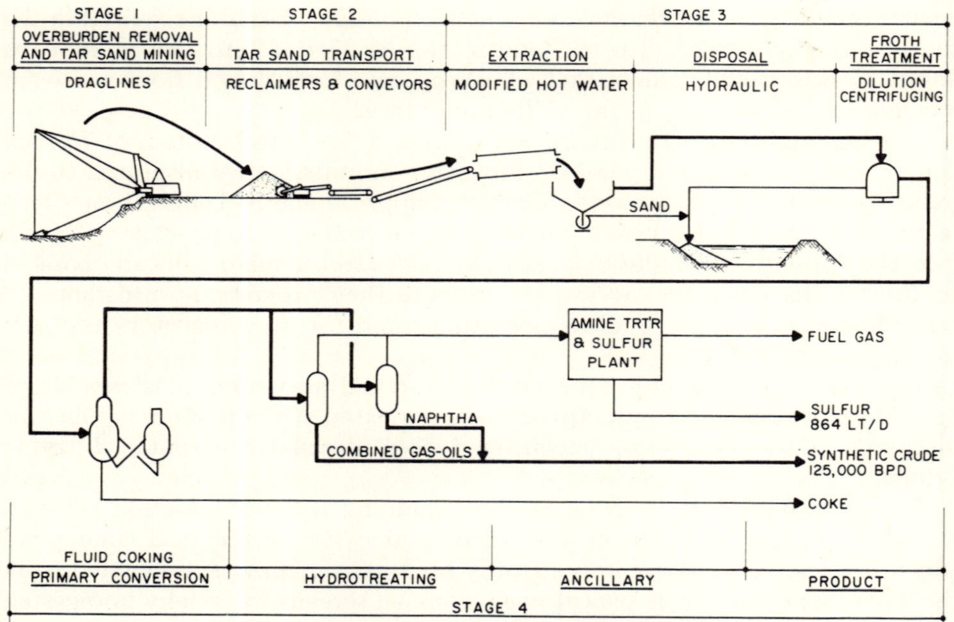

Fig. 5-6. Diagrammatic illustration of the Syncrude plant.

the project site, 25 miles north of Fort McMurray, was commenced in December 1973 and the plant is expected to start testing in the second half of 1977 with the official on-stream date expected about January 1, 1978. (By agreement with the provincial government, the official on-stream date is defined as that date on which production of synthetic crude oil will have totalled 5,000,000 bbl.) This date, however, will be strongly influenced by availability of skilled labor and delivery of equipment. The scheme to be employed by Syncrude, in principle, is the same as that employed by G.C.O.S., although there will be many variations in conceptual detail as to how each of the four separate stages is approached. The overall plan calls for development of two parallel streams, combined initial production of which will be 104,500 bbl/day. With operating experience and "bottleneck" removal, it is expected to bring production up to the rated level as quickly as possible.

In the Syncrude mining scheme two parallel open-pit mining faces will be opened and actively mined, advancing away from each other, throughout the life of the project. Each face will be served initially by two draglines of the 80- to 90-cubic-yard bucket capacity and utilizing 360-ft booms. As plant throughput improves and as overburden thickens, provision has been made for the addition of two additional draglines to each mining face. The draglines will operate from the top surface and from a bench about midway in the face. Overburden will be disposed of directly into the pit and all mined ore will be placed in a pile paralleling the pit edge on the top surface. No equipment will be placed in the bottom of the pit — a condition which allows earlier than normal replacement of sand tailings and sludges back into the mined-out area.

Unfortunately, the mined-out area will not accommodate all of the original material after it has been disturbed or processed. This condition results, for the most part, from the swelling of clays entrapped in the processed tar sand and the manner in which materials, including the overburden, are redeposited in the pit. As a result of the requirement for added pit volume, a 50-ft plus dyke will be required surrounding the entire mined-out area. This dyke will be easily constructed, mostly by the draglines as they remove the overburden around the perimeter of the pit.

Considerable concern developed during the design of the Syncrude mining pit over the stability of the mining face. This concern developed because of a lack of knowledge of the behavior of tar sand exposed in a high steep face, and further complicated by severe loading problems at the top edge due to machines and the tar-sand reclaim pile together with an unexpected serious formation hydrological problem. The economic drive in using draglines, however, was substantial and Syncrude carried out extensive field tests with a 195-ft boom—17-cubic-yard bucket dragline. These tests lasted over a period of many months and cost several million dollars. A board of internationally known soil experts was retained to guide and study the results of the tests. A set of conditions was eventually established whereby the board was prepared to recommend a procedure by which dragline mining of the tar sands could proceed.

Once the tar sands have been placed on the top surface parallel to the pit face, Syncrude expects to use reclaim wheels to load the mined tar sand onto a system of conveyor belts (or possibly trains) for transport to the extraction plant.

Syncrude's method of extraction will be quite similar to that employed by G.C.O.S. The tar sands will be pulped in tumblers with steam and hot water, placed in extraction vessels for primary recovery, followed by secondary recovery of oil from a middlings stream and final cleanup of the oily froth in a dilution centrifuge process.

In the upgrading step, Syncrude deviates quite sharply from the G.C.O.S. approach. Fluid coking is used in the primary conversion of the heavy oil. In this process two overhead streams will be taken off: naphtha and gas oil. Each will be hydrotreated separately to the extent necessary to make, when blended, a low-sulfur synthetic crude oil of approximately 33° API.

The residual material from fluid coking contains an excessive amount of solids and sulfur and is not suitable as a fuel under today's technology; however, the fluid-coking process itself is internally heat balanced. In this process, heat for the coking reaction is provided by burning the requisite amount of gross coke produced in the burner vessel. In addition, by limiting the amount of combustion air, a substantial quantity of CO is produced which is collected and used in CO boilers to generate substantial quantities of steam for use elsewhere in the plant.

Fluid coking results in the production of somewhat more saleable product than does delayed coking, as well as producing significant quantities of off gases. A portion of the butanes find their way into the final-product stream but the remainder of the butanes and the lighter gas end up in the plant fuel stream. In spite of the various sources of heat generated during the fluid-coking process, there will not be enough heat to put the overall plant in heat balance. In the neighborhood of 20 million cu ft/day of natural gas will be required to bring about this balance.

Shell mining project

Shell Canada Limited appeared before the Alberta Energy Resources Conservation Board with an application to build a 100,000-bbl/day synthetic crude oil plant based on a plan outlined in Fig. 5-7, which in many respects is similar to the approaches employed by G.C.O.S. and Syncrude. In its application to the Board [24], Shell summarized its plan as follows:

(1) Mining

"Four intermediate size (75—90 cubic yard) walking draglines will be used to remove

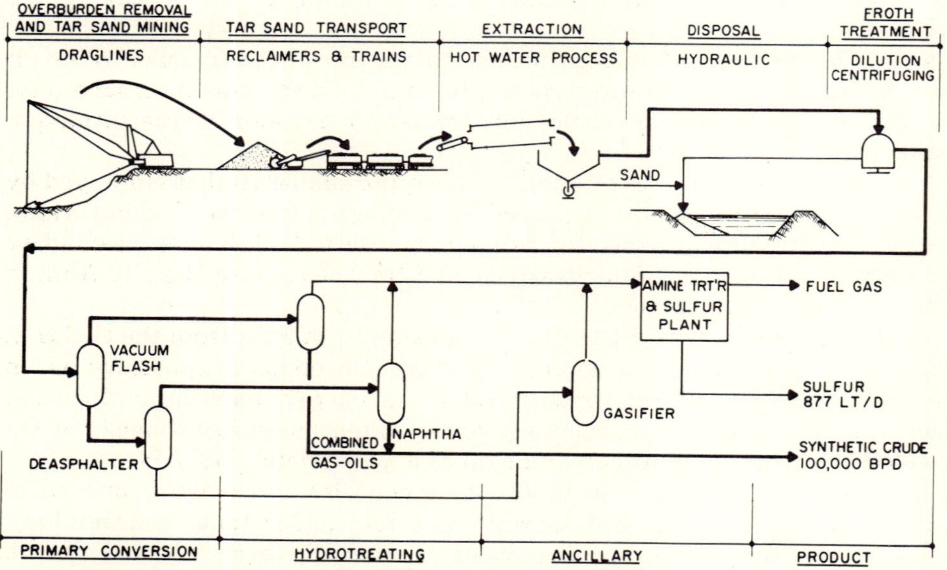

Fig. 5-7. Diagrammatic illustration of the proposed Shell plant.

the overburden and mine the tar sands. The ore will be cast onto piles, picked up by reclaimers and front-end loaders, and conveyed to rail-car loading stations."

(2) Ore transportation

"The ore will be transported from the mine to the extraction plant by locomotives hauling 100-ton side-dump cars."

(3) Extraction

"The K.A. Clark hot-water extraction process will be used to effect the bitumen—mineral separation. Basic components of the process are tar-sand conditioning, froth flotation, and naphtha dilution centrifuging."

(4) Diluent recovery

"The naphtha diluent and entrained water will be separated from the bitumen by conventional water flashing and hydrocarbon distillation."

(5) Primary separation

"Vacuum flashing and solvent deasphalting will be used to remove the lighter portion of the bitumen. The remaining residue will be fed to the utility plant."

(6) Hydroprocessing

"The lighter oil removed from bitumen will be hydrotreated and hydrocracked to produce a low-sulfur crude. Auxiliary units include hydrogen generation and sulfur removal."

(7) Tankage and pipeline

"Sufficient tankage for operating flexibility will be provided on-site. The synthetic crude will be moved by pipeline to Edmonton."

(8) Utilities plant

"Steam and electric power will be supplied by the utilities plant. The utilities process configuration will maximize utilization of residue and minimize air pollution. These facilities will be described in detail in a future application pursuant to Section 7 of The Hydro and Electric Energy Act."

Other mining projects

Fina, already mentioned, representing a group of companies, has submitted an application to the Energy Resources Conservation Board for a permit to produce 122,500 bbl/day of synthetic crude oil by means of a scheme based on mining wheels, hot-water extraction, and fluid coking. At the present time Fina is preparing additional materials for the Board. Until this is complete, a date for the formal hearing will not be announced. With formal announcement, technical and other details of the project will be made public.

Still other commercial tar-sand projects appear to be nearing the commercial stage. Hudson's Bay representing a group of oil companies is active in the mining area of the Athabasca tar sands, Amoco are significantly busy in the in-situ area of its Athabasca tar-sand deposit and, according to rumor in oil circles, Shell will be reconsidering its plans for an in-situ project, this time in the Peace River deposit.

Shell in-situ project

Although the application of Shell Oil Company of Canada,Ltd., to the Alberta Oil and Gas Conservation Board dated September 6, 1962, to produce 47,450,000 bbl/year of crude hydrocarbon product (about 100,000 bbl/day synthetic crude) was subsequently withdrawn, this application bears some significance to the commercial development of the Athabasca tar sands in that it represented the first major attempt to produce oil from the Athabasca deposit by means of an in-situ thermochemical-stimulation process. Reportedly, the application was withdrawn for reasons other than technical and presumably would still represent a viable approach to the commercial production of a major portion of the Athabasca tar-sand deposit.

Shell based its application on laboratory research, theoretical work, and the results of an extensive field test spread over the period from 1960 to 1962. During the field test [25], Shell was able to demonstrate that acceptable rates of oil production, in the form of low-viscosity emulsions, could be established by:

(1) Creating horizontal fracture by the injection of weak alkaline solution.

(2) Injection of additional amounts of heated aqueous alkaline solution to create a continued path.

(3) Injection of steam at desired rate and pressures to heat the formation.

In this production technique, Shell estimated that displacement efficiency could go as high as 87% and residue tar saturation could be as low as 7%.

In the commercial operation, Shell contemplated an expanding system of injection and production wells operating at first as a concentrated pattern to establish operating parameters for the particular characteristic of the tar sands at its commercial site with subsequent expansion into the full commercial development. During sustained rated production levels, Shell expected to bring a group of 84 wells on production every 58 days with a corresponding abandonment scheme carried out as production rates of individual wells fell below acceptable levels.

Shell proposed to dilute and then dehydrate the oil—water emulsion produced from the wells and visbreak this product in the field. Two overhead streams were to be collected from the visbreaking operation — one as a source of diluent to be added to the emulsion in the dewatering process, and the other as the main product stream which would be pipelined to Edmonton, about 275 miles away. Residual material from the visbreaking operation was to be used as fuel. In Edmonton, Shell proposed to separate the pipeline stream into a naphtha/LGO cut and a heavy gas-oil cut. The two streams were to be separately hydrotreated and blended to form the synthetic crude. When blended this would have resulted in 96,854 bbl/day of synthetic crude plus 408 LT/D of sulfur.

Muskeg in-situ project

The Muskeg Oil Company (wholly owned subsidiary of Pan American Petroleum Corporation — later Amoco) submitted an application to the Alberta Oil and Gas Conservation Board on October 25, 1968, for the commercial production of 15,000,000 bbl of crude bitumen (oil extracted from the tar sands) at rates up to 8000 bbl/day from the Athabasca tar sands. This requested production constituted phase one of a two-phase operation and was designed to optimize well spacing, formation heating, and well completion and operating techniques. The second phase, to be covered by a later application, was to have begun in the third year following project approval and was expected to result in a logical development leading to the production of at least 60,000 bbl/day of bitumen. Even though the application was withdrawn, Muskeg has continued with its field pilot testing and, in spite of the fact that commercial production has not materialized, the continued activities of this company represent a significant step in the commercial development of the Athabasca tar sands.

The in-situ recovery process proposed by Muskeg Oil Company was described as follows [26]:

> "The Applicant proposes to produce bitumen from the Athabasca oil sands by developing selected areas in 10-acre well patterns with 9-spot configuration. Four 3.5-acre five spots will be included in the initial block (group of patterns) to permit further optimization of well spacing and pattern geometry. In all cases, the McMurray Formation will be hydraulically fractured and preheated by a combustion process. The bitumen will then be recovered by a combustion displacement process utilizing air and water.
>
> The object of the fracturing and heating operations is to raise the average temperature within a pattern to about 200°F, a level at which the viscosity of the bitumen is sufficiently reduced so that it can be made to flow readily. Field tests have demonstrated that a combustion process can be utilized to obtain the required elevated temperatures. After such formation heating is accomplished, conventional oil displacement methods can be used to produce the bitumen.
>
> The displacement process to be used by the Applicant is a Combination of Forward Combustion and Waterflooding (hereafter referred to as the COFCAW process). In this forward-combustion process, the formation in the vicinity of an injection well is ignited. Injected air then moves a combustion front toward producing wells. The performance of the forward-combustion process is significantly improved by simultaneously injecting air and water. The injected water vaporizes and passes through the combustion front carrying heat from behind to ahead of the front. This results in rapid heat-front movement and efficient utilization of heat. From the combined effects of heat and gas and water drives, oil is displaced ahead of the combustion front.
>
> The Applicant has tested COFCAW experimentally in several conventional oil fields in addition to the Athabasca oil sands. A commercial-sized (480 acres) operation is currently underway in the Sloss field in Nebraska. The process details are discussed in a recent paper and United States Patents No's. 3,171,479 and 3,196,945 assigned to Pan American on March 2, 1965, and July 27, 1965, respectively. Corresponding Canadian Patent No's. 739,769 and 739,768 were issued on August 2, 1966."

Conclusions

The commercial development of the Athabasca tar sands is no longer a speculative possibility but a reality. The only questions that remain are the rate at which this source of oil can be developed and the extent of development. On the one hand there are proposals such as the one suggested by Herman Kahn [27], Director of the Hudson Institute, whereby massive development of the Athabasca tar sands would take place involving a $ 15—20 billion effort on the part of Japan, Europe, and the United States to get 2,000,000 bbl/day flowing within three years. Under existing conditions the problems of implementing such a plan would be astronomical. The political concerns related to provincial and federal conflicts in Canada would have to be dealt with, as well as environmental concerns and major concerns related to individual priorities, which would have to be set by each respective country participating. In spite of all the obvious difficulties, however, the proposed plan for a massive development did receive support from the Hon. Jean Pierre Goyer, Canadian Federal Minister of Supply and Services. Mr. Goyer made a number of speeches across Canada but widespread public acceptance of the plan has been lacking.

On the other hand a more likely approach to the development of the Athabasca tar sands would involve a plan based on (1) the time required to isolate and define specific ore bodies suitable for commercial development, (2) the limitation dictated by the availability of professional engineers and skilled labor, and (3) the ability of designers to optimize the technical configuration and operation of each plant in order that competition from existing and future sources of energy can be met.

In time, the practical extent to which the Athabasca tar sands can be developed, if approached in an intelligent manner, will be limited only by the imagination and determination of the operators.

Next, one may examine the present and anticipated results of efforts that are currently under way and that are planned for the foreseeable future and how these efforts might relate to the energy-supply problem.

Many people have examined the relationships between supply and demand for petroleum in Canada. Although the details of these studies differ, they each point to a pretty obvious fact: Canada will not be able to maintain its current production rates beyond the early 1980's.

Fig. 5-8, based on information included in a Gulf Oil Canada Limited submission [28], provides one example of a realistic Canadian supply—demand projection. From the early 1980's the gap between supply and demand will widen rapidly, even only if demand west of the Ottawa Valley is considered, unless alternate sources of supply can be produced in increasing volumes. There are several possibilities: (1) the Arctic mainland and islands, (2) the eastern Canadian offshore, and (3) the Athabasca tar sands.

The pros and cons of Arctic and eastern Canada offshore areas have been discussed by many experts and, no doubt, they do constitute prospective sources of oil; however, the extent of future production from these areas has not been demonstrated and it is not the intention here to hazard a guess as to what might be expected. By contrast, oil does not have to be discovered in the Athabasca tar sands (or any other deposit along the heavy-oil trend

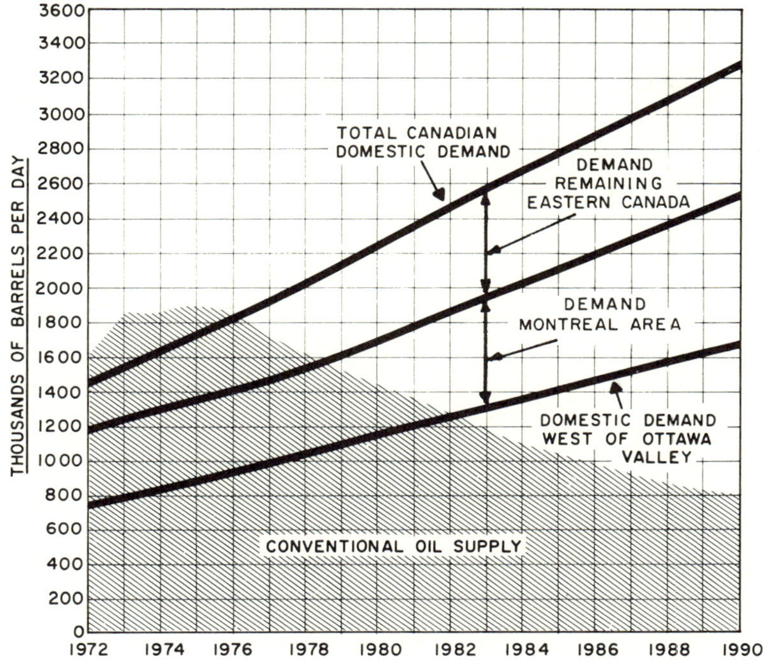

Fig. 5-8. Conventional domestic supply and overall demand for oil in Canada, 1972—1990.

in Alberta). Here the oil is known to exist in very large quantities and possibility of commercial production has been demonstrated. With a firm base from which to start, what realistic oil production can be expected from the Athabasca tar sands? An attempt to answer this question has been made in the chart illustrated in Fig. 5-9. The chart, which should be accurate to within plus or minus one year, reveals that production from the Athabasca tar sands will reach about 400,000 bbl/day of synthetic crude by 1985 and about 700,000 bbl/day by 1990.

In preparing the chart due recognition was given to the limited supply of professional help and skilled labor as well as slow delivery schedules of equipment and supplies. In addition, only those projects were included where planning and field work is already in progress. Highly speculative "big talk" ventures were not considered.

Projections after 1990 were not attempted; however, if work proceeds before 1990 as anticipated then after that date industry would, in all probability, be capable of building at least one new tar-sand plant per year sufficient in size to take care of the annual incremental increase in Canadian demand without any help from Arctic or offshore sources. If this were true, then, still without any help from Arctic or offshore productions, the

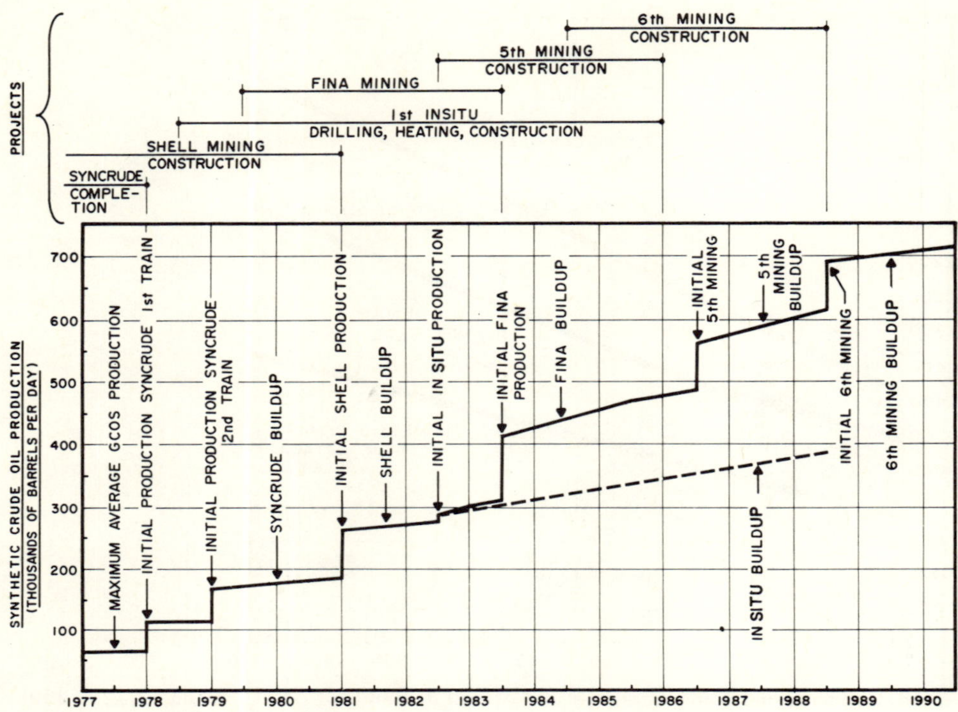

Fig. 5-9. Buildup of production from the Athabasca tar sands.

Athabasca tar sands could probably make Canada self-sufficient in energy sometime after the year 2000. The potential exists for a faster buildup, but it would take a far greater effort on the part of both industry and governments than has been experienced to date.

It would be unfair at this stage not to mention prospects in other deposits along the Alberta heavy-oil trend. Two significant pilot operations are in progress, one in the Cold Lake deposit and one in the Peace River deposit. It is very likely that both will be in production by 1990 with a combined output of synthetic crude of from 100,000 to 200,000 bbl/day. On an optimistic basis, then, production from the entire heavy-oil trend in Alberta could be as much as 900,000 bbl/day by about 1990 *. Combined Canadian conventional and synthetic production would meet total demand west of the Ottawa Valley by this time, but could not accommodate the total Canadian demand or even the demand of the Ottawa Valley plus the Montreal market.

---

* See Addendum.

References

1 The Energy Resources Conservation Board, *Reserves of Crude Oil, Gas, Natural Gas Liquids and Sulphur, Province of Alberta,* Table 3-1 (Dec. 31, 1972).

2 G.W. Govier, *"Oil Sands — Fuel of the Future"* Talk to the *Can. Soc. Pet. Oil Sands Geol. Symp.*, Calgary, Alberta, p. 9 (Sept. 7, 1973).

3 G.A. Stewart, "Geological Controls on the Distribution of Athabasca Oil Sand Reserves", in: *K.A. Clark Volume,* Research Council of Alberta, p. 18 (1963).

4 V.P. Kaminsky, "Selection of a Mining Scheme for a Tar Sands Extraction Plant", paper presented to the *Can. Soc. Pet. Geol., Oil Sands Symposium,* Calgary, Alberta, Fig. 2 (Sept. 7, 1973).

5 Walwyn, Stodgell and Co., Ltd., *A Look at World Energy, The Athabasca Tar Sands,* p. 8 (June, 1973).

6 S.M. Blair, *The Development of the Alberta Bituminous Sands,* a report commissioned by the Alberta Government, p. 10 (Dec., 1950).

7 Walwyn, Stodgell and Co., Ltd., *A Look at World Energy, The Athabasca Tar Sands,* p. 9 (June, 1973).

8 R.R. Goforth, Syncrude Canada Ltd., Research Director, personal communication.

9 The Oil and Gas Conservation Board, *Supplemental Report to the Lieutenant Governor in Council, with respect to Application of Great Canadian Oil Sands Limited,* p. 7 (Sept., 1962).

10 *Application to Oil and Gas Conservation Board by Atlantic Richfield Company, Cities Service Athabasca, Inc., Imperial Oil Limited, and Royalite Oil Company Limited,* p. II-3 (May 3, 1968).

11 The Energy Resources Conservation Board, *Reserves of Crude Oil, Gas, Natural Gas Liquids and Sulphur, Province of Alberta,* p. 3-2 (Dec. 31, 1972).

12 M.A. Carrigy and J.W. Kramers, Guide to the Athabasca Oil Sands Area, Alberta Research Council, prepared for *Can. Soc. Pet. Geol. Oil Sands Symposium, Inf. Ser.,* 65, 105—106 (1973).

13 *Application to the Oil and Gas Conservation Board Under Part VI-A of the Oil and Gas Conservation Act by Atlantic Richfield Company, Cities Service Athabasca, Inc., Imperial Oil Limited, Royalite Oil Company Limited,* p. II-5 (May 3, 1978).

14 Walwyn, Stodgell and Co., Ltd., *A Look at World Energy, The Athabasca Tar Sands,* p. 2 (June, 1973).

15 S. Lynett, "Digging for Oil", *Imp. Oil Rev., 4,* p. 18 (1973).

16 Walwyn, Stodgel and Co., Ltd., *A Look at World Energy, The Athabasca Tar Sands,* pp. 2—4 (June, 1973).

17 Walwyn, Stodgell and Co., Ltd., *A Look at World Energy, The Athabasca Tar Sands,* pp. 4—6 (June, 1973).

18 Walwyn, Stodgell and Co., Ltd., *A Look at World Energy, The Athabasca Tar Sands,* p. 6 (June, 1973).

19 S. Lynett, "Digging for Oil", *Imp. Oil Rev., 4,* p. 21 (1973).

20 Walwyn, Stodgell and Co., Ltd., *A Look at World Energy, The Athabasca Tar Sands,* p. 6 (June, 1973).

21 S. Lynett, "Cold Lake", *Imp. Oil Rev., 1,* p. 6 (1974).

22 M.L. Natland, "Project Oil Sand", in: *K.A. Clark Volume,* Research Council of Alberta, pp. 143—155 (Oct., 1963).

23 E.D. Innes and J.V.D. Fear, "Canada's First Commercial Tar Sand Development", *Proc. 7th World Pet. Congr.,* Mexico City, *3,* 633—650 (April, 1967).

24 *Application of Shell Canada Limited and Shell Explorer Limited to the Energy Resources Conservation Board,* p. 16 (June 28, 1973).

25 In the matter of the Oil and Gas Conservation Act being Chapter 63 of the *Statutes of*

Alberta, 1957, and in the matter of an application by Shell Oil Company of Canada, Limited, persuant to Part VI-A of the said Act for the approval of a Scheme or Operation for the Recovery of Oil for a Crude Hydrocarbon Product from Oil Sands, Section II, II-1—9 (Sept. 6, 1962).

26 Muskeg Oil Company, *Application to the Oil and Gas Conservation Board under Part VI-A of the Oil and Gas Conservation Act*, p. C-1 (Oct., 1968).

27 G.H. Wright, "Herman Kahn's Frightening Scheme to Exploit the Athabasca Oil Sands", *Edmonton J.*, p. 5 (Jan. 15, 1974).

28 Gulf Oil Canada Limited, *Submission to the National Energy Board in the Matter of Exploration of Oil*, VI-1—12 (Dec., 1973).

Bibliography

1 *An Energy Policy for Canada, Phase I and II*, issued under the authority of the Minister of Energy, Mines and Resources, Ottawa, Ontario (1973).

2 DeGolyer and MacNaughton, *Twentieth Century Petroleum Statistics* (1973).

3 G.W. Govier, *Alberta's Oil Sands in the Energy Supply Picture*, Talk to the Can. Soc. Pet. Geol. Symp., Calgary, Alberta, Canada (Sept. 7, 1973).

4 The Hon. Jean-Pierre, Goyer, *Athabasca — The Time is Now*, address to the Calgary Chamber of Commerce, Calgary, Alberta (March 25, 1974).

5 G.R. Gray, *Conversion of Athabasca Bitumen*, presented to the American Institute of Chemical Engineering, Dallas, Texas (Feb., 1972).

6 *Interim Report Alberta's Requirements of Energy and Energy Resources, 1972—2001*, Energy Resources Conservation Board (Nov., 1973).

7 V.P. Kaminsky, *Selection of a Mining Scheme for a Tar Sands Extraction Plant*, presented to The Canadian Society of Petroleum Geologists, Calgary, Alberta, Canada (Sept. 7, 1973).

8 F.K. Spragins, *A Second Generation Tar Sands Plant*, reprint from the Proc. 8th World Pet. Congr., Moscow, U.S.S.R. (June, 1971).

9 D.M. Winch, *The Political Economy of Canadian Oil Policy*, McMaster University (Dec., 1973).

Addendum

The sharp increase in the Middle East oil prices in the early 1970's and the subsequent inflationary rise in the price of all goods and services necessitated the preparation of a new definitive estimate of the cost of the Syncrude Athabasca tar-sand project. The results of this study, which became available in late 1974, indicated a cost increase from approximately \$ 1 billion to slightly over \$ 2 billion. Shortly thereafter, Atlantic Richfield withdrew from the project. The working interest of the remaining three participants was realigned and three governments joined the group to take up the 30% interest dropped by Atlantic Richfield. A new consortium was formed with the following participating interests: Imperial Oil Limited — 31.25%; Canada-Cities Service, Ltd. — 22.00%; Gulf Oil Canada Limited — 16.75%; Petro-Canada Exploration Inc. (Government of Canada) — 15.00%, Alberta Syncrude Equity (Government of Alberta) — 10.00%; and Ontario Energy Corporation (Government of Ontario) — 5.00%. Syncrude Canada Ltd. remained as operator.

Currently, the project is on schedule and within budgetary limitations. The first synthetic crude oil is expected to flow in May 1978.

Although the tempo of heavy-oil activity in Canada was slowed down by inflationary and other concerns during the past two or three years, the growing imbalance between supply and demand of oil in Canada is as real as ever. Also, the possibility of early production from the frontier areas is not encouraging. Consequently, many oil companies are reexamining heavy-oil prospects. At least three separate groups are seriously considering the possibilities of building 100,000-bbl/day heavy-oil upgrading plants at the extreme southeastern end of the Alberta heavy-oil trend, where oil can be produced from conventional wells; however it is so heavy that it cannot be transported by pipeline in the raw form. Imperial Oil is continuing the operation of its pilot thermal-stimulation plant at Cold Lake at a production rate of 5000 bbl/day and is studying the possibility of turning the project into a commercial venture at 100,000 bbl/day. Shell has carried out pilot mining operations on its lease in the Athabasca area and is looking at commercial possibilities.

Probably the most significant of all recent developments in heavy-oil activity is the establishment of the Alberta Oil Sands Technology and Research Authority. This group, an Alberta Government sponsored organization with an initial funding of $ 144 million, has been set up to assist in the research and development of in-situ type heavy-oil production. In addition to a number of small projects, at least four major field pilot tests are underway. In general, AOSTRA is sharing with industry the cost of such projects on a 50/50 basis.

Although circumstances have changed and the predictions suggested in Fig. 5-9 may not come to pass, interest in the heavy-oil belt, including Athabasca, remains high and most authorities believe that the projections suggested will be met by the indicated or other groups, but with some delay. Informed opinion now places overall heavy-oil production at a minimum of 450,000 bbl/day by 1990.

Chapter 6

THERMAL CRACKING OF ATHABASCA BITUMEN

JAMES G. SPEIGHT

Introduction

As pointed out in Chapter 5, the existence of bituminous sands along the Athabasca River was first reported in 1788 by Peter Pond, a fur trader with the Northwest Company, who observed the Indians in that area using the bitumen, washed out of the formation by water, as a caulking mixture for canoes [1]. Since that time the sands are mentioned in various literature reports [2,3], but serious mapping and exploration of the deposit did not ensue until S.C. Ells of the Canadian Department of Mines and Technical Surveys commenced his work in 1913 [4]. From then until ca. 1940, except for the continuing research program of the Research Council of Alberta [5], little interest was expressed in the deposit. During World War II, the possibility of an oil shortage in Canada and the United States became real, and an extensive program of exploitation of the tar sands was undertaken by the Canadian Government resulting in the construction of several pilot plants. During 1948—1949 the feasibility of the hot-water separation process was demonstrated on the pilot-plant scale under the technical supervision of Dr. K.A. Clark of the Research Council of Alberta [6] and was followed soon after by a technical study of the problem by S.M. Blair [7], who concluded that production of a synthetic crude oil from the Athabasca tar sands was economically feasible (see Chapter 5).

The Alberta Oil and Gas Conservation Board *[1] has estimated that the Athabasca deposit contains 626 billion *[2] bbl of oil in place (see Chapter 5). Of this total, over 400 billion bbl are considered as recoverable raw tar-sand oil from which about 300 billion bbl of upgraded synthetic crude are expected to be produced [8]. It is obvious that this reserve is enormous for, by comparison, of the conventional oil in place, 7 billion bbl are recoverable in Alberta [9]. On a world scale, the recoverable reserves of the Athabasca deposit elevate the Canadian reserves from about 2 to about 40% of world reserves [10,11]. It is not surprising, therefore, that attention has been turned to this vast deposit of raw material and that industrially sponsored research has accelerated considerably over the last two decades leading to formal proposals for the operation of several commercial ventures [12].

*[1] Now the Alberta Energy Resources Conservation Board.

*[2] The North American *billion* is used here as meaning *one thousand million*.

In general terms, the quality of the Athabasca bitumen compared to that of the majority of western Canadian crude oils is very low and, hence, the bitumen requires some degree of refining to upgrade it to the quality of a conventional crude oil. It is also to be remembered that, in any commercial operation where an upgraded synthetic crude oil is produced, a sizeable need for fuel at the processing site must also be satisfied. There are several ways whereby this can be accomplished and which involve either direct thermal treatment of raw tar sand or initial separation of the bitumen from the sand by a water-washing process [13] prior to thermal alteration of the material to an upgraded product. In any case, the object is to convert the charge to a distillate oil and coke; this latter, of course, fulfills the necessary fuel requirements.

This chapter is concerned primarily with work relating to the thermal alteration of the bitumen to produce a more saleable material which, for convenience, can be described as a synthetic crude oil; refining of the synthetic crude to useful consumer materials, such as gasoline, can then be accomplished by conventional means. For the purposes of this chapter, the literature is reviewed to May 1973 and reference to unpublished material is maintained at a minimum.

Thermal upgrading of the bitumen

The scientific literature abounds with material relevant to the thermal alteration of the Athabasca bitumen and it is from this work that today's commercial ventures have grown. The reported methods are outlined and discussed in turn below, but it must be remembered that, inasmuch as the scale of operations described varies from laboratory batch experiments to pilot-plant scale, comparisons of the relative merits of each particular process can only be made with extreme caution.

Thermal processing of the tar sand in situ

There are several reports in the literature which describe preliminary laboratory-scale investigations into recovering the bitumen, or a suitable derivative thereof, without disturbing the formation. The method employed usually involves combustion of the bitumen in the formation by means of ignition of an air—bitumen mix at the sand face in several adjacent production wells. The heat generated raises the temperature and increases the mobility (usually through combustive cracking) of the fluid. This is displaced into the production wells by the gas and vapors resulting from the injection of air or oxygen-enriched air for propagation of the combustion.

Elkins [14] claimed that recovery of oil from the Athabasca tar sands could be accomplished readily by an in-situ combustion, which involves drilling several wells into the deposit and burning the bitumen in the presence of injected air. As a result of the combustion process, heating of the deposit occurs near to the combustion zone and oil is driven into one or more of the wells for subsequent recovery and processing.

Salomonsson [15] proposed that heating elements be inserted into the formation and that the bitumen be pyrolyzed at 250—400°C to produce volatile lighter hydrocarbons which may be recovered from a collection zone. In the movement of these vapors towards this zone, they would contact the denser colder parts of the tar sand and the heavier components would condense to a liquid state and leach out more bitumen from the formation. The remaining coke would be more permeable than the tar sand itself and hence would not hinder liquid migration and could be employed in a later stage of the process as a combustion fuel for recovery of more hydrocarbon material from any residual bitumen.

The results of several laboratory tests on combustion in samples of Athabasca tar sands were presented briefly in 1960 by Perry [16], Reed [17,18], Berry [19], and their respective coworkers. The results showed that in-situ combustion of the bitumen would produce an upgraded oil having a 25° API gravity. Temperatures and combustion-zone velocities were a function of the rate at which air was fed into the zone. At air-flow rates above 20—40 cu ft/sq ft/hour, formation temperatures were above 650°F and thus in the coking range for the bitumen.

Watson [20] found that cracking of the Athabasca bitumen and recovery of a light oil from the formation could be achieved by injection of hot inert gases into the sand until a temperature of 700—1200°F is attained. This caused cracking of the bitumen and also established a zone of eroded and heated circulating sand from which oil could later be recovered.

Carpenter [21] claimed that good recovery of oil from tar sands could be achieved by heating the bitumen by means of a heat-transfer medium contained within a wellbore in the formation; the liquid heat-transfer medium was heated electrically. It is doubtful in this instance if any cracking of the bitumen occurred and it appears that the sole purpose was to alter the viscosity of the bitumen to such an extent that recovery from the borehole could be achieved quite easily.

Davis [22] conducted a series of investigations on the possibility of volatilizing the oil in situ using electrical energy as the source of heat. He concluded that complete volatilization and recovery of the oil using electricity would be uneconomical. He did suggest, however, that the process of volatilization need only be commenced with electrical energy and so produce a fissure pattern to provide passages for injection of compressed air to support the combustion. In this manner, a partial combustion of the bitumen in situ would provide a heat source for volatilization of the oil.

Schoeppel and Sarapuu [23] also described a laboratory investigation into the application of electrical energy as a means of recovering oil from tar sands. They found that the average composite liquid product obtained from the tar sands was equivalent to a 16° API-gravity oil, rich in aromatics and olefins but lean in paraffins, suggesting to them an end use as a raw material for chemical and solvent manufacture rather than as a gasoline-type fuel. The gas produced by the electrocarbonization process possessed an unusually high hydrogen content with a heating value as high as 758 Btu/cu ft.

An interesting proposal for in-situ thermal treatment of the bitumen was put forward by Natland in 1963 [24]. The proposed experiment involved the peaceful use of nuclear energy insofar as a 9-kiloton nuclear device would be detonated at a depth of 1250 ft — some 20 ft below the McMurray bitumen — at a remote undeveloped area. The explosion would create a cavity some 230 ft in diameter into which several million cubic feet of oil sand would fall. The bitumen would, thus, be subject to cracking temperatures, allowing it to be recovered by conventional oil-field methods. Additional oil would also be made available by the shock energy from the explosion. Inasmuch as the explosion would be completely contained underground, no fallout was predicted and most of the fission products would be effectively entrained in a vitreous slag concentrated at the bottom of the blast cavity. The remaining fission products would be confined to the immediate vicinity of the explosion and be retained, through an ion-exchange process, by the rock minerals, thereby becoming unavailable to contaminate oil or water. The politics of the era necessitated that the proposal should be shelved pending further study. In the meantime, Doscher [25] has estimated that heating the crude by the energy liberated from a nuclear detonation is not, at the moment, competitive with more conventional thermal processes, such as steam flooding. Doscher did suggest, however, that if the sole purpose of the nuclear detonation is to create a fractured zone, thereby facilitating application of an in-situ solvent, emulsion, or thermal process, the feasibility of using a nuclear detonation for development of a recovery procedure for the Athabasca bitumen must be considered. Indeed, the feasibility of such a process has been kindled by a similar proposal involving recovery of natural gas in the U.S.A. by means of a nuclear explosion [26,27], but serious thought must be given to the consequences of detonating such a device in the McMurray area.

Direct thermal processing of mined tar sand

An interesting attempt to separate the bitumen—sand mixture and at the same time crack the heavy crude oil involves a fluidized-bed technique. In this process, a sand bed is fluidized by a hot mixture containing about 85% nitrogen and light petroleum gas. For example, direct heating of tar sands using the fluidized-bed technique has been investigated and described in several publications by Gishler and his coworkers [28—33]. The method essentially involves feeding raw tar sand into a coker or still maintained at ca. 900° F, where the tar sand is heated by contact with a fluidized bed of clean sand; volatile portions of the bitumen distill from the tar sand as a heavy synthetic crude oil. Residual fractions of the bitumen are converted to coke resulting in the deposition of a layer of coke around each sand grain. Regeneration of clean sand is achieved in a second unit heated to ca. 1400° F, where the majority of the coke is burned from the sand, and clean hot sand is recirculated to the coker. Reaction off-gases are recirculated to fluidize the clean sand in the coker.

In a more recent investigation [34], Rammler described the application of the Lurgi—Ruhrgas process to the production of synthetic crude oil from the Athabasca oil sands. This process is similar in nature to those investigated by Gishler and his colleagues insofar as it involves a fluidized-bed technique using fine-grained heat carriers. Thus, tests using 45 tons of raw oil sand indicated that the process is suitable for the recovery of a synthetic crude oil from the sand.

The patent literature also contains examples of several licences granted to various companies where the major process is coking of raw tar sand. For example, Berg [35] claims that direct coking of raw tar sand produces a coker distillate, amounting to almost 73% of the bitumen contained in the sand, having a 19° API gravity. Coke production is low, i.e., <19% of the hydrocarbon feed of which 70% fulfills the heat requirements of the process. The remainder of the hydrocarbon feed appears as gases and a substantial desulfurization is noted with hydrogen sulfide appearing in the gaseous products leaving the coke with <3% w/w sulfur.

Improved conversion of bitumen to oil has been claimed by Haensel [36] using a process where oil sand is introduced into a cracking zone by means of a stream of a hydrocarbon carrier which simultaneously separates and thermally degrades bitumen in the tar sands to produce a synthetic crude oil and/or other lower-boiling products. In a similar process, Stewart and his colleagues [37] claim production of optimum quantities of a good-grade synthetic crude oil from the Athabasca tar sand. This involves thermal treatment of enriched fluidized tar sand which produces, at various stages of the process, coke, tar, gas oil and naphtha. Steinmetz [38] has described a similar process wherein bitumen previously extracted from the tar sands can be added as a froth to fresh tar sand and used as a feed for direct coking. The proportion of added bitumen in the feed can be adjusted so that, regardless of the amount of bitumen in the fresh sand, the mixed feed always contains the same amount of cokable material. This allows a better economic balance as the oil recovery is highly efficient and provides liquid hydrocarbon yields better than 90% of the theoretical.

On the other hand, Nathan and his colleagues [39—41] stated that maximum recovery of a synthetic crude oil, or other saleable products, can be obtained by first admixing tar sand and fluid bed materials at 250—600° F prior to introduction into a fluid-coking zone maintained at temperatures above 850° F. A similar claim has been made by Tse [42] who contends that increases in yield and quality of the synthetic crude oil and/or hydrocarbon products are noted when the tar sands are heated at 200—380° F to increase flowability of the material prior to introduction into the pyrolytic zone. In essence, the method provides some little modification to that of Nathan described above.

Bennett [43] has claimed that the system for retorting the bituminous sands can be refined by using a horizontally moving bed of the sands which are compacted and perforated prior to entry into the retort. This provides vertical flow channels which allow a controlled distribution of heat and burning within the retort. The method is claimed to produce excellent conversion of the bitumen to oil and allows efficient reuse of the heat generated during the burning.

The direct coking of tar sands is a simple direct method of recovering a synthetic crude oil from raw material and the technology involved is similar to that used in fluid cracking. The process has the elegant advantage over the hot-water separation in that feeds of high mineral content offer no particular problem and high yields of synthetic crude oil (>75% based on bitumen in the feed) are observed. Large amounts of sand *[1], relative to oil, however, must be circulated, and discharge of the hot sand from the unit may represent a significant heat loss. The process described by Rammler [34] is believed to overcome these problems by using it in conjunction with other processes whereby the tar sand employed (having undergone a beneficiation treatment) is comparatively rich in bitumen. Thus, other modifications such as preheating the air, required in the process, by passage through the hot sand are believed to present a more favorable economic picture. Rammler has estimated that the net processing cost for a barrel of synthetic crude oil is approximately 50c *[2].

Thermal processing of separated bitumen

The cracking of the bitumen from the Athabasca tar sands was the subject of an early report by Egloff and Morrell [44,45] who studied the effect of a temperature of 750° F and a pressure of 90 psi on the bitumen. They observed formation of a distillate, gases, and coke and noted that the bitumen lends itself to cracking at relatively low temperature and pressure to produce a stable motor fuel high in aromatics content. The quantity of gas formed was approximately 1000 cu ft/bbl (1 barrel = 42 gallons) of bitumen cracked, having a Btu value of approximately 1000 per cubic foot of gas. High coke yields were noted, there being approximately 139 pounds of coke produced per barrel of bitumen; the Btu per pound of coke approximated 15,000. Reference was also made to the thermal sensitivity of the bitumen in a lengthy paper by Ball [46]. In discussing the results of several tests

---

*[1] Yields are of the order of one barrel of heavy synthetic crude oil per ton of tar sand treated.

*[2] Prices as of 1970 for one barrel of synthetic crude oil were ca. $ 3.00 at Edmonton.

performed on the bitumen by commercial enterprises, Ball considered the bitumen to be remarkably sensitive to heat and, therefore, to refining processes.

Viscosity reduction by thermal methods was the subject of a study by Peterson [47] who reported considerable reduction in bitumen viscosity, without extensive coking, at 475—500° C at contact times of ca. 100 sec and pressures of the order of 200 psi. He noted that the conditions compared favorably with those of commercial visbreaking (460—485° C; 200—500 psi) although no data were available for process contact times. Peterson did note that at temperatures above 490° C, coke formation occurs readily but could be minimized under conditions of turbulent flow and 300—400 psi. The sulfur content of the bitumen decreased throughout these processes but the reduction was not great enough to solve the problem of its complete removal. Peterson went on to suggest that, under conditions of turbulent flow, a temperature of 485° C at 400 psi and a contact time of 150—200 sec would be sufficient to reduce the viscosity of the bitumen to ca. one-tenth of its original value without excessive coke formation.

Sterba [48] made a study of the thermal coking of the bitumen to yield a variety of products including gasoline, gas oil, coke and gaseous products. He concluded that the general character of the thermal decomposition is similar when coking bitumen in batch, fluidized-bed, or "once-through" types of operation, but the extent of the decomposition does vary from one process to another. The "once-through" coking technique, however, afforded product distributions similar to those obtained from other petroleums and, except for a higher ash content of the bitumen coke, there are marked resemblances between the bitumen coke and coke from high-sulfur petroleum feedstocks. The octane number of the gasoline from the bitumen is slightly higher than that of the gasoline from conventional oils, whereas sulfur contents of the bitumen gasoline are similar to those of gasoline from certain California crudes. The relative sulfur concentrations in the total distillates from bitumen and a petroleum pitch were approximately equal. Thus, Sterba concluded that there were no basic differences in the thermal decomposition of sulfur compounds found in these two types of raw material.

Gishler and his coworkers [49] reported the results of a study of the simultaneous dehydration and coking of the bitumen using a fluidized-solids technique. The method involved introduction of the warmed wet bitumen into the lower part of a fluidized bed at about 500° C causing a dry 16° API oil to be flash-distilled to a collecting system. Yields of oil were reportedly good (ca. 95%) and the properties were similar to those of the oil obtained by the fluidized processing of raw bituminous sand [50].

Warren and his colleagues [51] also studied the flash distillation of the bitumen. The process involved feeding a wet diluted bitumen mix (ca. 70% bitumen) into a flash coking unit maintained at temperatures >900° F. The reaction produced a synthetic crude oil, equivalent to about 75% of the bitumen in the feed, with less than 4% w/w sulfur. Coke yields amounted to 14—17% of the bitumen and had a calorific value of ca. 14,000 Btu/lb. The remainder occurred as hydrocarbon gases (C_1—C_5) and contained hydrogen, carbon monoxide and hydrogen sulfide.

Pasternack [52,53] proposed a partial coking, or thermal deasphalting process to provide minimal upgrading of the bitumen. The objective of the process was to remove minerals and water from a hot-water process wet product to yield a dehydrated mineral-free material containing the majority of the asphaltenes and exhibiting a near-original Ramsbottom carbon value. The method essentially involved distilling over a low proportion of the oil, with water, at temperatures up to 850°F. Coke formation occurred using entrained mineral particles as nuclei and filtration produced an essentially mineral-free product suitable for several applications, such as metallurgical coke, for which the original mineral content would have been a disqualifying factor.

In a following study, Pasternack [54—56] also observed the changes occurring in the bitumen and its components during thermal treatment. He observed changes which occur in molecular weight, ultimate composition, Btu value, specific gravity and penetration. In general, the molecular weight of the residual oil increased substantially with increased distillate removal. Maximum production of Ramsbottom carbon residue from the bitumen, or its coker distillate, occurred when the material was preheated at 800—825°F prior to carbonization. Residues, representing the high-boiling ends of the fluidized coker distillate, provide carbon residues with excellent cohesiveness and permit the production of hard low-temperature coke, which may be strong enough to be fired directly into a metallurgical furnace.

Henderson and Weber [57] studied the physical upgrading of the Athabasca bitumen together with several conventional crude oils. They noted that all oils showed permanent reductions in viscosity, specific gravity, and pour point as a result of mild thermal cracking. From their kinetic data, Henderson and Weber derived an activation energy of 49 kcal/mole for the bitumen and noted that only two of the six conventional crude oils had lower activation energies. McNab and his colleagues [58] had previously derived an activation energy of 49 kcal/mole for the Athabasca bitumen and used their data, along with that from other studies, to show that the bitumen was easier to crack than the average gas oil from a conventional crude which had an activation energy of ca. 55 kcal/mole. Consequently, it appears that the bitumen will crack considerably easier than certain conventional crudes; this is in agreement with the earlier observations made by Ball [46] on the sensitivity of the bitumen to thermal treatment.

The mild thermal alteration of the bitumen, along with several other crude oils and asphalts, was the subject of a study by Erdman and Dickie [59]. At the temperature of the investigations (350°C), it was concluded that the bitumen was quite heat sensitive relative to the remaining asphaltic oils.

More recent work by the writer [60—62] has shown that not only the bitumen and deasphalted heavy oil but also the asphaltenes can be cracked at ~460—470°C to produce a distillate, coke, and gaseous materials. The interesting point here is that considerable quantities of the nitrogen, oxygen, and sulfur are eliminated as their hydrogenated counterparts. Indications are that the nitrogen, oxygen, and sulfur which remain do so because of the stability conferred upon them by incorporation into ring systems. Application of a variety of spectroscopic techniques to investigate the nature of the near-aliphatic distillate (H/C = 1.65) shows that *n*-paraffins (up to at least C_{32}) are present and that an overall dealkylation of aromatics to methyl (predominantly) and ethyl (minority) groups has occurred. The hydrocarbon gases produced during the cracking varied from C_1 to C_6, but were predominantly C_3 and normal C_4 compounds, and the gaseous mix had excellent burning properties.

Hydrocracking of the separated bitumen

Products from thermal-coking investigations have been shown to contain organically bound sulfur distributed throughout their boiling-point ranges. Indeed, the amounts reported are usually several times the preferred amount designated suitable for further refining and, therefore, alternate treatment methods are required which include some form of sulfur removal. A partial, if not complete, solution to the problem has occurred with the application of a concurrent hydrogenation—cracking technique to the bitumen, thus allowing removal of the large majority of sulfur from the synthetic crude oil product. It is not surprising, therefore, that thermal processing of the bitumen in the presence of hydrogen was the subject of several earlier reports of investigations into the nature of the bitumen, as it appeared to be a natural extension of the conventional thermal degradative processes as well as the above-mentioned necessity to produce an essentially sulfur-free synthetic crude oil.

One of the earliest reports on the subject dates from the 1922—1926 period (cf. [63]). In the former year, a quantity of tar sand was sent to Bergin's laboratory at Rheinau and in 1926 a report of the test was forwarded to the Fuel Research Laboratories in Ottawa. There were no operating details, but it appears that cracking of the bitumen in an atmosphere of hydrogen produced gasoline (30% *), creosoting oil (5%), diesel oil (25%), and pitch (25%). The nature of the 15% not accounted for was not stated.

Boomer and Saddington [64] noted that, at temperatures up to 900°C, the bitumen was hydrogenated to form a light synthetic crude oil amounting to 80% or more of the original material. Optimum temperatures were in the

* Yields are given as percent by weight of bitumen charged.

region of 380° C and the best catalysts were ammonium molybdate and aluminum chloride. They also reported that the material consumed, with relative ease, an amount of hydrogen equivalent to 3% by weight of the bitumen. The synthetic crude oil had a sulfur content of about one-half that of the bitumen and was easily refined to produce stable, white gasoline. They concluded that the process was superior to the conventional cracking method by virtue of a greater oil yield and lower coke production.

In a subsequent publication, Boomer and Saddington [65] reported a second series of investigations on the batch hydrogenation of the bitumen. Again, they reported that hydrogenation of the majority of the bitumen occurred readily and at comparatively low temperatures. The rate of hydrogenation was dependent upon temperature and presence of a catalyst; for example, molybdic anhydride was reported to approximately double the rate of reaction. Furthermore, the extent and rate of hydrogenation decreased in repeated experiments with the same sample of bitumen due to the adverse effect of an increasing proportion of simple hydrocarbons. An increase in the relative amount of hydrogen to bitumen increased the absolute rate of reaction and also the degree of hydrogenation possible in one treatment. As a result of the shorter time required, coke formation was largely suppressed. The authors concluded that 75% of the bitumen could be easily converted to synthetic crude oil and, hence, gasoline; about 15% * was resistant to hydrogenation up to 500° C, and about 10% appeared as gaseous products.

Boomer and Edwards [66] also studied the destructive hydrogenation of Athabasca bitumen in tetralin at 400° C at pressures of the order of 2500 psi. They observed yields of synthetic crude oil greater than 78% with up to 7% of the bitumen occurring as a solid and the remainder as gases. It was interesting to note that 51% of the sulfur appeared as hydrogen sulfide, but without tetralin a substantial decrease in the desulfurization occurred and only 34% of the sulfur was released as hydrogen sulfide. The rate of the reaction was markedly increased by the presence of tetralin and the authors equated the role of the solvent to that of a catalyst as it substantially increased the reaction velocity without significantly altering the extent of the reaction. In the same report [66], Boomer and Edwards described an attempt to use tetralin alone as the hydrogenating agent and used natural gas (93% methane plus 5% nitrogen) to create the desired pressure (3510 psi). The formation of naphthalene suggested to the authors that the tetralin was, in fact, releasing hydrogen within the system, and they concluded that more than half of the hydrogen content of the tetralin reacted with the bitumen to produce the synthetic crude oil.

---

* This is presumably the majority of asphaltene fraction which constitutes 17% of the bitumen and can only be hydrogenated at relatively higher temperatures with a concurrent increase in coke formation.

At about the same time, Warren [63] reported the results of a series of investigations on the hydrogenation of the bitumen. He noted that coke formation is increased by high temperatures, low pressures, and long reaction times, and sulfur removal was enhanced by low temperature, long duration, and high pressure. The kerosene yield did not appear to be influenced by temperature or pressure and was only slightly affected by increased duration of the reaction. The amount of hydrogen absorbed was independent of the temperature and increased only slightly with duration of the reaction, but was directly proportional to the pressure. Warren also reported that the catalysts employed in some of his investigations did not greatly increase the yield of gasoline nor did they cause removal of sulfur as hydrogen sulfide. He did find, however, that sulfide could be completely eliminated from the charge by combination with iron oxide or calcium oxide and that, if preheated with hydrogen, these materials were capable of reducing coke formation. In the absence of hydrogen pretreatment, iron oxide produced the greatest amount of coke and nickel carbonate the least, with chromic oxide, zinc oxide, tin, ammonium molybdate, and copper oxide showing decreasing coke formation in that order. Warren concluded that on the basis of his experiments and those carried out to that date on conventional cracking [44,45] there appeared to be a balance in favor of hydrocracking.

Warren and Bowles [67] investigated the hydrogenation of the bitumen using a molybdic oxide catalyst supported on coke at temperatures in the range of 428—452° C and pressure of ca. 2700 psi. They noted yields of 73% by volume of the dry bitumen charge (or ca. 100% by volume of the net charge). The majority (ca. 90%) of the products boiled below 410° F and there were some (<10% w/w) olefins in the product. Sulfur content of the product was 0.6% and hydrogen consumption was 5.7% by weight of the total charge; coke formation was negligible. The authors concluded that some refining of the synthetic crude oil would be necessary to produce a marketable product.

A later publication by Warren and his colleagues [68] described the effects of process variables on sulfur content, yields, hydrogen consumption, and physical properties of the products. They noted that temperature exerted the greater influence on product properties and showed that a small decrease in operating temperature below 420° C (788° F), with other process variables remaining constant, caused a large increase in the sulfur content of the product. Above 425° C (797° F), however, a corresponding increase in temperature caused only a slight decrease in the sulfur content of the product. The volume yield of product increased with the temperature to a maximum of near 100% at 400° C (752° F), holding constant to about 440° C (824° F) and falling off rapidly above this temperature. Pressure had an important influence in the range below 1000 psi, with the sulfur content of the product increasing rapidly, but at higher pressures it only had a small effect on the product composition and properties. Thus at the optimum conditions suggested by the authors, i.e., 420° (788° F) and 1000 psi, the product contained 0.2% sulfur and the process consumed 600 cu ft/bbl of feed for a volume yield of near 100%. Perhaps it is noteworthy that at temperatures below 420° C (788° F) hydrogen consumption dropped considerably, but at higher temperatures hydrogen consumption increased drastically.

Montgomery and his coworkers [69,70] reported an investigation of the desulfurization of coker distillates from separated bitumen using a hydrogenation technique over a cobalt molybdate catalyst. They obtained similar results to those obtained by Warren and his colleagues [68] that optimum conditions for a particular feedstock were of the order of 420° C (788° F) and 1000 psi, with a hydrogen flow of 4500 cu ft/bbl of feed, and achieved a volume yield of oil close to 100% containing only 0.2% by weight sulfur. They also noted the rate of decline of catalyst activity and found that at a pressure of 1000 psi little coking occurred at temperatures below 430° C (806° F), but above this temperature coke deposition increased exponentially with temperature. They observed that the quality of the motor gasolines produced in the process was a little below that of commercial regular-grade gasoline and the gas oils were on the borderline of acceptability as diesel fuel. The heavier distillates were cited as usable for fuel oils or cracking and recycle stocks.

Other workers [71,72] pyrolytically distilled the bitumen and examined the hydrogenation of the distillate over a fixed bed of cobalt molybdate on alumina at temperatures in the range of 797—896° F and pressures of 1000—10,000 psi. They noted that the rate of catalyst deactivation due to coke deposition was approximately inversely proportional to the cube of the reaction pressure and that pressures above 2500 psi and temperatures of 840—860° F were required to remove 90% or more of the original sulfur; the calculated hydrogen consumption was 1700 cu ft/bbl of feed. Nitrogen was observed to be more difficult to remove than sulfur, but an increase in pressure produced a greater degree of nitrogen removal than sulfur removal. The overall volume recovery of synthetic crude oil was excellent (ca. 100%) at all conditions above 1000 psi, and good conversions of feedstock to gasoline-range boiling material was good at high temperatures and ca. 3000 psi. In an extension of this work to the Athabasca bitumen [73], the authors noted very high conversions (ca. 100%) to a synthetic crude oil of which less than 4% boiled above 915° F. There was a reasonably low hydrogen consumption (<1500 cu ft/bbl of bitumen feed), and product distribution could be varied by alterations in hydrogen recycle rate or operating pressure. For example, increasing the hydrogen recycle rate tended to produce heavier products, whereas production of lighter products accompanied increased pressure.

White [74,75] has claimed that bituminous emulsions from the hot-water separation process can be effectively and economically subjected to destructive hydrogenation at temperatures in the range of 700—900° F and pressures of 1000—3500 psi with or without a catalyst. A good-grade synthetic crude oil, having a low sulfur content, is produced and the process requires only 800—1000 cu ft of hydrogen per barrel of bitumen feed.

Parsons [76] reported a method of estimating the amount of hydrogen to upgrade the Athabasca bitumen together with a series of residual oils and tars. Several process factors were taken into account and Parsons concluded that surprisingly little hydrogen was required to upgrade the Athabasca bitumen together with a series of residual oils and tars, and the maximum cost per barrel would not be more than 30c (~1400 cu ft). If, in fact, only 600 cu ft of hydrogen was required per barrel [68] and recovery of the residual hydrogen was possible, the cost would of course be much lower (ca. 12c per barrel).

Presumably the low hydrogen requirement is due to the existence of a dehydrogenation—hydrogenation process within the system whereby coke is produced on the one hand (dehydrogenation) and a near-aliphatic synthetic crude oil is produced on the other hand (hydrogenation) [61]. In fact, experiments by the writer [62] during the conventional cracking process show that increased amounts of lower-molecular-weight hydrocarbons are produced using reduced bitumen, or fractions thereof, compared to cracking in the absence or presence of gaseous hydrogen. In fact, it is apparent that the increase in the proportions of the hydrocarbon gases bears a relationship to the hydrogen uptake by the material. It is apparent, therefore, that the predominant mechanism of hydrocracking appears to be one of hydrogen transfer within the system with added gaseous hydrogen contributing only slightly to the product mix.

In a later publication, Parsons and his coworkers [77,78] made a comparison of the thermal and catalytic hydrogen of the bitumen. The experiments were made in the liquid phase using conventional flow apparatus at pressures ranging from 500 to 3500 psi. The rate of accumulation of tar, coke, and mineral matter in the reaction vessel was greatest at high conversion levels and low pressures. Thus, they noted that continuous operation was not possible at 500 psi, but at 1000 psi the concentration of residuum (including clay) could be reduced to 18—20% without serious difficulty. Considerable gas formation occurred at all pressures in the thermal experiments and at low pressures in the catalytic system, but it was only at high pressures that the catalyst suppressed gasification relative to the conversion of the residuum, resulting in a marked increase in the yield of the liquid product. The authors concluded that the maximum permissible extra cost for catalytic processing, compared to thermal hydrocracking (capital cost plus catalyst), would be 25—30c per barrel.

In a separate report O'Grady and Parsons [79] noted the efficiencies of different forms of catalyst * during catalytic hydrotreating of the bitumen.

* Oxide form — cobalt molybdena on alumina.
Sulfide form — cobalt molybdena on alumina pretreated with H_2S at 850°F/15 psi/1 hr.

They speculated that the greater cracking activity of the sulfide form of the catalyst leads to more rapid coke buildup on the surface and rapid deactivation of the sulfide catalyst, whereas the oxide form had low-enough cracking activity for coke precursors to be washed away from the catalyst surface by the liquid-phase portion of the reaction medium.

Parsons and his colleagues [80], in agreement with the earlier work of Boomer and Edwards [66], have reported marked improvements in the conversion of the bitumen during hydrocracking by dilution of the feedstock with small quantities of low-boiling gas oils. They also noted that any factor which reduces the concentration of low-boiling hydrocarbon species in the reactor (such as topping the feed or using a high gas-flow rate) reduces conversion and enhances coke formation. They concluded, however, that (1) the hydrogen-transfer capability, or catalytic activity, of the low-boiling fractions is small compared with that of a good heterogeneous hydrogenation catalyst, and (2) success, or failure, with the dilution technique hinges, to some degree, on the thermal history of the preparation of the feedstock.

Product properties

The aforegoing section provides an account of the considerable amount of work relevant to conversion of the bitumen to a synthetic crude oil. Unfortunately, the patent literature gives very few, if any, detailed product inspections, and reliance, therefore, must be put on the more conventional scientific literature for product properties and illustrations of process performance.

Synthetic crude oil

Inspections of synthetic crude oils prepared from the bitumen are presented in Table 6-I. As mentioned previously, absolute comparisons of product properties, where the products arise from a variety of processes, must be made with caution as several process variables can account for considerable differences in the product. For the purpose of the table, an attempt has been made to bring together the available data to show if a general processing trend has, in fact, emerged as a result of the aforementioned thermal-treatment methods.

The data presented in Table 6-I show that at temperatures of $430 \pm 30°C$ a synthetic crude oil (API gravity $>15°$) can be produced from the bitumen in good yield. The sulfur content of the product, however, remains substantially higher than desired (ca. 4.0 wt%) and occurs throughout the boiling range of the synthetic crude oil. More significant results are obtained when the bitumen is subjected to destructive hydrogenation and marked reductions (to <1.6 wt%) are then noted in the sulfur content of the products as well as excellent conversion yields. Furthermore, the data generally indicate that destructive hydrogenation, especially of a coker distillate, is the preferred route from bitumen to a synthetic crude oil insofar as this approach brings about the most spectacular decreases in sulfur content. Indeed, on a commercial scale, it is possible to remove sulfur (and nitrogen) down to parts-per-million level by suitable choice of process conditions. As an illustration of the effect of process conditions on product properties, the variation of sulfur content with temperature during catalytic hydrodesulfurization is presented in Fig. 6-1.

Gases and coke

The synthetic crude oil, although by far the major product, is by no means the only product of the cracking processes. Other interrelated products are gaseous materials and coke (both are capable of being integrated into the process as fuel), and it is possible to vary the processing conditions depending upon current requirements. As an example of product interrelation, Fig. 6-2 shows the relationship between distillate yield, API gravity of the distillate, and coke yield.

The suitability of the coke and gases for fuels was recorded in an early publication by Egloff and Morrell [44,45], who obtained 1000 cu ft of gas and 139 lb of coke per barrel of bitumen treated and noted that the Btu per cu ft of gas approximated 1000, whereas the coke had a calorific value of ca. 15,000 Btu/lb. Later investigations (summarized in Tables 6-II, 6-III, and 6-IV) have confirmed this.

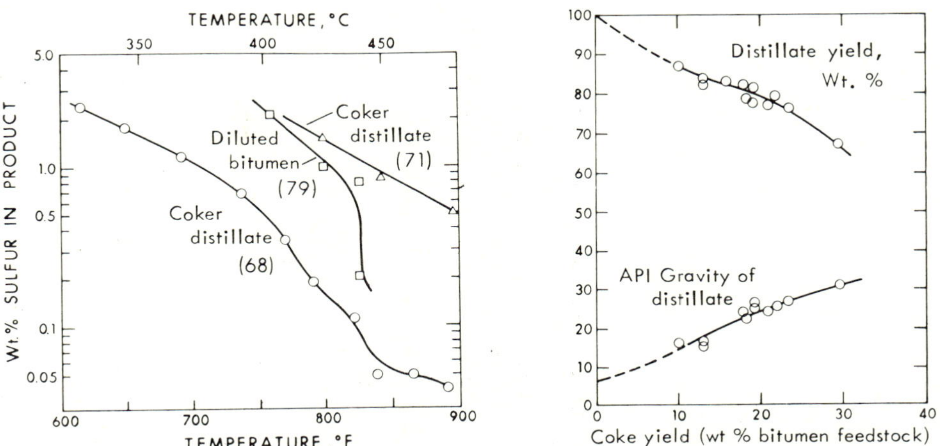

Fig. 6-1. Effect of temperature on catalytic hydrodesulfurization. (After Camp [81].)

Fig. 6-2. Yields and API gravities of coker distillates. (After Sterba [48].)

TABLE 6-I

Properties of synthetic crude oils prepared from Athabasca bitumen

Composition (wt% on d.a.f. basis)	Original bitu-men	In situ crack-ing	Process											
			destructive distillation						fluidized bed					
									tar sand				bitumen	
Carbon	80.8	84.0		84.4	84.4		86.0	85.9			83.8	85.0		
Hydrogen	9.9	10.0		11.3	11.2		11.8	11.5			11.2	11.7		
Oxygen	3.1	0.9		0.6	0.1		0.3	0.3			0.8	0.4		
Nitrogen	1.1	0.5		0.1	0.2		0.4	0.4			0.2	0.2		
Sulfur	5.1	4.5	4.0	3.6	4.1		1.6	1.9	4.0	3.8	4.0	2.7	4.0	<1.3
Molecular weight	532	~250		322	393		238	242		421	368			
Carbon residue (Conradson)	14.3			2.0	4.2					7.0		1.3	7.5	
Gravity (°API)	7.3	16.0	19.0	19.7	15.4	38.8			15.9	14.8	15.9	31.5	19.3	>15.0
Oil yield														
wt%			75.3		78.2	56.0	55.0	66.0					79.4	
vol%					84.0								85.5	
Reference	[61]	[23]	[51]	[69]	[71]	[44]	[61]	[61 [*1]]	[31]	[69]	[69]	[34]	[32]	[78]

	destructive hydrogenation											
Carbon		85.0										
Hydrogen		12.7										
Oxygen												
Nitrogen												
Sulfur	< 2.8	1.6	1.4	0.1	0.9	1.9				0.2	0.9	0.8
Molecular weight												
Carbon residue (Conradson)		9.0	12.5						2.8			
Gravity (°API)	> 14.0	20.2	22.3	32.1	28.4	32.5	17.3	17.9	20.0	30.0	25.5	25.9
Oil yield												
wt%		90.5	71.3	93.7	77.0	84.0	94.5	95.0	78.0	94.7	90.6	93.5
vol%				102.0						101.0	97.2	99.8
Reference	[78*2]	[65]	[65*2]	[68*2]	[79]	[79]	[80]	[80*4]	[66*3]	[69*3]	[69*3]	[69*3]

*1 Using deasphalted bitumen as the feedstock.

Notations for destructive hydrogenation processes:

*2 Without catalyst.

*3 Feed was coker distillate having 3.5—4.5 wt% sulfur.

*4 Hydrocarbon diluent present.

TABLE 6-II. Composition of gaseous products

	Composition (vol%)			
Component				
Methane	21.7	18.8	33.6	55.8
Ethylene	5.1			1.5
Ethane	13.3		6.3	
Propylene	8.7			2.3
Propane	5.6			
Iso-butane	1.3			1.9
n-Butane	5.6		4.8	1.6
Iso-pentane	2.8			0.8
n-Pentane	1.2			3.6
Hydrogen	17.1	61.5	41.8	25.7
Carbon monoxide	3.9	7.4	3.2	3.4
Hydrogen sulfide	13.7			
Carbon dioxide			1.0	
Other (CO_2, SO_2, etc.)		12.3	9.3	3.4
Calorific value				
Btu/lb	17,598			
Btu/cu ft		529	758	
Reference	[51]	[23]	[23]	[49 *]

* Normalized to exclude nitrogen used in the process.

TABLE 6-III. Product yields from the delayed coking of bitumen

Product	Scale of operation		
	pilot plant	pilot plant	commercial (G.C.O.S.)
Product yields (wt%)			
H_2S	1.2		
Hydrocarbon gases (C_1—C_4)	7.0		
Total gases (C_1—C_4)		8.3	7.9
Gasoline (C_{5+})	15.4	12.1	12.7
Kerosene		10.1	15.0
Gas oil	55.0	41.4	36.2
Fuel oil		4.2	6.0
Coke	21.0	22.7	22.2
Unaccounted	0.4	1.3	
Gas composition (mole%)			
H_2	11.0	24.2	
CH_4	47.1	32.9	
C_2H_4	1.8	2.3	
C_2H_6	14.2	11.6	
C_3H_6	6.1	3.4	
C_3H_8	11.7	7.6	
C_4H_8	3.6		
C_4H_{10}	4.5	0.6	
H_2S and others		17.4	
Reference	[48]	[82]	[83]

TABLE 6-IV

Properties of coke from Athabasca bitumen

Property	Sample number			
	1	2	3	4
Proximate analysis (wt%)				
Volatile	10.4	14.3	13.0	10.85
Fixed carbon	84.3	77.0	82.9	
Ash	5.3	8.7	4.1	2.95
Heating value (Btu/lb)	14,300	14,000	14,710	
Sulfur (wt%)	6.0	6.0	6.0	6.42
Reference	[81]	[81]	[81]	[48]

Furthermore, the fuel requirements for the production of coker distillate from the bitumen were estimated by Carson and Booth [84], who concluded that for a plant producing 20,000 bbl/day of hydrogenated synthetic crude oil, natural-gas requirements would vary in the range of 400—2000 cu ft/bbl of product (Table 6-V). It can be seen from the table that utilization of the coke for fuel brings about considerable reduction in the gas requirements.

TABLE 6-V

Natural-gas requirements for processing Alberta bitumen (After Carson and Booth [84])

Process	Natural-gas requirements			
	thousands of cu ft/hr	millions of cu ft/day	millions of cu ft/ million bbl of product	cu ft/bbl of product
Hot-water separation, fluidized coking	691	16.6	830	830
Hot-water separation, dehydration, conventional coking	397	9.5	475	475
As above — coke not used	1287	30.9	1545	1545
Cold-water separation, dehydration, conven-ventional coking	326	7.8	392	392
As above — coke not used	1536	36.8	1840	1840

Commercial upgrading of the bitumen to a synthetic crude oil

The aforementioned processes have shown that one of the main drawbacks to straight thermal treatment of the bitumen is the occurrence of organically bound sulfur throughout the range of products. This can be overcome to a great extent by thermal treatment in the presence of hydrogen with, or without, the aid of a catalyst. Indeed, it is as a result of this work and the work carried out in earlier pilot-plant studies and semicommercial operations that it is now possible to process the bitumen economically.

At present, recovery of the bitumen by mining methods is also an economic feasibility as there remain several large areas where the overburden is not deeper than 150 ft. Consequently, in-situ methods of recovery, or for that matter in-situ conversion, of the bitumen — and there are numerous claims to this effect [85,86] — are, for the moment, precluded. As the richer deposits are depleted, in-situ techniques will, no doubt, become prominent and eventually may become the sole method of bitumen, or product, recovery.

There have been numerous proposals for the commercial production of a synthetic crude oil from the Athabasca bitumen but, due to amalgamations or unforeseen circumstances, the field was narrowed considerably. At present one plant is in commercial operation in the tar-sand area with another holding the option to go into commercial production in the near future; a third proposal was recently (1968) withdrawn. A brief description of these ventures is given below (also see Chapter 5).

The Great Canadian Oil Sands (G.C.O.S.) plant which has been in operation for several years involves a delayed-coking technique followed by hydrogen treating of the distillates to produce the synthetic crude oil [87,88]. The selection of delayed coking over mild thermal-cracking processes, such as visbreaking, was based on the high yields of residuum produced in these processes, which would exceed the fuel requirements of the process, especially if the distillates had to be shipped elsewhere for hydrogen treating. The product distribution and properties were also more favorable than those encountered in the other thermal processes. A simplified process flowsheet is illustrated in Fig. 6-3 and a comparison of bitumen and synthetic crude oil properties is given in Table 6-VI.

At present, one other commercial plant is in the planning stages and is to be built by Syncrude Canada, Ltd. [12]. It will be substantially larger than the G.C.O.S. plant and is designed to produce a minimum of 125,000 bbl/day of 32° API synthetic crude oil and related products (Fig. 6-4) and is scheduled to commence production in 1977. Again, hydrotreating will be one of the key factors in producing low-sulfur products. Syncrude workers have also reported [90] that their hydrovisbreaking process consumes only modest amounts of hydrogen and has good economic balance when the residuum is 13—15% w/w of the charge, and they concluded that hydrovis-

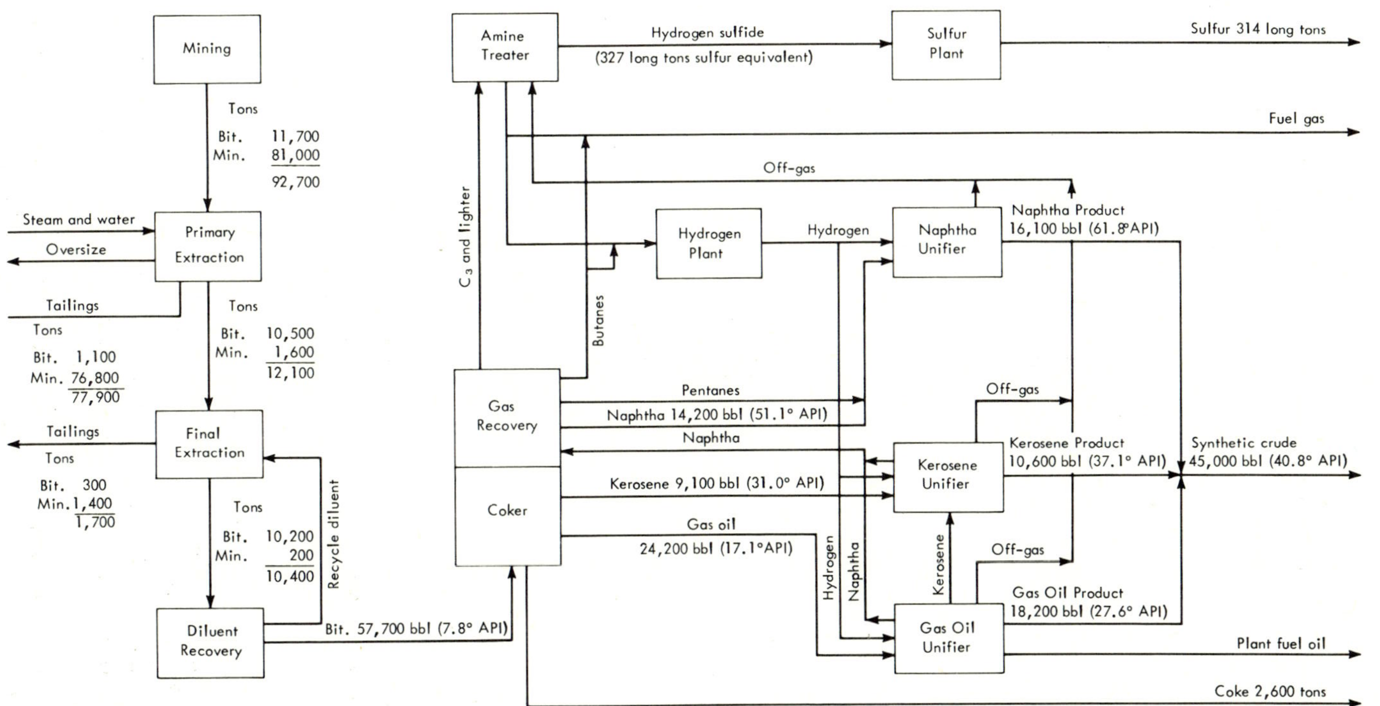

Fig. 6-3. Process flowsheet for the G.C.O.S. plant. (After Innes and Fear [87].)

TABLE 6-VI

Properties and analysis of G.C.O.S. bitumen and synthetic crude oil

	Bitumen	Synthetic crude (C_{6+})
Gravity (°API)	6.5	37.6
Distillation (°F)		
IBP	505	210
5%	544	277
10%	610	300
30%	795	379
50%	981	486
95%	—	738
EP	1030	833
% Rec.	50	99
Viscosity		
SUS @ 100°F	35,100	34.4
SUS @ 210°F	513	—
Pour point (°F)	+50	−30
Elemental analysis (wt%)		
Carbon	83.2	86.3
Hydrogen	10.4	13.4
Sulfur	4.2	0.03
Oxygen	0.94	—
Nitrogen	0.36	0.02
Hydrocarbon type (wt%)		
Aphaltenes	19	0
Polar aromatics	32	0
Aromatics	30	21
Saturates	19	79
Trace metals (ppm)		
V	250	0.01
Cu	5	0.02
Ash (wt% on ignition-free basis)	0.65	nil
Ramsbottom carbon (wt%)	10	0.04

breaking, as developed by them, is the preferred upgrading approach.

Shell Canada, Ltd. also put forward a proposal to produce 100,000 bbl/day of 33°API synthetic crude oil [91]. The bitumen refining scheme was analogous to that of G.C.O.S. with the exception that the thermal-cracking unit was to produce a liquid pitch, rather than coke, for use as a refinery fuel. For a variety of reasons, Shell withdrew the proposal in 1968.

An inspection of the products produced in a semicommercial and a commercial plant is presented in Table 6-VII. The results of other commercial and semi-

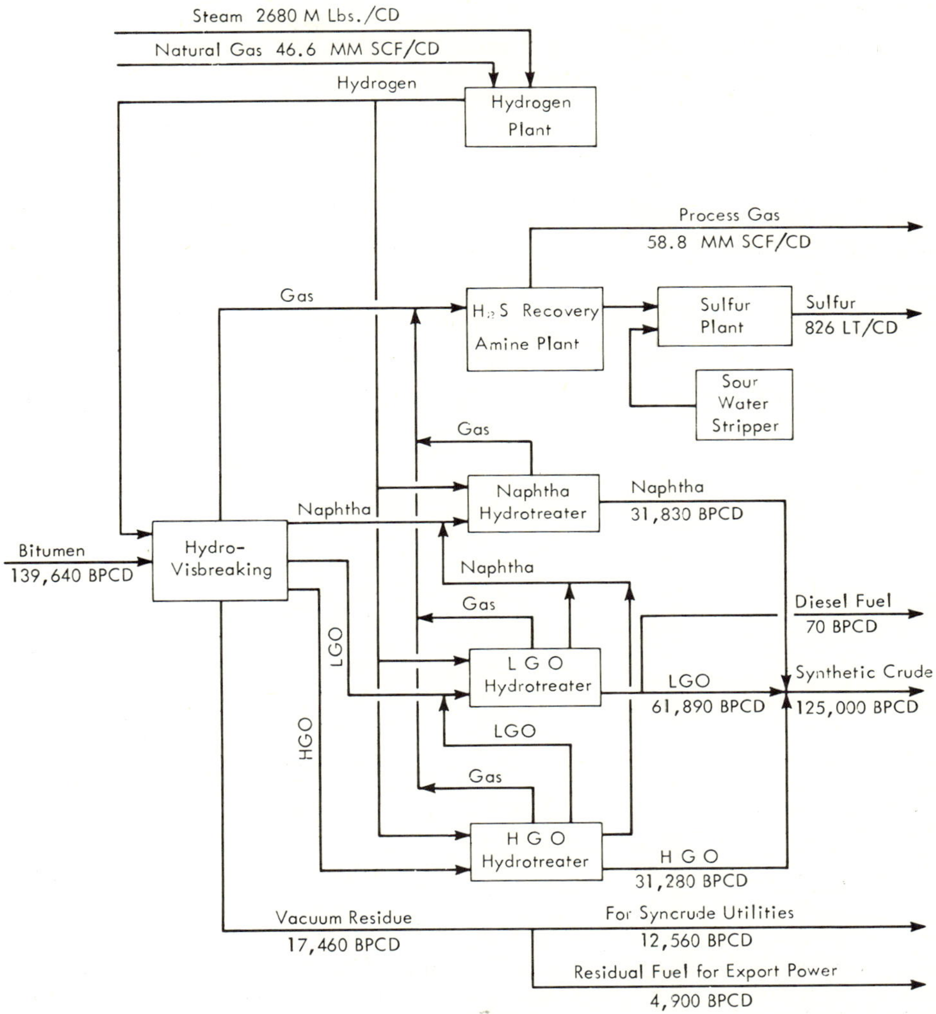

Fig. 6-4. Process flowsheet for the propose Syncrude Canada, Ltd. plant. (After Anonymous [89].)

commercial units have also been reported and products inspections are presented in Tables 6-VIII and 6-IX.

Great Canadian Oil Sands, Ltd., the commercial pioneer in tar-sand development, have increased their daily production of synthetic crude oil quite substantially over the past few years (Table 6-X). With the advent of the proposed Syncrude Canada, Ltd. plant, the flow of synthetic crude oil from the tar sands will make a significant contribution to the availability of fuel in Canada and, perhaps, even in the whole of North America.

TABLE 6-VII

Product inspections of delayed coking of bitumen

Inspection	Scale of operation				
	pilot plant		commercial design basis		
	gasoline	gas oil	naphtha	kerosene	gas oil
Coker distillate (wt%)	19.5	69.6	16.5	13.7	51.8
Gravity (°API)	51.9	16.6	46.8	32.9	18.3
Sulfur (wt%)	1.86	4.04	2.2	2.7	3.8
Bromine no.	80	47	61	36	20
Nitrogen, total (ppm)			150	400	2000
Vol%					
Aromatics			19	39.2	62.1
Olefins			32	14.4	2.0
Paraffins and naphthenes			49	46.4	35.9
Distillation temperature (°F)					
IBP	126	443	180	380	515
5%	165	466	202	396	530
10%	186	486	220	409	550
30%	232	550	268	428	600
50%	275	621	295	441	645
70%	315	690	314	458	697
90%	358	715	347	477	780
95%			360	490	807
EP (end pt.)	400	760+	400	535	850
Distillation recovery (%)	98.5	97.5			
Reference	[48]	[48]	[82]	[82]	[82]

The synthetic crude oil produced by Great Canadian Oil Sands, Ltd. has already been tested as a substitute diesel fuel by Canadian Pacific Rail on their Montreal—Windsor route, and it appears that the synthetic crude oil compares favorably with conventional diesel fuel.

In the aforegoing pages, the author has attempted to relate the development of synthetic crude oil production from the Alberta tar sands dating from early laboratory work to the more grandiose commercial-scale production. As several other companies are involved in feasibility studies for the construction of tar-sand processing plants, it is evident that development of the tar sands is only beginning and production of gasolines may even become a sideline. With the introduction of gas turbine, electric, and propane automobiles and the like, it is possible that the future of the tar sands lies in their being used as a chemical raw material. As examples, sulfonated and

TABLE 6-VIII

Properties of synthetic crude oil refined from Athabasca bitumen

Property	Source of samples							
	G.O.C.S.		Cities Service Atha-basca	Shell Canada, Ltd.	Direct fluid coking		UOP *1 pilot plant	Feed for hydro-treating tests
	pilot plant	commer-cial			Abasand	Bitu-mount		
Gravity (°API)	37.6	38.3	30.2	32.8	15.6	15.8	25.5	15.4
Distillation temperature (°F)								
IBP	210	162			176	182	118	
5%	277	221			338	412	182	
10%	300	254	265		460	482	245	
30%	379	408	455	400	630	600	438	
50%	486	507	585	540	682	650	550	
90%	680	615		885			750	
end pt.	833	715					760	
Distillation recovery (%)	99							
Elemental analysis (wt%)								
Carbon	86.3							84.4
Hydrogen	13.4							11.2
Nitrogen	0.02		0.096	0.09				0.24
Oxygen								0.10
Sulfur	0.03	0.022	0.182	0.29	4.0	3.9	3.6	4.1
Heavy metals *2 (ppm)	0.04							
Hydrocarbon type (wt%)								
Asphaltenes	0							
Resins	0							
Oils (aromatic)	21							
Oils (saturated)	79							

(*continued*)

TABLE 6-VIII (continued)

Properties of synthetic crude oil refined from Athabasca bitumen

Property	Source of samples							
	G.O.C.S.		Cities Service Athabasca	Shell Canada, Ltd.	Direct fluid coking		UOP *1 pilot plant	Feed for hydro treating tests
	pilot plant	commercial			Abasand	Bitumount		
UOP *1 "K"							11.0	
Process	delayed coking; hydrotreating		fluid coking; hydrotreating	thermal cracking; hydrotreating	direct coking of tar sand 977°F		delayed coking	destructive batch distillation
Reference	[87]	[83]	[92]	[91]	[30]	[30]	[48]	[71]

*1 Universal Oil Products Co.
*2 Vanadium, nickel, and copper.

TABLE 6-IX

Hydrodesulfurization of coker-distillate fractions

Inspection	G.C.O.S								
	naphtha			kerosene			gas oil		
	feed	product		feed	product		feed	product	
		pilot plt.	com. *		pilot plt.	com.		pilot plt.	com.
Coker distillate (wt%)	16.5			13.7			51.8		
Synthetic crude (vol%)		30.6	30.8		19.0	27.2		50.4	42.0
Gravity (°API)	46.8	50.9	55.3	32.9	39.7	38.6	18.3	28.7	27.5
Sulfur (ppm)	22,000	50	15	27,000	50	50	38,000	800	410
Nitrogen, total (ppm)	150		2	400	1	50	2000	trace	500
Vol%									
Aromatics	19	18		29.3	13.8	12.7	62.1	29.8	25.3
Olefins	32			14.4	<1.0		2.0		
Paraffins and naphthenes	49			46.4	85.7		35.9		
Distillation temperature (°F)									
IBP	180	174	162	380	388	358	515	499	498
5%	202	260	194	396	398	385	530	512	526
10%	220	274	206	409	402	398	550	522	540
30%	268	282	238	428	411	418	600	561	568
50%	295	296	278	441	415	438	645	611	588
70%	314	310	316	458	423	460	697	655	615
90%	347	334	369	477	433	496	780	740	675
95%	360	344	396	490	448	513	807	785	706
end pt.	400	366	462	535	468	533	850	869	715
Reference	[82]	[83]	[83]	[82]	[83, 93]	[83]	[82]	[83, 93]	[83]

* com. = commercial.

TABLE 6-X

Daily production of synthetic crude oil at the G.C.O.S. plant (After Humphreys [88])

1968	24,400 bbl/day
1969	27,300 bbl/day
1970	32,700 bbl/day
1971	42,200 bbl/day
1972	51,000 bbl/day

sulfomethylated derivatives have satisfactorily undergone tests as drilling mud thinners and the results are comparable to those obtained with commercial mud thinners. In addition, the compounds may also find use as emulsifiers for the in-situ recovery of the Athabasca bitumen. There are also indications that these materials and other similar derivatives of the bitumen, especially those containing functions such as carboxylic or hydroxyl will readily exchange cations and could well compete with synthetic zeolites. Other uses of the hydroxyl derivatives and/or the chloro-asphaltenes include high-temperature packings or heat-transfer media [94]. Reactions incorporating nitrogen and phosphorus into the bitumen are particularly significant at a time when the effects on the environment of many materials containing these elements are receiving considerable attention [95]. The resulting compounds may act as potential slow-release soil conditioners which will release the nitrogen or phosphorus only after considerable weathering or bacteriological action. One may proceed a step further and suggest that the carbonaceous residue remaining after release of the hetero-elements may be beneficial to humus-depleted soils, such as the gray-wooded and solonetzic soils found in Alberta and various other localities. It is also feasible that coating a conventional quick-release inorganic fertilizer with a water-soluble or water-dispersible derivative will provide a slower-release fertilizer and an organic humus-like residue. In fact, variations of this theme are multiple. Numerous patents deal with the uses of bitumen, bitumen coke, or bitumen derivatives, but only further work will reveal how practical these projected uses may be. None can be discounted as long as research continues and the need for new uses of petroleum remains.

Acknowledgment

The author wishes to thank Dr. N. Berkowitz for continual encouragement, Mr. J.F. Fryer for his comments on the manuscript, and Dr. S.E. Moschopedis for helpful discussions.

References

1 H.A. Innis, *R. Soc. Can. Trans., Sect. 2*, p. 131 (1928).
2 R.G. McConnell, in: *Can. Geol. Surv. Annu. Rep., 5A*, p. 21 (1890—91).

3 R.G. McConnell, in: *Can. Geol. Surv. Annu. Rep., 12A*, p. 11 (1899).

4 S.C. Ells, *Dep. Mines Tech. Surv., Mines Branch, Inf. Circ., IC-139*, Ottawa (1962).

5 Bibliography of K.A. Clark, in: M.A. Carrigy (Editor), *Athabasca Oil Sands, Res. Counc. Alberta, Inf. Ser., 45*, xi—xii (1963).

6 D.S. Pasternack, *Report on Operations at Bitumount during 1949*, Research Council of Alberta, Edmonton, Alta. (1949).

7 S.M. Blair, *Report on the Alberta Bituminous Sands*, Govt. of Province of Alberta, Edmonton, Alta. (1950).

8 Oil and Gas Conservation Board, *A Description and Reserve Estimate of the Oil Sands of Alberta*, Calgary, Alta. (1963).

9 G.W. Govier, "Alberta's Synthetic Crude Oil Development Policy", *Colo. Sch. Mines Q., 65*, 227—241 (1970).

10 C.W. Bowman, "Athabasca Oil Sands Development — 50 Years in Preparation", *Sci. J., 5A*, p. 28 (1969).

11 C.W. Bowman, "The Athabasca Oil Sands Development — 50 Years in Preparation", in: *Origin and Refining of Petroleum, Adv. Chem. Ser., 103*, 81—93 (1971).

12 F.K. Spragins, "Second Generation Tar Sands Plant", *Proc. 8th World Pet. Cong., 4*, 35—41 (1971).

13 F.W. Camp, Personal communication.

14 L.E. Elkins, "Oil Production from Bituminous Sands", *U.S. Patent 2,734,579* (Feb. 14, 1956).

15 G.J.W. Salomonsson, "Underground Gasification of Precarbonized Fuel Deposits", *U.S. Patent 2,914,309* (Nov. 24, 1959).

16 R.H. Perry, D.W. Green and J.M. Campbell, "Reverse Combustion — A New Oil Recovery Technique", *J. Pet. Technol., 12*, 11—12 (1960).

17 R.L. Reed, D.W. Reed and J.H. Tracht, "Experimental Aspects of Reverse Combustion in Tar Sands", *J. Pet. Technol., 12*, 14—15 (1960).

18 J.E. Warren, R.L. Reed and H.S. Price, "Theoretical Considerations of Reverse Combustion in Tar Sands", *J. Pet. Technol., 12*, 14—15 (1960).

19 V.J. Berry Jr. and D.R. Parrish, "A Theoretical Analysis of Heat Flow in Reverse Combustion", *J. Pet. Technol., 12*, 15—16 (1960).

20 K.M. Watson, "Oil Recovery by Subsurface Thermal Processing", *Can. Patent 621,230* (May 30, 1961).

21 C.A. Carpenter, "Methods and Apparatus for Heating Oil Sands", *U.S. Patent 3,113,622* (Dec. 10, 1963).

22 C.M. Davis, *Proc. Athabasca Oil Sands Conf.*, Research Council of Alberta, Edmonton, Alta., p. 141 (1951).

23 R.J. Schoeppel and E. Sarapuu, *Soc. Pet. Eng., AIME, Prod. Res. Symp.*, Tulsa, Paper SPE-281 (1962).

24 M.L. Natland, "Project Oil Sand", in: M.A. Carrigy (Editor), *Athabasca Oil Sands, Res. Counc. Alberta, Inf. Ser., 45*, 143—155 (1963).

25 T.M. Doscher, "Technical Problems in In Situ Methods for Recovery of Bitumen from Tar Sands", *Proc. 7th World Pet. Congr., 3*, 625—632 (1967).

26 D.C. Ward, C.H. Atkinson and J.W. Watkins, "Technical Problems in In situ Methods for Recovery of Bitumen from Tar Sands", *J. Pet. Technol., 18*, 139—145 (1966).

27 M.D. Nordyke, "Project Gasbuggy — A Nuclear Fracturing Experiment", *J. Pet. Technol., 11*, p. 303 (1971).

28 P.E. Gishler, "The Fluidization Technique Applied to Direct Distillation of Oil from Alberta Bituminous Sands", *Can. J. Res., F28*, 62—70 (1950).

29 W.S. Peterson and P.E. Gishler, "Small Fluidized Solids Pilot Plant for Direct Distillation of Oil from Alberta Bituminous Sand", *Can. J. Res., F28*, 62—70 (1950).

30 W.S. Peterson and P.E. Gishler, "Small Fluidized Solids Pilot Plant for Direct Distillation of Oil from Alberta Bituminous Sands", *Proc. Athabasca Oil Sands Conf.*,

Research Council of Alberta, Edmonton, Alta., p. 207 (1951).
31 B. Matchen and P.E. Gishler, "Small Fluidized Solids Pilot Plant for Direct Distillation of Oil from Alberta Bituminous Sands", *Natl. Res. Counc. Rep., C51-515*, Ottawa (1951).
32 W.S. Peterson, H. Keller and P.E. Gishler, "Fluidized Solids Coking of Canadian Heavy Crude Oils", *Can. J. Technol., 34,* 104—120 (1956).
33 P.E. Gishler and W.S. Peterson, "Treatment of Bituminous Sand", *Can. Patent 530,920* (Sept. 25, 1956).
34 R.W. Rammler, "Production of Synthetic Crude Oil from Oil Sand by Application of the Lurgi Ruhrgas Process", *Can. J. Chem. Eng., 48,* 552—560 (1970).
35 C.H.O. Berg, "Fuel Production from Oil Shale", *U.S. Patent 2,905,595* (Sept. 22, 1956).
36 V. Haensel, "Controlling Temperatures in Fluidized Solid Systems for Conversion of Hydrocarbons", *U.S. Patent 2,733,193* (Jan. 31, 1956).
37 J. Stewart, S.C. Fulton and A.W. Langer Jr., "Recovery of Oil from Bituminous Sands", *U.S. Patent 2,772,209* (Nov. 27, 1956).
38 I. Steinmetz, "Coking of Mix of Tar Sands Froth Product", *U.S. Patent 3,466,240* (Sept. 9, 1969).
39 M.F. Nathan, G.T Skaperdas and G.C. Grubb, "Fluid Coking of Tar Sands", *U.S. Patent 3,320,152* (May 16, 1967).
40 M.F. Nathan, G.T. Skaperdas and G.C. Grubb, "Fluid Coking of Tar Sands", *Can. Patent 823,186* (Sept. 16, 1969).
41 M.F. Nathan and G.C. Grubb, *Can. Patent 819,100* (July 29, 1969).
42 H.F. Tse, "Pyrolytic Methods of Treating Bituminous Tar Sands and Preheating of Sand", *U.S. Patent 3,518,181* (June 30, 1970).
43 J.D. Bennett, "Method of Damping Vibration and Article", *U.S. Patent 3,623,972* (Nov. 30, 1971).
44 G. Egloff and J.C. Morrell, *Trans. Am. Inst. Chem. Eng., 18,* p. 347 (1926).
45 G. Egloff and J.C. Morrell, "Cracking of Bitumen Derived from Alberta Tar Sands", *Can. Chem. Metall., 11,* p. 33 (1927).
46 M.W. Ball, "Development of Athabasca Oil Sand", *Trans. Can. Inst. Min. Metall., 44,* 58—91 (1941).
47 W.S. Peterson, Thesis, University Alberta, Edmonton, Alta. (1944).
48 M.J. Sterba, *Proc. Athabasca Oil Sands Conf.*, Research Council of Alberta, Edmonton, Alta., p. 257 (1951).
49 G.W. Hodgson, B. Matchen, W.S. Peterson and P.E. Gishler, "Oil from Alberta Bitumen — Simultaneous Dehydration and Coking Using Fluidized Solids", *Ind. Eng. Chem., 44,* 1492—1496 (1952).
50 W.S. Peterson and P.E. Gishler, "Oil from Alberta Bituminous Sand", *Pet. Engr., 23,* C66—74 (April, 1951).
51 F.L. Booth, R.E. Carson, E.J. Burrough and T.E. Warren, *Can. Dep. Mines Tech. Surv., Mines Branch, Tech. Mem., 208-58,* Ottawa (1958).
52 D.S. Pasternack, "Low-Ash Asphalt and Coke from Athabasca Oil-Sands Oil", in M.A. Carrigy (Editor), *Athabasca Oil Sands, Res. Counc. Alberta, Inf. Ser., 45:* 207—229 (1963).
53 D.S. Pasternack, "Removal of Suspended Mineral Materials from Asphaltic Petroleum Oil", *U.S. Patent 3,365,384* (Jan. 23, 1968).
54 D.S. Pasternack, "Thermal Cracking of Athabasca Oil Sands Oil. I. Changes in Some Properties of the Oil and Its Components", *J. Can. Pet. Technol., 3(2),* 39—45 (1964).
55 D.S. Pasternack, "Thermal Cracking of Athabasca Oil-Sands Oil. II. Cohesiveness of Carbon Residue from Fluidized-Coker Distillate", *J. Can. Pet. Technol., 3(3),* 91—94 (1964).
56 D.S. Pasternack, "Thermal Cracking of Athabasca Oil Sands Oil. III. A Study of

Asphaltic Components of Alberta Petroleum Oils and Asphalts", *J. Can. Pet. Technol.*, *3(4)*, 141—144 (1964—1965).
57 J.H. Henderson and L. Weber, "Physical Upgrading of Heavy Crude Oils by the Application of Heat", *J. Can. Pet. Technol.*, *4(4)*, 206—212 (1965).
58 J.G. McNab, P.V. Smith and R.L. Betts, "The Evolution of Petroleum", *Ind. Eng. Chem.*, *44*, 2556—2563 (1952).
59 J.G. Erdman and J.P. Dickie, "Mild Thermal Alteration of Asphaltic Crude Oils", *Am. Chem. Soc., Div. Pet. Chem.*, *9*, B69 (1964).
60 J.G. Speight, *Proc. 19th Can. Chem. Eng. Conf.*, Edmonton, Alta. (1969).
61 J.G. Speight, "Thermal Cracking of Athabasca Bitumens, Athabasca Asphaltenes and Athabasca Heavy Oil", *Fuel*, *49(2)*, 134—145 (1970).
62 J.G. Speight, "High Temperature Mass Spectroscopy of Athabasca Asphaltenes and the Relationship to Cracking Processes", *Am. Chem. Soc., Div. Fuel Pet. Chem., Prepr.*, *15(1)*, 57—61 (1971).
63 T.E. Warren, in: *Can, Dep. Mines Branch Rep.*, *725*, p. 115 (1930—31).
64 E.H. Boomer and A.W. Saddington, "Hydrogenation of Bitumen from Bituminous Sands of Alberta", *Can. J. Res. 2*, 376—383 (1930).
65 E.H. Boomer and A.W. Saddington, "Hydrogenation of Bitumen from Bituminous Sands of Alberta", *Can. J. Res.*, *4*, 517—539 (1931).
66 E.H. Boomer and J. Edwards, "Hydrogenation in Tetralin Medium. I. Destructive Hydrogenation of Bitumen and Pitch", *Can. J. Res..*, *13B* 323—330 (1935).
67 T.E. Warren and K.W. Bowles, *Can., Dep. Mines Branch*, *Rep. 737*, p. 86 (1932).
68 T.E. Warren, F.L. Booth, R.E. Carson and K.W. Bowles, *Proc. Athabasca Oil Sands Conf.*, Research Council of Alberta, Edmonton, Alta., p. 289 (1951).
69 F.L. Booth, R.E. Carson, K.W. Bowles and D.S. Montgomery, *Can. Mines Branch Res. Rep.*, *R30*, (1958).
70 D.S. Mongomery, *Can. Dep. Mines Tech. Surv.*, *Mines Branch*, *Rep. FRL-237*, Ottawa (1956).
71 A.R. Aitken, W.H. Merrill and M.P. Pleet, "Hydrogenation of a Coker Distillate Derived from Athabasca Bitumen", *Can. J. Chem. Eng.*, *42(5)*, 234—238 (1964).
72 W.H. Merrill, D.H. Quinsey, M.P. Pleet and W.H. Merrill, *Am. Inst. Chem. Eng.*, *Symp. Advances High Pressure Technol.*, New Orleans (March 16, 1969).
73 D.H. Quinsey, R.B. Logie, M.P. Pleet and W.H. Merrill, *Proc. 19th Can. Chem. Eng. Conf.*, Edmonton, Alta. (1969).
74 E.W. White, "Hydrogenation of Bituminous Emulsion Obtained from Tar Sands", *U.S. Patent 3,291,717* (Dec. 13, 1977).
75 E.W. White, "Process for Hydrogenation for Bituminous Emulsions", *Can. Patent 792,733* (Aug. 20, 1968).
76 B.I. Parsons, *Dep. Energy, Mines Resour.*, *Mines Branch Tech. Bull.*, *TB100*, Ottawa (1965).
77 B.I. Parsons, *Proc. 19th Can. Chem. Eng. Conf.*, Edmonton, Alta. (1969).
78 J.J. Cameron, M.A. O'Grady and B.I. Parsons, *Dep. Energy, Mines and Resour.*, *Mines Branch Rep.*, *R217*, Ottawa (1969).
79 M.A. O'Grady and B.I. Parsons, *Dep. Energy, Mines and Resour.*, *Mines Branch Rep.*, *R194*, Ottawa (1967).
80 E.C. MaColgan, R.G. Draper and B.I. Parsons, *Can. Dep. Energy, Mines and Resour.*, *Mines Branch, Rep.*, *R246*, Ottawa (1971).
81 F.W. Camp, "Tar Sands", in: *Encyclopedia of Chemical Terminology*, *19*, Interscience, New York, N.Y., pp. 682—732 (1963).
82 W.A. Bachman, "Plant Starts, Athabasca Now Yielding Its Hydrocarbons", *Oil Gas J.*, *65*, 69—72 (Oct. 23, 1967).
83 G.F. Andrews, E.W. Dobson and H.M. Lewis, "Great Canadian Oil Sands Experience

in the Commercial Processing of Athabasca Tar Sands", *Preprints, Am. Chem. Soc., Div. Petrol. Chem., 13(2),* F5—F18 (1968).

84 R.E. Carson and F.L. Booth, *Natural Gas Requirements for Processing Alberta Bituminous Sands,* Canada Mines Branch, Fuel Res. Lab., Ottawa, unpublished manuscripts (1952).

85 J.F. Fryer and J.G. Speight, *Athabasca Oil Sands Index, Selected Abstracts, I, Bitumen Recovery,* Research Council of Alberta, Edmonton, Alta. (1973).

86 J.F. Fryer and J.G. Speight, *Athabasca Oil Sands Index, Selected Abstracts, II. Bitumen Processing, Properties and Utilization,* Research Council of Alberta, Edmonton, Alta. (1973).

87 E.D. Innes and J.V.D. Fear, "Canada's First Commercial Tar Sand Development", *Proc. 7th World Pet. Congr., 3,* 633—650 (1967).

88 R.D. Humphreys, Paper presented at the *Pet. Soc. Can. Inst. Min. Metall.,* Edmonton, Alta. (May, 1973).

89 *Applications to the Alberta Oil and Gas Conservation Board by Atlantic Richfield Co., Cities Service Athabasca Inc., Imperial Oil Ltd., Royalite Oil Co. Ltd.* (Aug. 7, 1971).

90 G.R. Gray and J.A. Haston, *Proc. 19th Can. Chem. Eng. Conf.,* Edmonton, Alta. (1969).

91 A.D. Hoskins, "How Hydrogen Will Be Used to Upgrade Athabasca Tar to Sweet Crude Oils", *Proc. Am. Pet. Inst., Sect. III 44,* 426; *Oil Gas J.,* pp. 122—124 (May 18, 1964).

92 R.A. Given, *Natl. Pet. Refiners Assoc. Meet.,* Paper 63-15 (April, 1963).

93 P.F. Lovell, H.E. Reif, R.O. Burke, P.H. Hertel and B.T. Abbott, "First Commercial Tar-Sands Crude", *Oil Gas J., 64,* 66—72 (Aug. 15, 1966).

94 S.E. Moschopedis and J.G. Speight, "Oxidative Degradation of Athabasca Asphaltenes", *Fuel, 50(2),* 211—217 (1971), and references cited therein.

95 J.G. Speight, "The Structures of Athabasca Asphaltenes and the Conversion of These Materials to Useful By-Products", *Prepr., Am. Chem. Soc., Div. Fuel Pet. Chem., 15,* 76—82 (1971).

Chapter 7

PROPERTIES AND STRUCTURE OF BITUMENS

CHAUR SHYONG WEN, GEORGE V. CHILINGARIAN and TEH FU YEN

Introduction

Numerous organic-rich fossil materials of widely different geologic ages represent an important potential source of fuels. Although a number of physicochemical methods have been used in the structural elucidation of natural bitumens, the knowledge of the composition of these substances is still very limited. The understanding of the fundamental properties and structure of bitumens may aid in devising new methods and improving existing technologies for converting the organic matter into useful liquid and gaseous products.

Organic matter in bituminous substances has been regarded as the residue of organic life that occurs as viscous impregnations in sandstones, siltstones, shales, and carbonates (limestones and dolomites). These deposits were probably derived from a saline-lacustrine sapropel and owe their variable properties to differences in environment of deposition. A variety of naturally occurring bituminous materials is present among the organic constituents of oil shales. These substances range from liquid petroleums through heavy, viscous asphalts to solid asphaltites and asphaltic pyrobitumens. The types of organic compounds that can be extracted by solvents, particularly when carbon disulfide is used as a solvent, are commonly called bitumens. Actually the amount of these bitumens is very small when compared with the huge amounts of insoluble kerogenic material remaining behind in the sedimentary rocks. At present, the identity of the chemical source materials of bitumens, as well as the reactions involved in their genesis, are largely unknown. A general understanding of the properties and the chemical structure of these substances is important.

The following discussion is a summary of studies of properties of naturally occurring bituminous substances, as well as an elucidation of the chemical structure of bitumen fractions of these substances. The principal physical properties and some of the chemical characteristics of bitumens are presented. Structural parameters based on the chemical and physical determinations are collected in order to investigate the average structure of bitumens. Although discussion is intended to be general, oil-shale bitumens are described as type examples, because data are still lacking on most of the bituminous substances except those associated with the Green River Formation, which have been studied in considerable detail. The chapter also includes a brief survey of the bituminous-material classification and bitumen fractionation.

Properties of bitumens

Different bitumens extracted from different bituminous substances vary considerably in their physical and chemical properties. This indicates that the composition of the organic material in different source rocks of bitumens was a major factor in determining the kind of bitumens formed and can be used for characterizing and identifying them. The principal physical properties and chemical compositions of bituminous substances are discussed in the following sections.

Classification of bituminous substances

Much confusion exists in the use and interpretation of the terms "bitumens", "asphaltic bitumens", or "native asphalts". During the last three decades, efforts were made to develop a uniform nomenclature for bituminous materials all over the world [1—3]. No agreement could be reached in this field and various schemes have been proposed by different investigators for the classification of bituminous substances. The authors prefer the classification presented in Fig. 7-1 which is a modified version of the proposed scheme by Abraham [2].

Vassoevich [41] presented an excellent discussion on the state of confu-

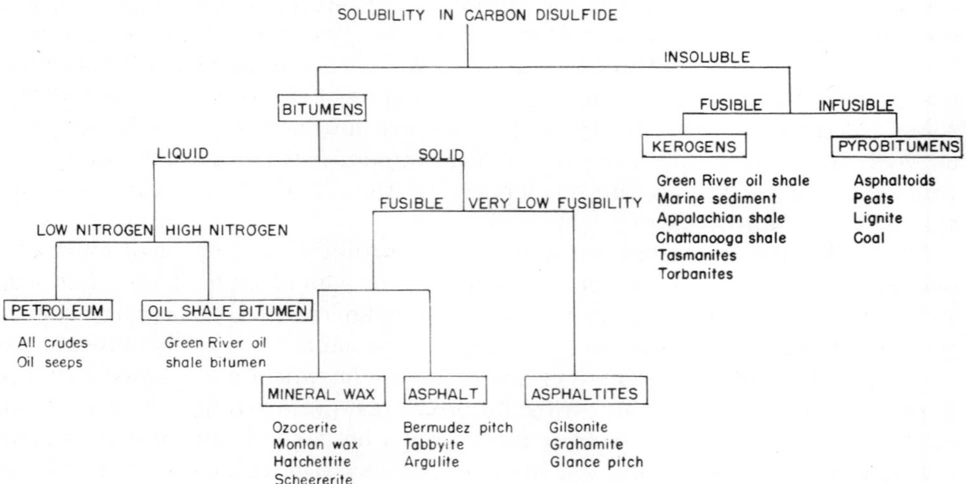

Fig. 7-1. Terminology and classification of naturally occurring bitumens and related materials. (Modified after Abraham [2].)

sion which exists in using the terms "bitumen" and "bituminous". He proposed to use the term "bitumoid" (meaning "bitumen-like") to refer to all those substances which can be extracted (1) without destruction (or almost without it) by nonpolar and weakly polar organic solvents and (2) with SO_2, from soils, sediments, rocks, and other materials, in general. Bitumoids can be differentiated on the basis of solvent type, e.g., ether, chloroform, and benzol, and conditions of extraction, i.e., temperature, pressure, sample preparation, etc. The value of this chemical-analytical term becomes apparent when one realizes that one should not use the same term to designate (1) "bitumens" and naturally occurring substances, such as oil, asphalt, and ozocerite, and (2) extracts from peats, marine sediments, soils, etc., obtained on using various organic solvents. According to Bell and Hunt [5], bituminous substances which decompose before melting are called pyrobitumens. These authors differentiated between the asphaltic pyrobitumens (asphaltoids) and nonasphaltic pyrobitumens (peats, coals), with the former containing less than 5% oxygen.

The writers have not discussed bitumens which are extracted from coal here. The benzene-extractable materials from peat and lignites consisting of esters of the montan-wax type, together with heavy hydrocarbons and resins, are also called bitumens. According to Francis [68], the bitumen present in coal is very complex and the complexity increases with rank. Pontonie et al. [69] have classified coal bitumens according to bituminization in five stages: (1) parent substance — probitumen; (2) youngest sapropelic stage of bituminization — from labile protobitumen into stable protobitumen; (3) stabilization phase of bituminization — from metabitumens into kerobitumens and polymerbitumens; (4) enbituminization — protobitumen can bypass stage (3) and change directly into enbitumen; and (5) coalification phase — enbitumen can change into asphaltite and kerobitumen can change into nigritite of the following three forms: exinonigritite, keronigritite and polynigritite. Most of these are beyond the scope of the present chapter.

Mineral wax is composed chiefly of straight-chain paraffin hydrocarbons, i.e., ozocerite, or consists of a mixture of fatty acids and esters, i.e., montan wax. Scheererite and hatchettite are local names for varieties of ozocerite found in Scotland and Switzerland, respectively.

Viscous liquids or solid, low-melting bitumens are classified as asphalts, including Bermudez pitch, tabbyite and argulite. Bituminous substances with fusing points higher than 230°F [4] are classified as asphaltites. Gilsonite, grahamite and glance pitch belong to this class.

Other naturally occurring organic substances are classified as kerogens and pyrobitumens. The term "kerogen" is applied to the insoluble organic matter occurring in sedimentary rocks which yields oil by pyrolysis. As mentioned previously, bituminous substances that decompose before melting are called pyrobitumens and include asphaltoids, peat, lignite and coals. No attempt is made in this chapter to discuss all of the asphaltoids and other pyrobitumens.

Pyrobituminous substances which have distinct chemical and physical properties are not discussed in this chapter. Probable properties of the various bitumens based on published data are presented below.

Mineral waxes

Earth waxes (ozocerite, hatchettite and scheererite) vary in color from dark yellow-brown to black waxy and occur either in veins and nodules or as films on brecciated rocks [5]. The main constituents of earth waxes are hydrocarbons, including straight- and branched-chain saturated paraffins with some napthene groups. The sum of oxygen, nitrogen, and sulfur contents averages less than 0.5%. A typical Uinta Basin ozocerite contains paraffin mixture usually up to 80—95% [5,6]. There is also a high content of normal alkanes in which C_{29} or C_{31} are the most abundant ones. The H/C ratio of 1.96 is approximately that of the $-CH_2$-group. Other physical properties of ozocerite include a low refractive index of 1.53, low carbon residue of around 4.1%, a specific gravity of 0.9 (at 77° F), and fusing point of about 140—176° F.

Montan wax is an extract obtained from some kinds of brown coals or lignites. The crude montan wax is dark brown in color and melts at about 180° F [7]. In contrast to the high paraffinic content of ozocerite wax, montan wax is primarily an ester—alcohol wax, having nearly 60% esters and alcohols in its waxy fraction [8,9]. In addition to the high content of esters and alcohols, free acids contained in montan wax usually exceed 15% of the total waxy fraction. The carbon-chain lengths of the major portions of both acids and alcohols seem to lie in the range of even carbon numbers of C_{26} to C_{28} (C_{30} is also quite abundant in acids). Specific gravity of montan wax lies in the 0.9—1.0 range.

Asphalts

Many native asphalts occur as viscous impregnations in sandstones and carbonates. Native asphalts are found in profusion at various locations in the world [5,10—12]. Although asphalt seeps have been known for a long time in many countries, the best known and largest are those of the Bermudez pitch lake in Venezuela and the Trinidad asphalt lake. Some of the native asphalts found in the United States include tabbyite (solid) in the Tabby Canyon, Uinta Basin, and argulite, an asphalt-saturated sandstone in the Argyle Creek Canyon, south of Ouray, Uinta Basin.

Most native asphalts are of marine origin [13,14] and consist of high-molecular-weight compounds. The native-asphalt deposits contain from 5 to 20% of carbon-disulfide-soluble bitumens in rock asphalts [3,5] to about 65% bitumen in Bermudez pitch in the crude state. Refined Bermudez asphalt is soluble to the extent of about 95% in carbon disulfide [10]. Infrared and elementary analysis indicate that few oxygenated compounds are present in asphalts. Additionally, asphalts contain unusually high percentages of vanadium and occur with or without appreciable percentages of mineral matter. In all other properties, such as refractive index (~1.57), sulfur content (~1—6%), and atomic H/C ratio (~1.62), asphalts resemble other bitumens (see Fig. 7-1). The properties of several different native asphalts are summarized in Table 7-I.

Asphaltites

The asphaltites were probably derived from a saline-lacustrine sapropel. They are commonly associated with oil shales [5]. These substances are classified into three groups: gilsonite, glance pitch, and grahamite. The differentiation is made primarily on the basis of specific gravity and their high softening points (above 270° F) [13]. All three substances are almost completely soluble in carbon disulfide and their typical properties are presented in Table 7-II. Asphaltites are natural minerals and not manufactured products; therefore, similarly categorized substances (e.g., gilsonite) from various areas have somewhat different origins and variable compositions.

The largest deposits of gilsonite in the United States are found in the Uinta Basin. The high-carbon residue (~16%) and refractive index (~1.62) indicate presence of condensed rings [5]. Elemental analysis shows that gilsonite is high in nitrogen content (2.0—2.8%) as compared with the other

TABLE 7-I

Selected properties of native asphalts (After Abraham [2] and Miller [10])

Property	Bermudez Pitch	Trinidad	Tabbyite
Specific gravity (77° F)	1.07	1.40	1.01
Fusing point (° F)			
(ring and ball method)	145—160	210	178
Streak	Black	Blue	Brown
Fixed carbon (wt%)	12.9—14.0	10.8—12.0	8.1—9.2
Mineral matter (wt%)	1.5—6.5	27	4.8—6.0
Bitumen (soluble in CS_2)	92—97	56—57	92—95
Specific heat (at 300°F)	0.41	0.35	—
Hardness (Moh's scale)	<1	2	1
Penetration (77° F)	20—30	1.5—4.0	0
Ultimate analysis (wt%)			
Carbon	82.8	82.3	81.3
Hydrogen	10.7	10.7	11.2
Sulfur	5.8	6.2	1.2
Nitrogen	0.8	0.8	2.1
Oxygen (by difference)	0	0	4.2

TABLE 7-II

Typical properties of asphaltites (After Ervin [4] and Yen [59])

Property	Gilsonite	Grahamite	Glance pitch
Specific gravity (77° F)	1.01—1.10	1.15—1.20	1.10—1.15
Fusing point (° F)			
(ring and ball method)	270—375	375—625	270—375
Streak	brown	black	black
Fixed carbon (wt%)	10—20	35—55	20—35
Mineral matter (wt%)	<1	up to 50	0.3
Bitumen (soluble in CS_2)	98	45—100	99
Heating value (Btu/lb)	18,000	—	—
Specific heat (300° F)	0.52	—	—
Hardness (Moh's scale)	2	2	1
Penetration (77° F)	0—3	0	0
Ultimate analysis (wt%)			
Carbon	85.5	86.6	—
Hydrogen	10.0	8.7	—
Sulfur	0.3	1.8	—
Nitrogen	2.5	2.2	—
Oxygen	1.5	0.7 (by difference)	—

Uinta Basin bitumens. Tabbyite contains 2.1% and wurtzilite 1.8—2.2% of nitrogen [5]. Porphyrins in gilsonite were investigated by McGee [15], who isolated porphyrin fractions with yields of around 0.004—0.03%. The entire porphyrin content is present as a nickel complex. The presence of this porphyrin suggests that gilsonite is essentially or totally of plant origin.

Oil-shale bitumen

The amount of bitumen associated with the kerogen of oil shale is small, ranging from 0.5 to 5% of the total weight of oil shale. The extractable material (bitumen) increases with increasing extraction temperature and with higher polarity and chemical reactivity of the solvent. It was found that the amount of total extract is slightly higher by using the polar solvent than by using the nonpolar solvent [16]. The main compounds in the benzene-soluble materials of oil-shale bitumen include normal, branched, and cyclic alkanes, aromatic hydrocarbons, resins, and polar constituents [17—19]. The relative amounts of these constituents are not completely uniform from one deposit to another, or even within the same deposit, and minor deviations are not uncommon. Based on literature survey, no special attempt has been made to determine the physical property of natural oil-shale bitumen, except for its H/C ratio (~1.59) and its molecular weight (~1200) [20]. A typical oil-shale bitumen of the Green River Formation is very viscous, brown in

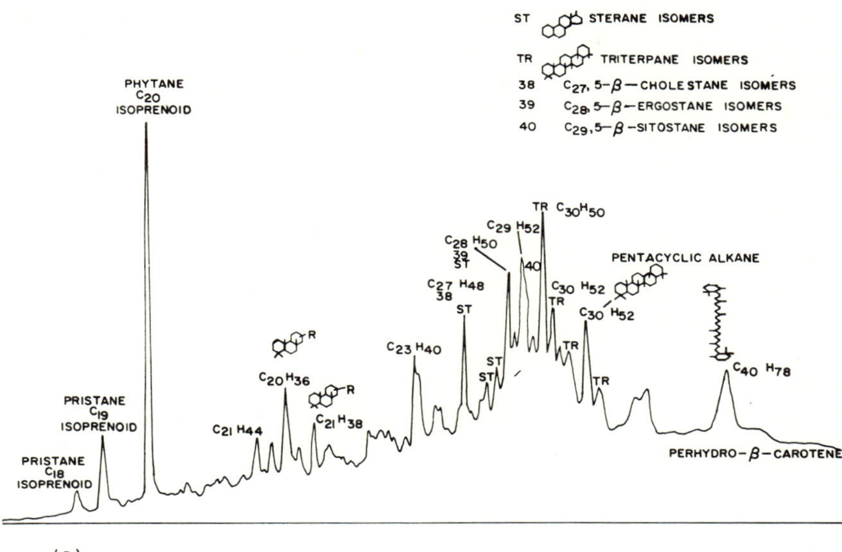

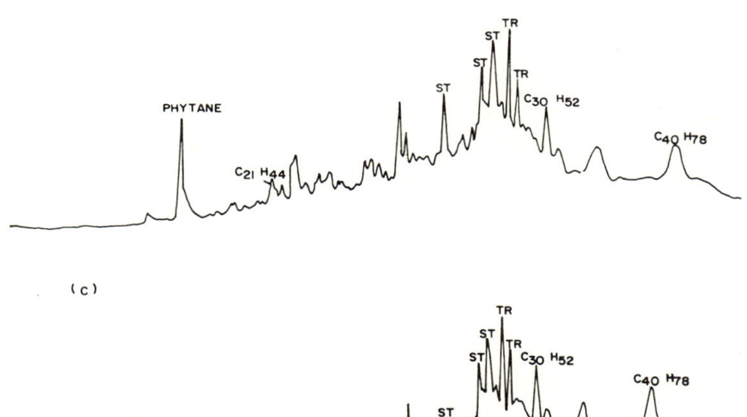

C21 H44

ST

ST

ST

TR

TR

C30 H52

C40 H78

(D)

Fig. 7-2. Gas chromatography analyses of Green River oil-shale bitumen: original (A), and heat treated at 320°C (B), 410°C (C), and 500°C (D).

color, and contains a considerable amount of nitrogen heterocyclic compounds [17,19]. An empirical formula for Green River bitumen based on ultimate analysis can be calculated as being $C_{83}H_{132}ON_4$. This formula indicates that oil-shale bitumen has a molecular weight of about 1200 [20]. The effect of heat treatment on oil-shale bitumen has been determined by heating samples at different temperatures (320, 410, and 500° C) for fifteen minutes [21]. After the heat treatment at these three different temperatures, samples were analyzed by gas chromatography (Figs. 7-2). The analyses show that the benzene-soluble oil-shale bitumen can endure temperatures of up to 410° C, except for the phytane and pristane fractions which disappear and become benzene-insoluble coke formations (Figs. 7-2b and 7-2c). After heating at 500° C, there is no significant peak in the gas chromatogram (Fig. 7-2d), although the benzene-extract fraction of this sample still has a light brown color. Inasmuch as bitumen has been expected to act as the heat-transfer agent [22—24] during the oil-shale retorting, the relation and interconversion among bitumen, shale oil and gas, and carbon residue (coke) are complicated. At high temperatures, bitumen seems to increase the tendency of coke formation during pyrolysis.

Structure of bitumens

Fractionation of bitumens

The bitumens listed in Fig. 7-1 can be separated into fractions on the basis of their solubility in various organic solvents. A complete fractionation scheme is given in Fig. 7-3: (1) the oily constituents are propane soluble; (2) resins are *n*-pentane soluble but propane insoluble; (3) asphaltenes are benzene soluble but *n*-pentane insoluble; and (4) carbenes are carbon disulfide

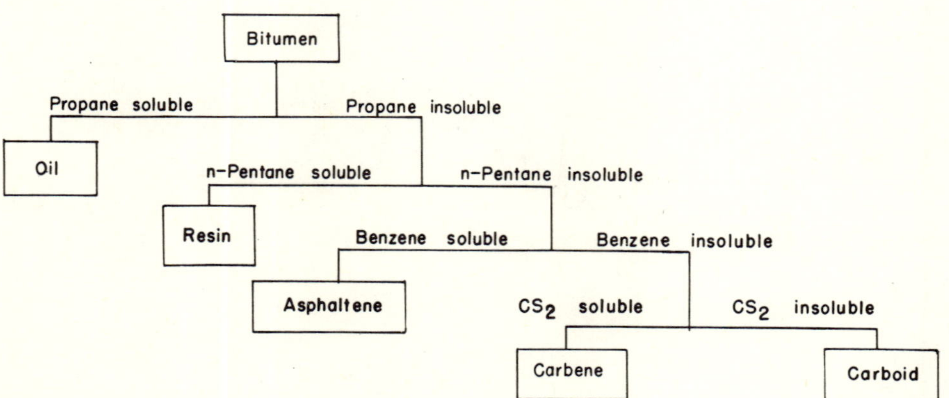

Fig. 7-3. Fractionation scheme of bitumens.

soluble but benzene insoluble. The sequential progression from oils-to-resins-to-asphaltenes-to-carbenes is associated with an increase in macromolecular size. Jewell et al. [70] have developed the separation procedures for petroleum crudes and residues. The SARA (saturate—aromatic—resin—asphaltene) column, consisting of four discreetly packed zones with a sequence of H^+ cation-exchange resin, OH^- ion-exchange resin, ferric chloride on clay, and OH^- ion-exchange resin, has been used to separate the material into oils, resins, and *n*-pentane-insoluble residue. The residue is further separated into asphaltenes and benzene insolubles by extraction with benzene. The compounds obtained on using the SARA method can be handled individually for further analysis and subdivision. More detailed discussions on the structure of bitumens is given in the following sections based on this fractionation of bitumens.

Structural parameters of bitumens

The structure of bitumens can be investigated by using the previously described fractionation of bitumens. Various methods for the structural elucidation of high-molecular-weight compounds can be applied to each fraction of bitumens. These include simple physical assays (specific gravity, refractive index, molecular weight, etc.), chemical analysis (elemental analysis, oxidation rate, etc.), and spectral instrumental measurements (X-ray diffraction, nuclear magnetic resonance, infrared, electron spin resonance, etc.). The major structural parameters of bitumens, based on all available data, are described below:

Molecular weight. The molecular weights of bitumens have been the subject of many studies, which yielded different results [25—27] depending upon the method employed (Table 7-III). Inasmuch as complex asphaltene molecules of bitumens tend to associate strongly, even in dilute solutions, the usual analytical procedures fail to differentiate between the true molecular weight and the particle weight. The number-average particle weight is a measure of the ease of dissociation of various asphaltenes in a given solvent. For example, in petroleum asphaltene, some particles retain their unit cell structures essentially intact, whereas others separate into smaller units with loss of one or more unit sheets [26]. Some petroleum asphaltenes even approach the dimensions of the unit sheet as determined from X-ray or mass-spectrometric analyses [28].

Data obtained by vapor-pressure osmometry (VPO) studies of molecular weights of asphaltene fractions in various solvents, which have different dielectric constants, suggest that the high molecular weights of the asphaltene molecules could be due to the electrostatic associations in the asphaltene individual units and the association between the unit and the bulk solvent [29]. The data presented in Table 7-III show that the degree of

TABLE 7-III

Molecular weights of bitumens

Sample	Solvent	Molecular weight	Method	Reference
Bermudez pitch	ØH[*1]	620	freezing point	[11]
Trinidad asphalt	ØH	1132	freezing point	[11]
Gilsonite	ØH	4252	freezing point	[11]
Gilsonite (aromatic fraction)	ØH	228	VPO[*2]	[67]
Gilsonite (naphthenic fraction)	ØH	267	VPO	[67]
Gilsonite (asphaltene)	ØH	8130	VPO	[26]
	THF[*3]	10,900	VPO	
	THF	4550 (M_N)[*4]	GPC[*5]	
	THF	12,300 (M_W)[*6]	GPC	
Boscan asphaltene	ØH	6220	VPO	[26]
	THF	6360	VPO	
	THF	3160 (M_N)	GPC	
	THF	4000 (M_W)	GPC	
Boscan resin	ØH	1270	VPO	[26]
	THF	1170	VPO	
	THF	3390 (M_N)	GPC	
	THF	4500 (M_W)	GPC	
Shale-oil asphaltene	ØH	749	VPO	[38]
Residue shale-oil asphaltene	ØH	630	VPO	
Shale oil	—	306—328	—	[78]

*1 ØH = benzene; *2 VPO = vapor pressure osmometry; *3 THF = tetrahydrofuran;
*4 M_N = number average molecular weight; *5 GPC = gel permeation chromatography;
*6 M_W = weight average molecular weight.

association of asphaltenes decreases with increasing polarity of the solvent. Although gel permeation chromatography (GPC) offers a possibility for the separation of individual sheets, because it is carried out in a dilute solution under high pressure, few investigators used it for the molecular-weight determination of bitumen fractions [26,30,31].

Indeed the molecular weight determined by the conventional methods is of little value for directly calculating molecular parameters of bitumen fractions. Particle weight (determined by using the VPO method) or a single unit-sheet weight (by the GPC method), however, could help in the determination of the average structure of bitumens.

X-ray diffraction. The X-ray diffraction method is the most direct for studying the structure of bitumens. A number of parameters can be obtained by resolving the γ and (002) bands in the X-ray scattering pattern [32,33]. The (002) spacing which appears in graphite, carbon blacks, pitch, coke, and coal is accepted as representing the spacing between the layers of a condensed aromatic ring [32—35]. The γ band centered around $(\sin\theta)/\lambda$ = 0.10—0.11 Å^{-1} has been shown to be associated with spacing in the saturated portion of the bitumen molecule [32,36]. Accordingly, a comparative estimate of the aromaticity, f_a, can be made from the areas of the resolved peaks corresponding to the γ and the (002) bands [32]:

$$f_a = \frac{C_A}{C} = \frac{C_A}{C_A + C_S} = \frac{A_{(002)}}{A_{(002)} + A_{(\gamma)}} \tag{7-1}$$

where A represents the areas under the respective peaks; and C_S, C_A and C, the number of saturated, aromatic, and total carbon atoms per structural unit. Fig. 7-4 shows the resolution of X-ray diffraction patterns for several petroleum and gilsonite asphaltenes in which the reduced intensity is modified from the coherent and incoherent scattering [32].

X-ray diffraction reveals considerable information concerning the structure of bitumens. The layer distance between aromatic sheets, d_m, by using the maximum of the (002) band is obtained from the following equation:

$$d_m = \frac{\lambda}{2 \sin\theta} \tag{7-2}$$

where λ = wavelength of the X-ray radiation and θ = Bragg angle. Similarly, the distance between the saturated portions of the molecules, d_γ, is obtained from a slightly modified version of the same equation by using values from the γ band:

$$d_\gamma = \frac{5\lambda}{8 \sin\theta} \tag{7-3}$$

X-ray diffraction also permits determination of the average diameter of the aromatic sheets, L_a, on the basis of Scherrer crystallite-size formula:

$$L_a = \frac{1.84\lambda}{\omega \cos\theta} = \frac{0.92}{B_{1/2}} \tag{7-4}$$

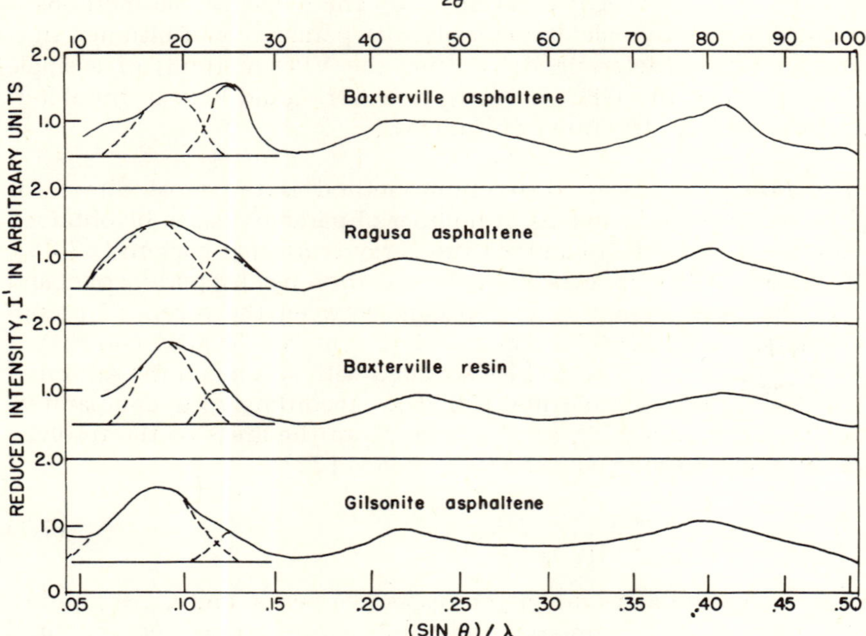

Fig. 7-4. X-ray diffraction patterns plotted on the basis of reduced intensities. The dashed lines show the resolved γ and (002) bands. (Modified after Yen et al. [32].)

where $B_{1/2}$ = the width at half maximum using the (11) band and ω = band width. This requires only the determination of the width of the (11) band at half maximum in units of $(\sin\theta)/\lambda$ [37].

In addition, the average height of the stack of aromatic sheets can be calculated by using the following equation:

$$L_c = \frac{0.9\lambda}{\omega\cos\theta} = \frac{0.45}{B_{1/2}} \tag{7-5}$$

From the values of L_c and d_m, it is possible to calculate the effective number of aromatic sheets, M_e, associated in a stacked cluster:

$$M_e = \frac{L_c}{d_m} \tag{7-6}$$

The results obtained by calculating all the parameters from X-ray measurements of three shale-oil asphaltenes [38] are summarized in Table 7-IV. The sharp doublet of (100) and (200) bands at 4.15 and 3.70 Å have been found for residual shale-oil asphaltene. In his study of petroleum asphaltenes, Yen

TABLE 7-IV

Aromaticity and crystallite parameters for shale-oil asphaltenes determined by X-ray diffraction

Asphaltene	Shale-oil asphaltene I*1	Shale-oil asphaltene II*2	Residue shale-oil asphaltene *3
f_a	0.39	0.43	0.42
d_m *4	3.57	3.57	3.57
d_γ *4	4.17	4.41	4.41
L_c *4	15	15	16
M_e	5	5	5
L_a *4,5	19	13	20
L_a *4,6	12	7	12

*1 Asphaltenes isolated from Don's Welding Retort Process.
*2 Asphaltene isolated from Paraho Retorting Process.
*3 Asphaltene isolated from residual shale oil (Paraho).
*4 All expressed in Å.
*5 From Scherrer's Equation.
*6 From Diamond's curve.

[39] has reported these two peaks as wax-like, long-chain alkyl compounds.

A number of X-ray diffraction patterns of naturally occurring bituminous materials and their derivatives have been examined [40]. Fig. 7-5 shows X-ray patterns of oil-shale kerogen, trona acids, and a kerogen derivative (asphaltine), which was obtained by extracting kerogen in tetralin at 350° C, followed by precipitation with pentane. The positions of the prominent peaks of three samples centered at $(\sin \theta)/\lambda$ = 0.10—0.11 Å^{-1} show strong saturated structure of γ bands. The slight shift of low-angle peaks (from 0.10 to 0.11 Å^{-1}) in kerogen and its derivatives indicates the presence of some isolated double bonds in their saturated structure [20]. Fig. 7-6 shows the X-ray diagrams of three grahamite samples. These three samples have very similar X-ray patterns, containing peaks corresponding to a spacing of (1) 0.10 Å^{-1} (γ band) and (2) 0.24 Å^{-1} (002 band).

Nuclear magnetic resonance. Nuclear magnetic resonance (NMR) spectroscopy has wide application in the elucidation of petroleum asphaltene [42,43]. The hydrogen-distribution data obtained by the NMR method has been applied to the analysis of carbon structure, with particular emphasis on estimating several important structural parameters: the degree of aromatic-ring substitution, the size of the condensed aromatic-ring sheets, and the aromaticity. Inasmuch as in proton NMR the area under a peak is directly proportional to the number of contributing hydrogens, the fraction of the total number of hydrogens which are saturated, i.e., H_S/H or h_s, can be obtained from the areas under the curve in the respective regions of resonance [42]. A typical

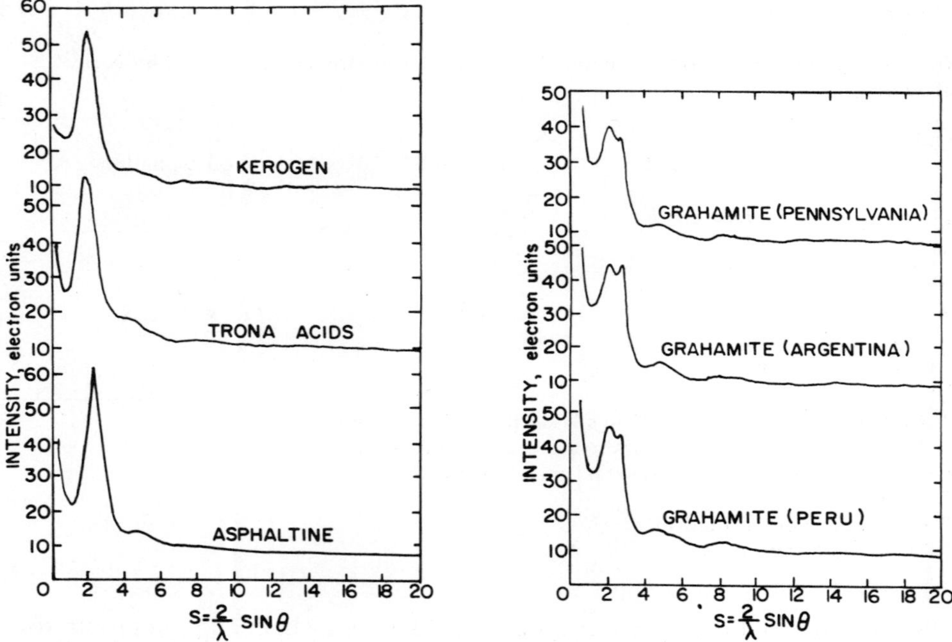

Fig. 7-5. X-ray scattering intensities of kerogen, trona acids, and asphaltine. (Modified after Ergun [40].)

Fig. 7-6. X-ray scattering intensities of some grahamites. (Modified after Ergun [40].)

NMR spectrum of shale-oil asphaltene is shown in Fig. 7-7 in which:

$$h_s = \frac{H_\alpha}{H} + \frac{H_N}{H} + \frac{H_R}{H} + \frac{H_{sMe}}{H} \tag{7-7}$$

where h_s = the fraction of saturated hydrogen atoms. The other assignments of the parameters of the above equation are given in Fig. 7-7.

From the area under the component peaks, it is possible to estimate the degree of substitution of the aromatic sheets expressed as C_{su}/H_I ratio, i.e., the average number of carbon atoms directly attached to an aromatic sheet, C_{su}, divided by the number of hydrogens, H_I, which would be attached if the aromatic sheet was totally unsubstituted. The equation is given by the following relation based on the assumption that for a unit of structure, C_{su} is approximately equal to one-half of the number of α-hydrogen:

$$\frac{C_{su}}{H_I} = \frac{H_\alpha/H}{2H_A/H + H_\alpha/H} \tag{7-8}$$

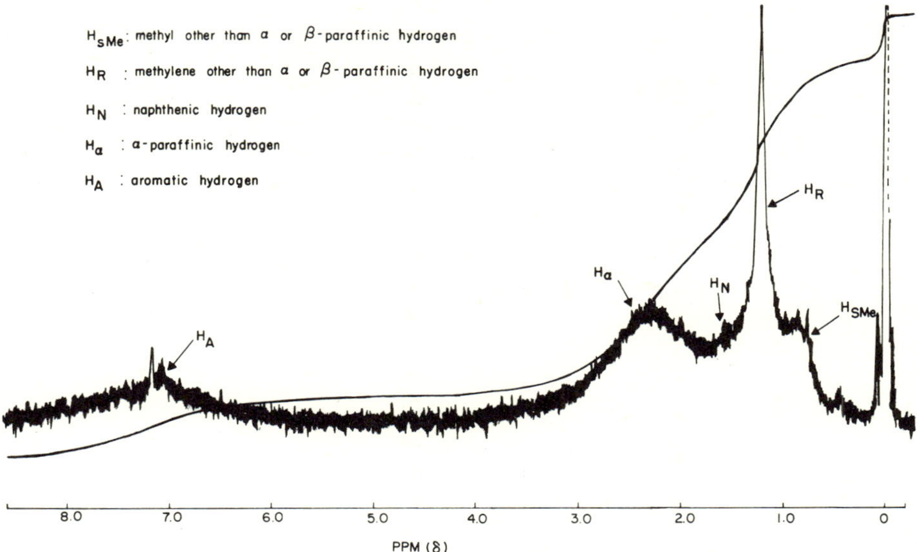

Fig. 7-7. NMR spectrum of asphaltene isolated from Green River shale oil.

The degrees of substitution obtained for the shale-oil asphaltenes, petroleum asphaltene and resin, and gilsonite asphaltene are presented in Table 7-V. The shale-oil asphaltenes show lower degree of substitution than other bitumens. The resin cut of the Baxterville crude oil shows a higher degree of substitution than does the corresponding asphaltene, i.e., 0.70 versus 0.62. The degree of substitution for the asphaltene fraction of gilsonite lies within the range (central portion) of the native petroleum asphaltenes.

The aromaticity, f_a, can be determined from NMR data in the following manner [44]:

$$f_a = \frac{C_A}{C} = \frac{\dfrac{C}{H} - \dfrac{1}{x}\dfrac{H_\alpha}{H} - \dfrac{1}{y}\dfrac{H_0}{H}}{\dfrac{C}{H}} \qquad (7\text{-}9)$$

where H_0 refers to all nonaromatic hydrogens other than those in α-paraffinic position. The parameter x is defined as H_α/C_α, the ratio of hydrogen to carbon for the α-paraffinic groupings, whereas y refers to H_0/C_0 ratio for the other attached groupings. Values for C/H ratio are obtained from the elementary analysis. Aromaticity calculated from NMR data depends largely on parameters x and y (Table 7-V).

Data obtained from NMR analysis also enable us to determine the shape

TABLE 7-V

Values for structural parameters of bitumens from NMR data

Bitumens	Shale-oil asphaltene I *1	Shale-oil asphaltene II *1	Residual shale oil asphaltene *1	Petroleum asphaltene I *2	Petroleum asphaltene II *3	Petroleum resin *4	Gilsonite asphaltene *5
f_a	0.49	0.50	0.47	0.48	0.54	0.41	0.38
C_{su}/H_I	0.58	0.62	0.56	0.48	0.62	0.70	0.62
H_I/C_A	1.00 *9	0.95 *9	0.86	0.38	0.34	0.61	0.60
C_{AR} *6	20.4	20.4	20.7	—	—	31.8	—
C_{SR} *7	11.8	11.9	8.6	—	—	13.6	—
N_{AR} *8	1.03	1.52	3.64	—	—	8.79	—
C_s/C_{su}	2.10	2.23	3.13	5.6	4.0	3.4	4.3
C_{sMe}/C_{Me}	0.36	0.35	0.45	0.71	0.50	0.48	0.55

*1 As Table 7-IV.
*2 Asphaltene from Mara, Venezuela [26].
*3 Asphaltene from Baxterville, Miss. [26].
*4 Resin from Baxterville, Miss. [26].
*5 Asphaltene from Uinta Basin, Utah [26].
*6 Total number of aromatic-ring carbons per molecule.
*7 Total number of substituted aromatic-ring carbons per molecule.
*8 Total number of aromatic rings per molecule.
*9 Based on coefficients $x(=H\alpha/C\alpha)$ and $y(=H_0/C_0)$ being equal to 2; if higher than 2, the H_I/C_A ratio will be considerably lower.

of the condensed aromatic sheets by estimating the value of the ratio of H_I to the number of aromatic carbon atoms, C_A:

$$\frac{H_I}{C_A} = \frac{\frac{1}{x}\frac{H_\alpha}{H} + \frac{H_A}{H}}{\frac{C}{H} - \frac{1}{x}\frac{H_\alpha}{H} - \frac{1}{y}\frac{H_0}{H}} \tag{7-10}$$

The NMR results for bitumens are presented in Table 7-V. The number of rings in the condensed aromatic-ring system per molecule varies from one, as in the case of shale-oil asphaltene, to 2—3 in the case of gilsonite asphaltene, to 3—4 in the case of the residue shale-oil asphaltene, and, finally, to 6 or more in the case of petroleum asphaltenes.

Another parameter, the substituent chain length expressed as C_s/C_{su}, i.e., the number of total carbon atoms attached to a position on the edge of an aromatic sheet divided by the number of carbon atoms directly attached to an aromatic sheet, can be determined from the following equation:

$$\frac{C_s}{C_{su}} = \frac{h_s - H_{sMe}/3H}{H_\alpha/H} \tag{7-11}$$

Table 7-V shows that C_s/C_{su} values for the shale asphaltene indicate medium-size chain lengths, i.e., between 2 and 3 carbons. For comparison, the gilsonite asphaltene contains about 4 carbon atoms, whereas the petroleum asphaltene has about 4—6 carbon atoms.

The other parameter to be investigated by using NMR technique is the C_{sMe}/C_{Me} ratio, i.e., ratio of the total number of methyl groups, excluding the α and β positions of an aromatic ring, to the total number of methyl groups. This ratio can be evaluated by means of the following equation:

$$\frac{C_{sMe}}{C_{Me}} = \frac{H_{sMe}/H}{(H_{Me}/H_s)(1 - H_A/H)} \tag{7-12}$$

where H_{Me}/H_s can be determined from infrared absorption data (see next section). The values listed in Table 7-V for the native petroleum asphaltenes, the petroleum resin, and the gilsonite asphaltene suggest that a third to more than two-thirds of the methyl groups are more remote than β position to the aromatic sheets. The low value of 0.36 for H_{Me}/H_s ratio of the shale-oil asphaltene compared with 0.45 for the residue shale-oil asphaltene again suggests that thermal cracking results in (1) the breaking off from the aromatic sheets of relatively long-chain groups and/or (2) the dehydrogenation of naphthenic groups to which methyl or ethyl groups are attached. This is also true for the petroleum asphaltenes [42].

Infrared analysis. The infrared method enables us to detect the presence of chemical groups, preclude or limit the existence of proposed structures, and demonstrate similarities and differences in chemical structure of bitumens and related materials in situ. Besides the nuclear magnetic resonance method, infrared spectroscopy provides information concerning substitution at the edges of the aromatic sheets. A composite infrared spectrum of a typical oil-shale asphaltene is shown in Fig. 7-8. The broad absorption band centered at 3220 cm^{-1} is attributed to O—H or N—H bonds; this interpretation has been reported for coal [45,46] and coal asphaltenes [47]. Two low-intensity bands at 3010 and 3050 cm^{-1} represent aromatic C—H stretching. Wiberley [48] has shown that the aromatic C—H band shifts to values as high as 3052 cm^{-1} as ring number decreases. The strong absorption bands falling between 2840 and 2950 cm^{-1} are due to naphthenic and/or aliphatic C—H vibration. The two vibrations of the methyl group are located at 2950 and 2885 cm^{-1}, whereas the methylene band, which is the strongest, is located at 2915 cm^{-1}. Another methylene absorption peak occurs at 2840 cm^{-1}. An intense absorption band located at 1600 cm^{-1} has been attributed partly to a conjugated C=C bond and partly to carbonyl, C=O group. Two very intense bands at 1455 and 1375 cm^{-1} are due to bending frequencies of (1) asymmetric C—CH_3 bond and/or methylene, and (2) symmetric C—CH_3 bond, respectively. The group bands located at 1255, 1090 and 1030 cm^{-1} are closely related to the aromatic oxygenated compounds, such as aromatic ethers [45]. The long wavelength bands at 860, 800, and 750 cm^{-1} are considered aromatic out-of-plane frequencies and are important with regard to the nature of the structure of aromatic clusters. The four-adjacent C—H bending band at 720 cm^{-1} and the five-adjacent band at 695 cm^{-1} are very weak in spectrum. The long-chain alkyl band at 735 cm^{-1} is only present in the spectrum of residue shale-oil asphaltene. This long-chain methylene

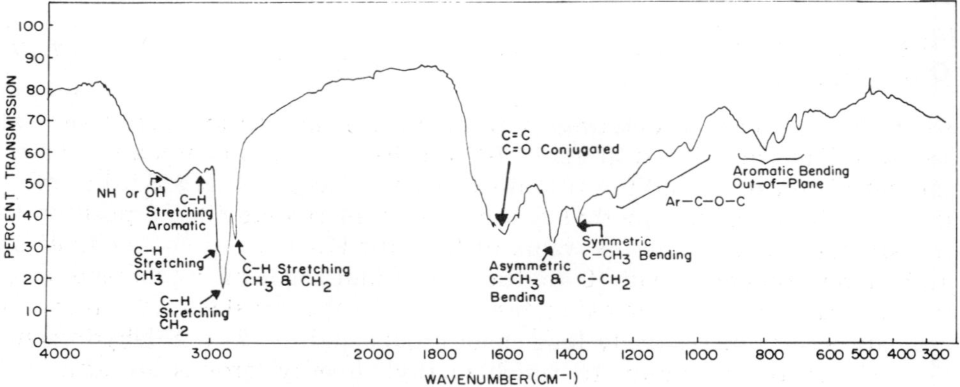

Fig. 7-8. Infrared spectrum of asphaltene isolated from Green River shale oil.

structure in residue shale-oil asphaltene has been confirmed by the X-ray diffraction analysis.

For the quantitative determination by infrared spectrum, the fraction of the hydrogen atoms which are attached to aromatic carbons, h_a, can be obtained by utilizing the infrared data as follows:

$$h_a = \frac{E_{3030}}{E_{3030} + E_{2920}} k \tag{7-13}$$

where E is the extinction coefficient at i wavelength and k is a constant.

It is also possible to determine the ratio of the number of hydrogen atoms present in methyl groups, H_{Me}, to the total number of hydrogen atoms attached to saturated carbon atoms, H_S, using the extinction coefficients of the absorption bands at 1380 and 2920 cm^{-1}:

$$\frac{H_{Me}}{H_S} = k' \frac{E_{1380}}{E_{2920}} \tag{7-14}$$

where k' is a constant [49]. The h_a and H_{Me}/H_S ratio values are presented in Table 7-VI for various bitumens.

Color intensity. The color intensity, I, is arbitrarily defined as the integrated absorption of a species between 750 and 400 nm. This relative measure is usually expressed as:

$$I = \int_{400\ nm}^{750\ nm} A d\lambda\ (mm^2 \cdot cm^{-1} \cdot ml/mg)\ 10^{-3} \tag{7-15}$$

where A* = intensity of absorbtion of the visible range of spectra; and λ = wavelength in nm. I is easily determined from visible range in a spectrophotometer. This parameter is a direct indicator for the extent of aromatic rings in the molecule. For example, refinery asphaltene has a higher value of color intensity than the native asphaltene [50]. The color intensity of carboid is higher than that of the carbene, that of carbene is higher than that of asphaltene, that of asphaltene is higher than that of resin, and that of resin is higher than that of gas oil [75,76].

Electron spin resonance. Electron spin resonance (ESR) provides a convenient method for the structural elucidation of complex macromolecules such as asphaltenes [50—52]. Among the basic parameters obtained by ESR measurements are N_g, the number of unpaired spins per gram of material, and g, the effective Landé g factor for the unpaired electron spins.

The g values determined for resin and asphaltene fractions are listed in Table 7-VII; bitumens exhibit a g value of about 2.00 [50,51]. In theory, the

* A = Area/(concentration × cell width).

TABLE 7-VI

Values for structural parameters of bitumens from infrared data

Bitumens [*1]	Shale-oil asphaltene I	Shale-oil asphaltene II	Residual shale-oil asphaltene	Petroleum asphaltene I	Petroleum asphaltene II	Petroleum resin	Gilsonite asphaltene
h_a	0.05	0.06	0.04	0.07	0.08	0.03	0.02
h_s [*2]	0.95	0.94	0.96	0.93	0.92	0.97	0.98
H_{Me}/H_s	0.46	0.41	0.38	0.34	0.42	0.31	0.41

[*1] As in Tables 7-V and 7-VI.
[*2] h_s is the fraction of saturated hydrogen atoms.

TABLE 7-VII

Values for structural parameters of bitumens from ESR data (After Yen et al. [50], Yen and Sprang [51], and Baedecker et al. [53])

Bitumens	*g* value	Spins/gram, $N_g \cdot 10^{-18}$
Petroleum asphaltenes		
Baxterville, Miss., U.S.A.	2.0029	6.4
Boscan, Venezuela	2.0034	1.8
Mara, Venezuela	2.0030	1.7
Petroleum resin		
Baxterville, Miss., U.S.A.	—	0.3
Ozocerite, Utah	2.0029	2.0
Tabbyite, Utah	2.0031	1.7
Gilsonite, Utah	2.0029	3.3
Grahamite, Utah	2.0029	3.5
Kerogen, marine sediment		
Tanner Basin, U.S.A.	2.0032	0.1

g value is a direct function of the degree of spin-orbital interaction. Variation of *g* values could be indirectly evaluated by a measurement of the heteroatomic contribution and the aromaticity. A recent investigation [51] involving the ESR *g* values of bitumen fractions yielded a more convenient classification scheme of bituminous materials. When the heterocyclic atomic contents of these bitumens are corrected for the spin-orbital coupling and correlated with *g* values, two distinct series of bitumens are obtained [51]. The classification scheme is shown in Fig. 7-9. A number of typical native petroleum asphaltenes tend to reflect a decrease in *g* values with a corresponding decrease in the percentage of heteroatoms (Series B). On the other hand, bitumens of asphaltites and asphaltoids fall into a constant *g* value (2.0030 ± 0.0003) zone. These *g* values, which are independent of the amount of organic heteroatom matter present, did not show any marked change among bitumens. Recently, a study of kerogens of marine sediments [53] showed *g* values of 2.0030 ± 0.0002 among unheated and heat-treated samples. The *g* values of coals and kerogens decrease with an increase in aromaticity during maturation.

ESR measurements on petroleum asphaltenes and native gilsonite asphaltenes [54] reveal a linear relation between aromaticity and the logarithm of spin concentration (Fig. 7-10). The data indicate that the free radical sites of these materials are associated with the aromatic rings of the bitumen skeleton. In addition, the variation of spin concentration with aromaticity could serve as an indication of the degree of localization of the free electron at the heteroatom sites of bitumens.

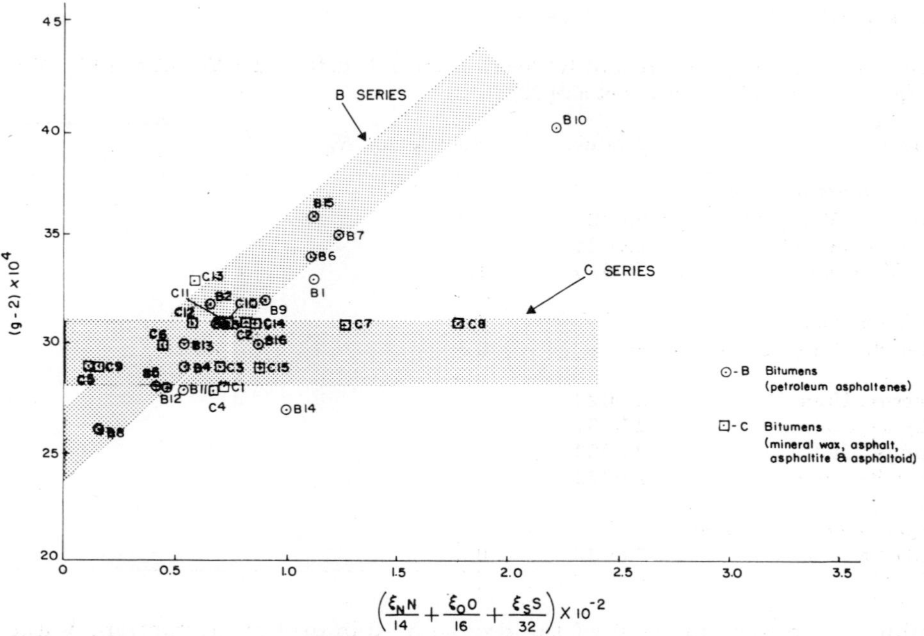

Fig. 7-9. Plot of g values versus heteroatom contribution. ξ_k = the spin-orbital coupling constant for k atom having p electrons (e.g., $\xi_N = 76$, $\xi_O = 151$, and $\xi_S = 382$); N = wt % of nitrogen; O = wt % of oxygen; and S = wt % of sulfur. (Modified after Yen and Sprang [51].)

ESR also can be used as a tool to gain more information concerning the arrangement of aromatic systems within the macrostructure of bitumens. The free electrons (or free radicals) are located within the planar polynuclear aromatic systems of bituminous crystallites (mesophase) in such a manner as to optimize their stabilization by the resonance of delocalized π-electrons. Under thermal conditions, these free electrons could exist as both singlet—triplet transitions as well as doublet states [74].

Oxidation rate. The slow rate of permanganate oxidation of petroleum asphaltenes and many other bituminous substances [55,56], as shown in Table 7-VIII, leads to a conclusion that most of the heteroatoms, particularly nitrogen and sulfur, are bound into ring structures. The oxidation rate of oil-shale kerogen is higher than that of petroleum asphaltenes. The heterocyclic cross-linkage of saturated naphthenics in kerogen structure results in a rapid decomposition during the oxidation [56,57]. In contrast to kerogen structure, the asphaltenes in petroleum have been pictured as having a two-dimensional fabric of condensed aromatic rings, short aliphatic chains, and some naphthenic-ring structures [58—60]. The stacked aromatic sheets in

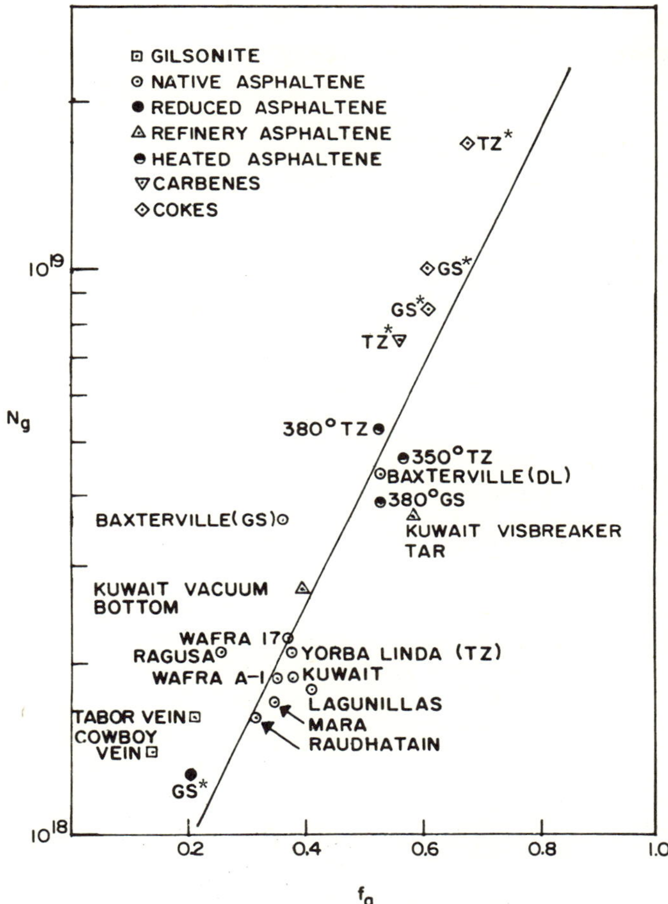

Fig. 7-10. Dependence of spin concentration on aromaticity; asterisks (*) indicate fractions from the corresponding asphaltene. (Modified after Yen et al. [54].)

asphaltene structure, linked with aliphatic and/or naphthenic rings, are extremely resistant to oxidation.

Refractive index. The refractive index of a bitumen is useful as a correlation index and as an indicator of molecular structure. Within any particular molecular-weight range, the order of increasing refractive indices is from paraffins, to naphthenes, to aromatics. Although in the high-molecular-weight ranges there is some overlapping, the rule does apply in general. Fig. 7-11 shows the relationship between the refractive indices of bitumens measured by the oil-immersion method [61] and the C/H ratio. There is a straight-line relationship between these two variables. Ozocerite, with a high

TABLE 7-VIII

Oxidation rates of bitumens (After Erdman and Ramsey [55])

Sample	Oxidation rate *
Shale (nonmarine)	
Insoluble carbonaceous material (Pa., U.S.A.)	20
Oil shale	
Green River, Colorado, U.S.A.	1.5
Petroleum asphaltenes	
Yorba Linda, Calif., U.S.A.	0.182
Santiago, Calif., U.S.A.	0.136
Boscan, Venezuela	0.074
Athabasca, Canada	0.073
Baxterville, Miss., U.S.A.	0.045
Petroleum asphaltenes from refinery fractions	
Straight-run residuum, California type	0.300
Straight-run residuum, Venezuelan type	0.078
Asphaltites	
Gilsonite (Uinta Basin, U.S.A.)	0.11
Gilsonite (Cowboy Vein, Uinta Basin, Utah, U.S.A.)	0.11
Grahamite (Okla., U.S.A.)	0.042
Manjak (Pinar del Rio, Cuba)	0.078
Asphaltoids	
Anthrapolite (Quebec, Canada)	0.302
Ingramite (Ingram mine, Utah, U.S.A.)	0.015
Albertite (Uinta Basin, U.S.A.)	0.013
Wurtzilite (Strawberry Canyon mine, Utah, U.S.A.)	0.003
Graphite (natural, Ceylon)	0.045

* Slope of plot of volume of oxidant (ml of 0.195 *M* $KMnO_4$ solution) versus time (min), corrected for residual pyrite where present.

paraffin content, has low refractive index and C/H ratio. With increasing indices and carbon content, there is a decrease in the paraffin content and an increase in the content of aromatic and condensed-ring structures. Inasmuch as paraffin and naphthene structures contain more hydrogen than aromatics, it is evident that gilsonite contains less of the former structures than ozocerite.

Density. Van Nes and van Western [71] developed a structural group analysis by measuring refractive index (n), density (d), and molecular weight (m). The so-called *n-d-m* method has been used widely in characterizing the

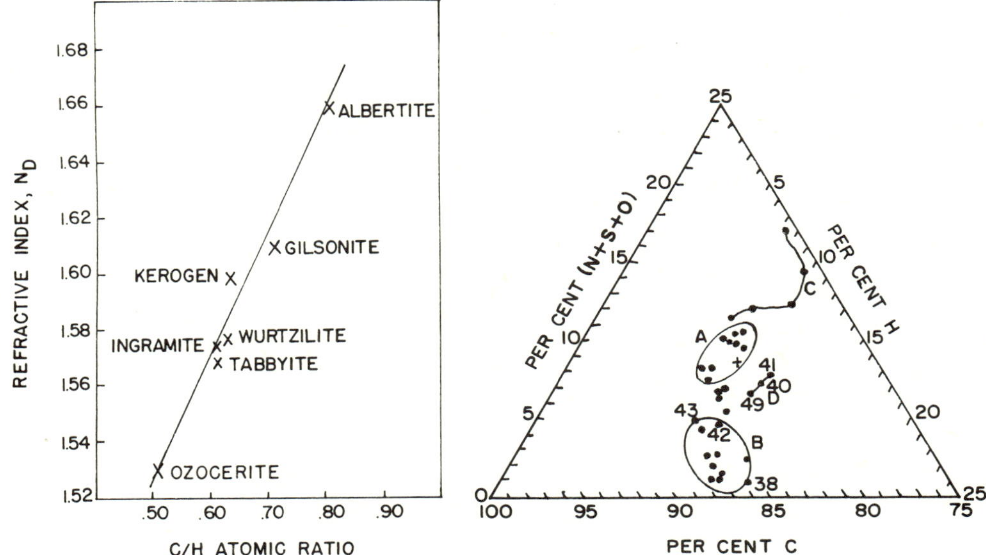

Fig. 7-11. Relationship between the refractive index and C/H atomic ratio for various bitumens.

Fig. 7-12. Compositions (weight percents of C, H, and heteroatoms) of four types of bituminous compounds. *A* = petroleum asphaltenes; *B* = thermal diffusion cuts from pentane-soluble fraction of 85 to 100 penetration asphalt (petrolenes); *C* = shale-oil asphaltenes and oil-shale derivatives (Green River Formation); *D* = asphalts (Bermudez pitch — point No. 49; Trinidad — point No. 40, and Tabbyite — point No. 41). Ozocerite (point No. 38), gilsonite (point No. 42), and grahamite (point No. 43) fall into B series. (Modified after T.F. Yen [79].)

mineral-oil fractions. Density is directly related to molar volume and can be easily measured by the helium displacement method. This densimetric method can be used to measure the ring condensation index for asphaltenes and related materials [72]. The compactness of the aromatic systems within the materials can be assessed [73]. For example, the aromatic systems in manjak and anthraxolite are peri-condensed and are close to those present in most petroleum asphaltenes. On the other hand, grahamite is only kata-condensed or linked.

Elemental composition. The major elements of bitumens are carbon, hydrogen, and oxygen. Nitrogen and sulfur are relatively minor constituents of bitumens on either atomic or weight basis. Some information concerning the molecular structures of bitumens can be obtained from the results of analyses for carbon, hydrogen, sulfur, and nitrogen contents. The ratio of carbon to hydrogen varies in different hydrocarbon structures. Naphthene

TABLE 7-IX

Elemental analysis of bitumens

Bitumens	C (wt%)	H (wt%)	C/H	N (wt%)	S (wt%)	O (wt%) *1
Petroleum asphaltenes						
Baxterville, Miss., U.S.A.	84.5	7.4	0.95	0.80	5.6	1.7
Mara, Venezuela	84.7	8.0	0.88	1.0	2.8	1.6
Petroleum resin						
Baxterville, Miss., U.S.A.	84.4	9.1	0.77		—	—
Green River oil shale						
Kerogen	77.7	10.1	0.64	2.6	1.1	8.5
Trona acids	75.7	9.8	0.64	1.9	1.6	11.0
Asphaltine	83.0	10.8	0.64	3.1	0.8	2.3
Shale-oil asphaltene I *2	79.7	8.2	0.81	3.5	1.2	4.6
Shale-oil asphaltene II *2	75.6	7.5	0.84	3.4	0.7	6.7
Residual shale-oil asphaltene *2	81.3	8.2	0.83	4.3	0.8	3.0
Kerogen, marine sediments *3	56.0	6.6	0.71	—	—	—
Ozocerite	85.4	13.9	0.51	0.4	0.3	0
Bermudez pitch	82.8	10.7	0.65	0.8	5.8	0
Trinidad	82.3	10.7	0.64	0.8	6.2	0
Tabbyite	81.3	11.2	0.61	2.1	1.2	4.2
Gilsonite	85.5	10.0	0.71	2.5	0.3	1.5
Grahamite	86.6	8.7	0.83	2.2	1.8	0.7

*1 By difference.
*2 As described in Tables 7-V and 7-VI.
*3 After Erdman and Ramsey [55].

and paraffin hydrocarbons have low C/H ratios, whereas condensed aromatics have high C/H ratios. Consequently, bituminous substances rich in the former structures have low C/H ratios, whereas carbonaceous substances rich in the latter structures have high C/H ratios. This interpretation based on C/H ratios, however, is valid only if the total quantity of nitrogen, sulfur, and oxygen in the substances being compared does not differ appreciably. Among bitumens, ozocerite (paraffin wax) has the lowest C/H atomic ratio (0.51), whereas grahamite (asphaltite) has the highest ratio (0.83) (Table 7-IX). The elemental composition of various bituminous substances is illustrated in Fig. 7-12 by means of a ternary graph, based on the weight percent of C, H, and heteroatoms.

Average structure of bitumens

The bitumens of naturally occurring bituminous substances are considered to be complex mixtures of high-molecular-weight hydrocarbons and nonhydrocarbons which can be separated into fractions consisting of oily material (soluble in propane), resins, asphaltenes, and carbenes.

These fractions merge into one another and their atomic C/H ratio increases with each succeeding member (see Table 7-IX), except for the carbenes which differ mainly in having more oxygen than the asphaltenes (see Figs. 7-8 and 7-13). Three types of hydrocarbons are present in bitumens: paraffinic, naphthenic, and aromatic. Nonhydrocarbons in bitumens have heterocyclic atoms consisting of sulfur, nitrogen, and oxygen. The asphaltenes usually contain more aromatic compounds than do the resins and the oily fraction. The resins contain aromatic or naphthenic hydrocarbons and the components of oily fraction may have naphthenic or paraffinic structure. The naphthenic plus paraffinic contents in bitumens increase with decreasing content of aromatic hydrocarbons, the oily fraction having the lowest aromatic content. The molecular weights of various bitumen fractions vary from about 300 to 1500 for the oily fraction and resins, and from about 600 to 10,000 for the asphaltenes, and to probably higher values for the carbenes.

Distinct differences between resin and asphaltene in bitumens are due to the differences in the structural units forming the constituent groups. The skeletal structure of bitumens, based on present-day knowledge of heavy-petroleum fractions, is pictured as a crumpled, largely two-dimensional fabric of condensed aromatic rings, intermingled with short aliphatic chains and fused naphthenic-ring systems [58,32]. The distinguishing disparity in solubility of resin and asphaltene is due to the differences in the contents of aromatic, naphthenic, paraffinic, and heterocyclic compounds. Inasmuch as the solubility and molecular weight of resins and asphaltenes do differ

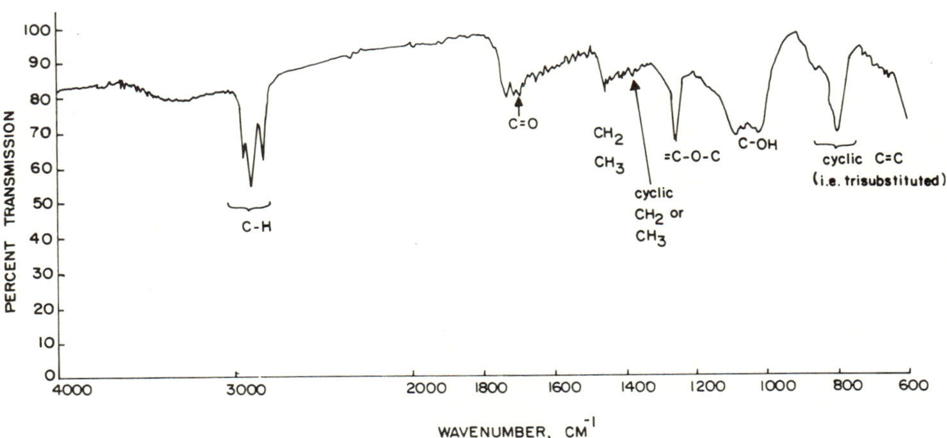

Fig. 7-13. Infrared spectrum of carbene isolated from the Green River shale oil.

greatly, it is inferred that the differences of structural units of their molecules lies mainly in the average size and shape of the fabric sheets. The material from the oil-shale bitumen (Green River Formation), which is insoluble in pentane, has an average molecular weight of 1320 and contains about 0.2% nitrogen, 1.0% sulfur, and 7.4% oxygen [62]. The polar portion of the pentane-soluble material, eluted from the alumina column by benzene and benzene—methanol mixtures, had an average molecular weight of 625 and contained 0.9% nitrogen, 1.4% sulfur, and 7.8% oxygen [62]. A series of heterocyclic compounds present in this material is shown in Table 7-X. A compound having an empirical formula of maleimides, $C_nH_{2n-5}NO_2$, has also been found in oil-shale retort water [63]. Doolittle et al. [64] obtained the exact mass and other spectral data for six oxygen- and oxygen—nitrogen-containing compounds which were removed from the resin fraction (eluted from the alumina column with benzene) of oil-shale bitumen by molecular distillation. Four of the compounds were of a homologous series with empirical formulas of $C_{11}H_{12}O$ to $C_{14}H_{18}O$. Spectral data suggest that these compounds may be either indanones, tetralones, or acetyl indans (Fig. 7-14).

A true understanding of resin fraction will probably not be reached through the study of the bitumens in isolation. It is proposed [77] that in the natural colloidal state, the resin is adsorbed in the asphaltene nuclear matrices in micellar form. Probably through aliphatic side chains, there is a transition of the aromatic resins to less polar, aliphatic and naphthenic components and a stable system in which the intermolecular adsorption forces are balanced.

The structural picture of bitumens, as known today, is based largely on the present-day knowledge of petroleum-asphaltene structure. The literature on structure of the asphaltenes has been reviewed recently by Yen [58]. Spectroscopic techniques (mass, infrared, NMR, electron spin resonance

TABLE 7-X

Water-miscible polar constituents from Green River oil shale (After Anders et al. [19], Wen [21], and Wen et al. [63])

General types of constituents	Specific types
$C_nH_{2n}O$	substituted cyclohexanols isoprenoid ketones
$C_nH_{2n-10}O$	tetralones substituted indanones
$C_nH_{2n-2}O_2$	gamma lactones
$C_nH_{2n-7}N$	substituted tetrahyroquinolines
$C_nH_{2n-11}N$	quinolines
$C_nH_{2n-1}NO$	alkoxypyrrolines
$C_nH_{2n-5}NO_2$	maleimides
$C_nH_{2n-3}NO_2$	succinimides

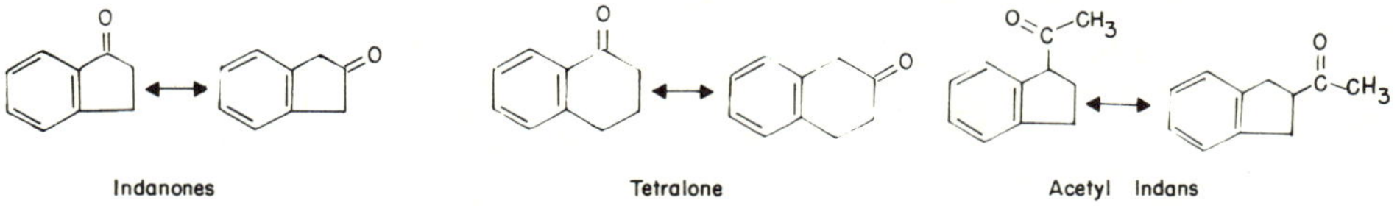

Indanones
Tetralone
Acetyl Indans

Fig. 7-14. Molecules isolated from resin of Green River shale-oil bitumen.

(ESR) and X-ray diffraction) and densimetric method have all been used to define the indistinct outlines of the asphaltene molecules. The data suggest that the asphaltene molecule appears to carry a core of stacked flat sheets of condensed aromatic rings. Approximately five of these sheets are associated by π—π interaction. They are stacked one above the other with a distance between sheets ranging from 3.55 to 3.70 Å giving an overall height for the stack of 16—20 Å; the average sheet diameter appears to be equal to 8.5—15 Å, as shown in Fig. 7-15. The stacked aromatic sheets show some disorder, probably induced by chains of aliphatic- and/or naphthenic-ring systems, which link the edges of the aromatic sheets and tend to hold the sheets apart. The condensed aromatic sheets contain oxygen, sulfur and nitrogen atoms, which are probably associated with the free-radical sites detected by ESR [50]. These sites may be the centers at which complexed vanadium and nickel are located. Vanadium and nickel are known to concentrate in the petroleum-asphaltene fraction and their content ranges from a few parts per million to a few thousand parts per million [65,66]. There are two types of bonding of metals to asphaltenes. The first type is the association of metallo-porphyrins and nonporphyrins with asphaltene molecules in which the metals are chelated or complexed to ligands. The other type is the bonding

Cluster
z
L_a 8.5-15 Å
d_M 3.55-3.70 Å
d_γ 5.5-6.0 Å
M_e = 5
L_c 16-20 Å

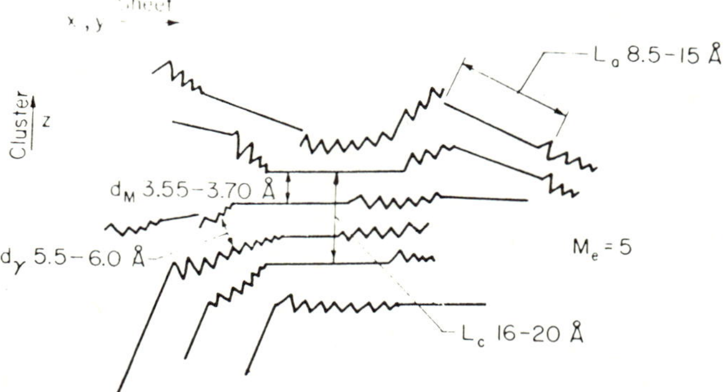

Fig. 7-15. Cross-sectional view of a petroleum asphaltene micelle; ~~~~ represents the zig-zag configuration of a saturated carbon chain or loose net of naphthenic rings; ——— represents the edges of flat sheets of condensed aromatic rings. (After Yen et al. [32]; courtesy of *Anal. Chem.*)

of metal ions to the defect centers of an aromatic sheet of asphaltene or resin as illustrated in Fig. 7-16.

Various asphaltene fractions from different bitumens appear to have similar structures with some differences: the degree of development of the fabric, i.e., the molecular weight, and the structure of heterocyclic compound, i.e., the heteroatomic positions. In contrast to the molecular weight of petroleum asphaltenes which ranges from 1000 to 10,000, the recently isolated asphaltene from shale oil had a molecular weight ranging from about 600 to 750 [38]. Structurally (the number of stacked flat sheets and the distance and diameter of condensed aromatic sheets) petroleum and shale-oil asphaltenes are similar. The distance between the sheets of saturated aliphatic and/or naphthenic-ring systems (connecting groups), however, appears to be 4.2—4.4 Å in shale-oil asphaltenes and 5.5—6.0 Å in petroleum asphaltenes. In addition, the average number of carbon atoms per saturated substituent appears to be 2—3 in shale-oil asphaltene as compared to 4—6 in petroleum asphaltenes.

The high contents of nitrogen and oxygen (heteroatoms) in shale-oil asphaltenes in comparison with those in petroleum asphaltenes suggest that the constitutions of heterocyclic compounds is quite different in these two asphaltenes. Based on infrared analysis [49], there is no quinone or semi-

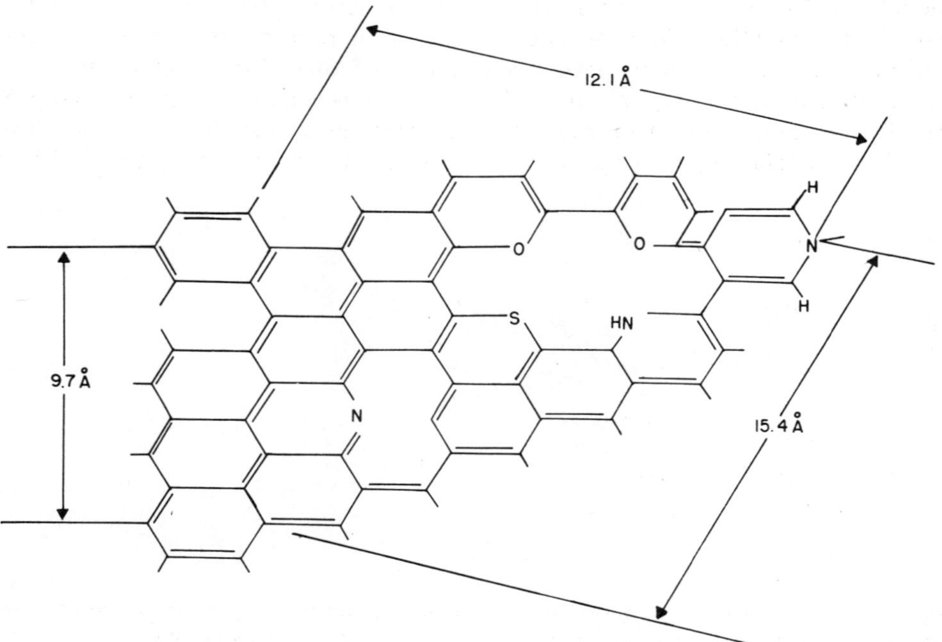

Fig. 7-16. Ligands derived from a defect center of an asphaltene sheet.

quinone type of oxygen in petroleum asphaltenes, although ether-type oxygen and quinoline-type nitrogen sites are detected. The slow rate of oxidation of petroleum asphaltenes suggests that most of the heteroatoms, particularly nitrogen and sulfur, are protected (tied up) in chains or rings. Shale-oil asphaltenes and many other asphaltene fractions of bitumens, however, could contain many reactive functional groups as compared to petroleum asphaltenes. The variation of ESR *g* values for native bitumens, compared to petroleum asphaltenes, indicates different heteroatomic contributions. Usually, structures containing hetero-nitrogenous components in cross-linked bridge are oxidized most rapidly. In the case of electrochemical oxidation [21], however, oxidation rate of shale-oil asphaltene has been found comparable to that of its other derivatives. Furthermore, the ultimate analysis indicated that high proportions of nitrogen and sulfur had been removed after the electrochemical oxidation.

Based on the present-day knowledge, bitumens can be visualized as multipolymers consisting largely of aromatic rings, linked by aliphatic and/or naphthenic chains. Cross-sectional views of the structural components of

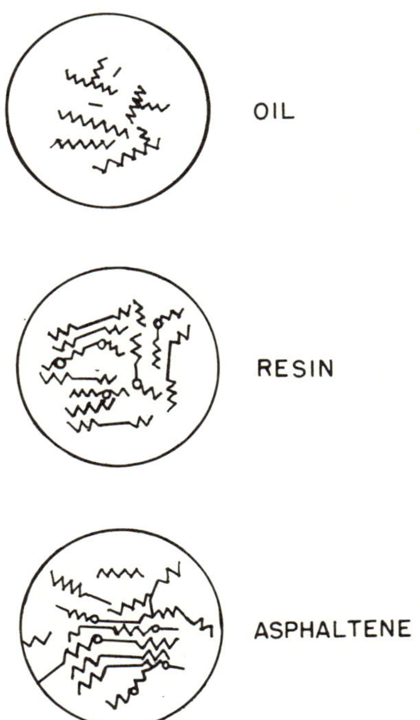

Fig. 7-17. Structural representation of bitumens; ~~~~ represents the zig-zag configuration of a saturated carbon chain or loose net of naphthenic rings; ——— represents the edges of flat sheets of condensed aromatic rings; and ○ represents the heteroatoms.

bitumens are shown in Fig. 7-17. The three major components of bitumens can be illustrated with the following symbols: straight line — aromatic; open zig-zag line — paraffinic; and closed zig-zag lines — naphthenic. Resins in bituminous materials can be treated as cross-linked analogs of asphaltene, except that size and shape of configuration is slightly different.

There are a large number of native bitumens, e.g., gilsonite asphaltene, which fall into the petroleum-asphaltene series. Both petroleum and gilsonite asphaltenes seem to have a common structural link as indicated by permanganate oxidation and physical spectroscopic investigations. As discussed above, other natural bitumens exhibit a macrostructure similar to that of petroleum asphaltenes. Although the emphasis of present-day structural studies of bitumens has been placed primarily on asphaltic fractions of petroleum, now there is a better understanding of structure of all natural bitumens. Obviously, the structures envisioned by various investigators represent only some of the possible combinations of unit sheets. Specific features may vary among the fractions, particularly molecular association in the resin fraction. None of the proposed structures by various investigators can represent the whole picture of an actual structure and reflect the properties of the actual substance. Structural parameters should be investigated further in order to approximate the actual structure of bitumens.

Conclusions

The properties and structures described in this chapter for bitumens extracted from the naturally occurring bituminous substances can only be regarded as "average" properties and structures. From the data presented, it is apparent that bitumens consist of a complex mixture of multipolymers which cannot be represented by a single structural formula. The various fractions (resins, asphaltenes, and others) represent part of a system of multipolymers, the chemical properties of which change systematically with increasing molecular weight. Statistical constitution analysis, based on spectroscopic and chemical studies, indicates that bitumens from various naturally occurring bituminous materials consist of molecules having similar structure, but having a wide molecular-weight range. The similarity in the properties of fractions dissolved in solvents suggests that characteristic structural groups tend to reoccur in the large molecules.

At the present time, data are still lacking on the chemical nature of many of the bitumen constituents and particularly as to how their structures are controlled by different depositional and postdepositional (diagenetic, catagenetic and metamorphic) environments. Direct chemical proof of structural modification is generally not available. Many major problems remaining to be solved include: (1) positive identification of the major source materials and elucidation of the principal reactions; (2) a configuration of a representative microstructure of bitumens based on the nature of bridge units, structural arrangement of reactive groups, and estimated structural parameters; and (3) genetic relationships among resins, asphaltenes, carbenes, and carboids, and their role in the formation of kerogen and other naturally occurring carbonaceous substances.

Although the resin and asphaltene fractions of bitumens and, in fact, the bituminous substances themselves, are minor in amount compared to the vast volume of insoluble kerogen material present in the sedimentary rocks, bitumens still represent an enormous energy source. Knowledge of the properties and structure of the substances contained in bitumens makes possible not only a better chemical utilization of these resources, but can also provide much information on the conditions and chemical changes necessary to convert these fossil raw materials into clean fuels. New methods should be developed for the economic utilization of these resources in view of the growing demand for energy.

Acknowledgment

Partial support from A.G.A. BR-18-12, PRF 6272AC-2, NSF GI-35683, AER-74-23797, and ERDA, E(29-2)-3619, E(49-118)-2031 and E(29-2)-3758 is acknowledged.

References

1 J.P. Pfeiffer, *The Properties of Asphaltic Bitumen*, Elsevier, New York, N.Y., 285 pp. (1950).

2 H. Abraham, *Asphalts and Allied Substances, 1*, Van Nostrand, Princeton, N.J., 6th ed., 370 pp. (1960).

3 A.J. Hoiberg, L.W. Corbett and R.B. Lewis, "Asphalt", in: *Kirt-Othmer Encyclopedia of Chemical Technology,2*, Wiley, New York, N.Y., 2nd ed., pp. 762—806 (1966).

4 H.O. Ervin, "Gilsonite", in: *Kirt-Othmer Encyclopedia of Chemical Technology, 10*, Wiley, New York, N.Y., 2nd ed., pp. 527—529 (1966).

5 G.K. Bell and J.M. Hunt, "Native Bitumens Associated with Oil Shale", in: I.A. Breger (Editor), *Organic Geochemistry*, Pergamon, Oxford, pp. 363—366 (1963).

6 B.F. Marschner and J.C. Winters, "Differences among Ozocerites", *Adv. Chem. Ser., 151*, 166—178 (1976).

7 E.S. McLoud, "Waxes", in: *Kirt-Othmer Encylopedia of Chemical Technology, 22*, Wiley, New York, N.Y., 2nd ed., pp. 156—173 (1966).

8 V. Wollrab and M. Streibl, "Earth Waxes, Peat, Montan Wax and Other Organic Brown Coal Constituents", in: G. Eglinton and M.T.J. Murphy (Editors) *Organic Geochemistry*, Springer, New York, N.Y., pp. 576—598 (1969).

9 W. Bergman, "Geochemistry of Lipids", in: I.A. Breger (Editor), *Organic Geochemistry*, Pergamon, Oxford, pp. 503—542 (1963).

10 J.S. Miller, "Native Asphalts and Bitumens", in: A.E. Dunstan, A.W. Nash, B.T. Brooks and H. Tizard (Editors), *The Science of Petroleum*, Oxford University Press, London, pp. 2710—2727 (1938).

11 R.N. Traxler, "The Physical Chemistry of Asphaltic Bitumen", *Chem. Rev., 19(2)*, 119—143 (1936).

12 C. Ponnamperuma and K.L. Pering, "Aliphatic and Alicyclic Hydrocarbons Isolated from Trinidad Lake Asphalt", *Geochim. Cosmochim. Acta, 31*, 1350—1353 (1967).

13 I.A. Breger and T.K. Miles, "Asphalt and Asphaltite", in: *McGraw-Hill Encyclopedia of Science and Technology,2*, McGraw-Hill, New York, N.Y., pp. 637—638 (1971).

14 W.H. Shearon and A.J. Hoiberg, "Petroleum-Base Protective Coatings", *Ind. Eng. Chem., 41*, 2672—2679 (1949).

15 L.R. McGee, *Porphyrins in Gilsonite*, Thesis, Univ. of Utah, Salt Lake City, Utah, 63 pp. (1956).

16 D.K. Vitorovic and P.A. Pfendt, "Investigation of the Kerogen of a Yugoslav (Aleksinac) Oil Shale", *Proc. 7th World Pet. Congr., 3*, 691—697 (1967).

17 W.E. Robinson, "Origin and Characteristics of Green River Oil Shale", in: T.F. Yen and G.V. Chilingarian (Editors), *Oil Shale*, Elsevier, Amsterdam, pp. 61—79 (1976).

18 W.E. Robinson and G.L. Cook, "Compositional Variations of the Organic Material of Green River Oil Shale — Colorado No. 1 Core", *U.S. Bur. Mines, Rep. Invest., 7492*, 32 pp. (1971).

19 D.E. Anders, F.G. Doolittle and W.E. Robinson, "Polar Constituents Isolated from Green River Oil Shale", *Geochim. Cosmochim. Acta, 39*, 1423—1430 (1975).

20 T.F. Yen, "Structural Investigations on Green River Oil Shale Kerogen", in: *Science and Technology of Oil Shale*, Ann Arbor Science, Ann Arbor, Mich., pp. 193—203 (1976).

21 C.S. Wen, *Electrolytic Processes of Oil Shale and Its Derivatives*, Thesis, Univ. of Southern California, Los Angeles, Calif., 239 pp. (1976).

22 V.D. Allred, "Kinetics of Oil Shale Pyrolysis", *Chem. Eng. Prog., 62(8)*, 55—60 (1966).

23 D.W. Fausett, J.H. George and H.C. Carpenter, "Secondary-Order Effects in the Kinetics of Oil Shale Pyrolysis", *U.S. Bur. Mines, Rep. Invest., 7889*, 21 pp. (1974).

24 C.S. Wen and T.F. Yen, "Optimization of Oil Shale Pyrolysis", *Chem. Eng. Sci., 32(3)*, 346—349 (1977).

25 S.R. Sergienko, *High-Molecular Compounds in Petroleum*, translated by J. Schmorak, Israel Program for Scientific Translations, pp. 401—437 (1965).

26 J.P. Dickie and T.F. Yen, "Macrostructures of the Asphaltic Fractions by Various Instrumental Methods", *Anal. Chem., 39(14)*, 1847—1952 (1967).

27 J.G. Speight, "The Structure of Athabasca Asphaltenes and the Conversion of These Materials to Useful By-Products", *Am. Chem. Soc., Div. Fuel Chem., Prepr., 15(1)*, 76—82 (1971).

28 J.P. Dickie and T.F. Yen, "High Resolution Mass Spectral Examination of the Resin Fraction of Petroleum Asphaltenes", *Am. Chem. Soc., Div. Pet. Chem., Prepr., 13(2)*, F140—F143 (1968).

29 R.S. Winniford, "Evidence for Association of Asphaltenes in Dilute Solution", *J. Inst. Pet., 49*, 215—221 (1963).

30 K.H. Altgelt and O.L. Harle, "The Effect of Asphaltenes on Asphalt Viscosity", *Ind. Eng. Chem., Prod. Res. Dev., 14(4)*, 240—246 (1975).

31 K.H. Altgelt, "Asphaltene Molecular Weights by Vapor Pressure Osmometry", *Am. Chem. Soc., Div. Pet. Chem., Prepr., 13(3)*, 37—44 (1968).

32 T.F. Yen, J.G. Erdman and S.S. Pollack, "Investigation of the Structure of Petroleum Asphaltenes by X-Ray Diffraction", *Anal. Chem., 33(11)*, 1587—1594 (1961).

33 S.S. Pollack and L.E. Alexander, "X-Ray Analysis of Electrode Binder Pitches and Their Cokes", *J. Chem. Eng. Data, 5*, 88—93 (1960).

34 A.R. Ubbelohde and F.A. Lewis, *Graphite and Its Crystal Compounds*, Oxford University Press, London, 217 pp. (1960).

35 S. Ergun and V.H. Tiensuu, "Interpretation of the Intensities of X-Ray Scattered by Coals", *Fuel, 38*, 64—78 (1959).

36 T.F. Yen and J.G. Erdman, "Asphaltenes (Petroleum) and Related Substances: X-

Ray Diffraction", in: G.L. Clark (Editor), *Encyclopedia of X-Rays and Gamma Rays*, Reinhold, New York, N.Y., pp. 65—69 (1963).

37 R. Diamond, "X-Ray Diffraction Data for Large Aromatic Molecules", *Acta Crystallogr., 10*, 359—364 (1957).

38 T.F. Yen, C.S. Wen, J.T. Kwan and E. Chow, "The Role of Asphaltenes in Shale Oil", *CIC—ACS Meet.*, Montreal (May, 1977).

39 T.F. Yen, "Long-Chain Alkyl Substituents in Native Asphaltic Molecules", *Nature Phys. Sci., 233(37)*, 36—37 (1971).

40 S. Ergun, "X-Ray Studies of Coals and Carbonaceous Materials", *U.S. Bur. Mines, Bull., 648*, 35 pp. (1968).

41 N.B. Vassoevich and P.P. Timofeev (Editors), *Investigation of Organic Matter of Recent and Ancient Sediments*, Nauka, Moscow, 411 pp. (1976).

42 T.F. Yen and J.G. Erdman, "Investigation of the Structure of Petroleum Asphaltenes and Related Substances by High Resolution Proton Nuclear Magnetic Resonance", *Am. Chem. Soc., Div. Pet. Chem., Prepr., 7(3)*, B99—B111 (1962).

43 J.W. Ramsey, F.R. McDonald and J.C. Petersen, "Structural Study of Asphalts by Nuclear Magnetic Resonance", *Ind. Eng. Chem., Prod. Res. Dev., 6(4)*, 231—236 (1967).

44 D.W. Van Krevelen, *Coal*, Elsevier, Amsterdam, 514 pp. (1961).

45 R.A. Friedel and J.A. Queiser, "Infrared Analysis of Bituminous Coals and Other Carbonaceous Materials", *Anal. Chem., 28*, 22—30 (1956).

46 R.A. Friedel and H.L. Retcofsky, "Spectrometry of Chars-Structure Studies", in R.A. Friedel (Editor), *Spectrometry of Fuels*, Plenum, New York, N.Y., pp. 46—69 (1970).

47 S.R. Taylor, L.G. Galya, B.J. Brown and N.C. Li, "Hydrogen Bonding Study of Quinoline and Coal-Derived Asphaltene Components with o-Phenylphenol by Proton Magnetic Resonance", *Spectrosc. Lett., 9(11)*, 733—741 (1976).

48 S.E. Wiberley and R.D. Gonzalez, "Infrared Spectra of Polynuclear Aromatic Compounds in the C-H Structures and Out-of-Plane Bending Regions", *Appl. Spectrosc., 15*, 174—177 (1961).

49 T.F. Yen and J.G. Erdman, "Investigation of the Structure of Petroleum Asphaltenes and Related Substances by Infrared Analysis", *Am. Chem. Soc., Div. Pet. Chem., Prepr., 7(1)*, 5—18 (1962).

50 T.F. Yen, J.G. Erdman and A.J. Saraceno, "Investigation of the Nature of Free Radicals in Petroleum Asphaltenes and Related Substances by Electron Spin Resonance", *Anal. Chem., 34*, 694—700 (1962).

51 T.F. Yen and S.R. Sprang, "Contribution of ESR Analysis Toward Diagenetic Mechanisms in Bituminous Deposits", *Geochim. Cosmochim. Acta*, in press (1977).

52 E.C. Tynan and T.F. Yen, "Association of Vanadium Chelates in Petroleum Asphaltenes as Studied by ESR", *Fuel, 48*, 191—208 (1969).

53 M.J. Baedecker, R. Ikan, R. Ishiwatani and I.R. Kaplan, "Thermal Alteration Experiments on Organic Matter in Recent Marine Sediments as a model of Petroleum Genesis", in: T.F. Yen (Editor), *Chemistry of Marine Sediments*, Ann Arbor Science, Ann Arbor, Mich., pp. 55—72 (1977).

54 T.F. Yen, E.C. Tynan and G.B. Vaughan, "Electron Spin Resonance Studies of Petroleum Asphaltics", in: R.A. Friedel (Editor), *Spectrometry of Fuels*, Plenum, New York, N.Y., pp. 187—201 (1970).

55 J.G. Erdman and V.G. Ramsey, "Rates of Oxidation of Petroleum Asphaltenes and Other Bitumens by Alkaline Permanganate", *Geochim. Cosmochim. Acta, 25*, 175—188 (1961).

56 W.E. Robinson, H.H. Heady and A.B. Hubbard, "Alkaline Permanganate Oxidation of Oil Shale Kerogen", *Ind. Eng. Chem., 45(4)*, 788—791 (1953).

57 D.K. Young, S. Shih and T.F. Yen, "Mild Oxidation of Bioleached Oil Shale", *Am. Chem. Soc., Div. Fuel Chem., Prepr., 19(2)*, 169—174 (1974).

58 T.F. Yen, "Structure of Petroleum Asphaltene and Its Significance", *Energy Sources, 1(4)*, 447—463 (1974).

59 T.F. Yen, "Chemical Aspects of Interfuel Conversion", *Energy Sources, 1(1)*, 117—136 (1973).

60 J.G. Erdman, "The Molecular Complex Comprising Heavy Petroleum Fractions", in: *Hydrocarbon Analysis, Am. Soc. Test. Mater., Spec. Tech. Publ., 389*, 259—300 (1965).

61 I.P. Forsman and J.M. Hunt, "Insoluble Organic Matter (Kerogen) in Sedimentary Rocks of Marine Origin", in: *Habitat of Oil, Am. Assoc. Pet. Geol., Spec. Publ.*, pp. 747—778 (1958).

62 W.E. Robinson and J.J. Cummins, "Composition of Low-Temperature Thermal Extracts from Colorado Oil Shale", *J. Chem. Eng. Data, 5*, 74—80 (1960).

63 C.S. Wen, T.F. Yen, J.B. Knight and R.E. Poulson, "Studies of Soluble Organics in Simulated In-Situ Oil Shale Retort Water by a Combined GC-MS System", *Am. Chem. Soc., Div. Fuel Chem., Prepr., 21(6)*, 290—297 (1976).

64 F.G. Doolittle, D.E. Anders and W.E. Robinson, "Spectral Characteristics of Heteroatom Compounds in Oil Shale", *Pittsburgh Conference on Analytical Chemistry and Applied Spectroscopy*, Cleveland, Ohio, Paper 167 (1973).

65 T.F. Yen, L.J. Boucher, J.P. Dickie, E.C. Tynan and G.B. Vaughan, "Vanadium Complexes and Porphyrins in Asphaltenes", *J. Inst. Pet., 55*, 87—99 (1969).

66 G.B. Vaughan, E.C. Tynan and T.F. Yen, "Vanadium Complexes and Porphyrins in Asphaltenes, II, The Nature of Highly Aromatic Substituted Porphyrins and Their Vanadyl Chelates", *Chem. Geol., 6(3)*, 203—219 (1970).

67 J.M. Sugihara and T.F. McCullough, "Analysis of Hydrocarbon Fraction of Gilsonite", *Anal. Chem., 28*, 370—372 (1956).

68 W. Francis, *Coal, Its Formation and Composition*, Edward Arnold, London, 806 pp. (1961).

69 R. Pontonie, P.W. Thomson and Fr. Thiergart, "Zur Nomenklatur und Klassifikation der Neogenen Sporomorphae (Pollen und Sporen), (Zur Geologie der Rheinischen Braunkohle, pp. 35—69)", *Geol. Jahrb., 65*, 551 (1949).

70 D.M. Jewell, E.W. Albaugh, B.E. Davis and R.G. Ruberto, "Integration of Chromatographic and Spectroscopic Techniques for the Characterization of Residual Oils", *Ind. Eng. Chem., Fundam., 13(3)*, 278—282 (1974).

71 K. van Nes and H.A. van Western, *Aspects of the Constitution of Mineral Oils*, Elsevier, Amserdam, 484 pp. (1951).

72 T.F. Yen, J.G. Erdman and W.E. Hanson, "Reinvestigation of Densimetric Methods of Ring Analysis", *J. Chem. Eng. Data, 6*, 443—448 (1961).

73 T.F. Yen and J.P. Dickie, "The Compactness of the Aromatic Systems in Petroleum Asphaltics", *J. Inst. Pet., 54*, 50—53 (1968).

74 T.F. Yen and D.K. Young, "Spin Excitations of Bitumens", *Carbon, 11*, 33—41 (1973).

75 I. Schwager and T.F. Yen, "Separation of Coal Liquids from Major Liquefaction Processes to Meaningful Fractions", in: T.T. Ellington (Editor), *Liquid Fuel from Coal*, Academic Press, New York, N.Y. (in press) (1977).

76 T.F. Yen, I. Schwager, J.T. Kwan and P.A. Farmanian, "Characterization of Coal-Derived Asphaltene", *Pittsburgh Conference on Analytical Chemistry and Applied Spectroscopy*, Cleveland, Ohio, Paper 69 (1977).

77 J.G. Erdman, "Geochemistry of the High Molecular Weight Non-Hydrocarbon Fraction of Petroleum", in: U. Colombo and G.D. Hobson (Editors), *Advances in Organic Geochemistry*, MacMillan, New York, N.Y., pp. 215—237 (1964).

78 T.A. Hendrickson, *The Synthetic Fuels Data Handbook*, Cameron Engineering, Inc., Denver, Colo., 308 pp. (1975).

79 T.F. Yen, "Terrestrial and Extraterrestrial Stable Organic Molecules", in: R.F. Landel and A. Rembaum (Editors), *Chemistry in Space Research*, Elsevier, Amsterdam, pp. 105—153 (1972).

Chapter 8

IN-SITU RECOVERY OF OIL FROM OIL SANDS

JACQUES G. BURGER

Introduction

It has been shown in Chapter 2 that tar sands represent very abundant fuel resources for the world. Major deposits occur in Canada (919 billion bbl of oil in the tar sands of Alberta), Venezuela (700 billion bbl of oil in the Orinoco tar belt) and the United States (25 billion bbl of oil in the tar sands of Utah). Technical and economic problems, however, have maintained the production of these deposits at a very low level. As a matter of fact, tar sands contain practically immobile hydrocarbons, and the productivity of wells drilled in such deposits is either nil or extremely poor. It is necessary, therefore, either to mine the sand and to separate the bitumen in a surface plant, as described in Chapter 5, or to use sophisticated in-situ techniques.

Very serious environmental problems are associated with mining techniques; furthermore, mining is inconceivable in deep deposits or in formations with a too high overburden/thickness ratio [1]. In considering the tar-sand deposits of Canada and Venezuela, only 38 billion bbl of the in-place reserves (crude bitumen) in the Athabasca deposit [1] are currently considered as proved resources recoverable by surface-mining techniques. This corresponds to 6% of the reserves in the Athabasca deposit and 2.35% of the tar-sand resources of Canada and Venezuela. Furthermore, the viability of underground mining of oil sands is still unproved in the field [6]. Hence, recovery of these huge reserves is potentially dependent on in-situ techniques.

During the past twenty years, a number of pilot projects have been aimed at developing various in-situ processes. The proposed techniques, however, involve injection of a fluid into the formation; therefore, it is required that the existing overburden be thick enough for the formation to contain the injected fluids. This constraint is particularly important for the Athabasca tar sands, characterized by a low-quality overburden: it is generally admitted that a minimum overburden of 500 ft is essential in that region. The Athabasca deposits occurring under 150—500 ft of overburden are currently estimated as poor candidates for both surface-mining and in-situ techniques.

The main problems when applying an in-situ method to tar-sand reservoirs are the low effective permeability, the lack of primary energy in the formation, and the lack of mobility of bitumen. Furthermore, specific well problems may be involved.

Various in-situ methods are briefly presented here. In addition, a number of field projects are described from the information available in the published literature.

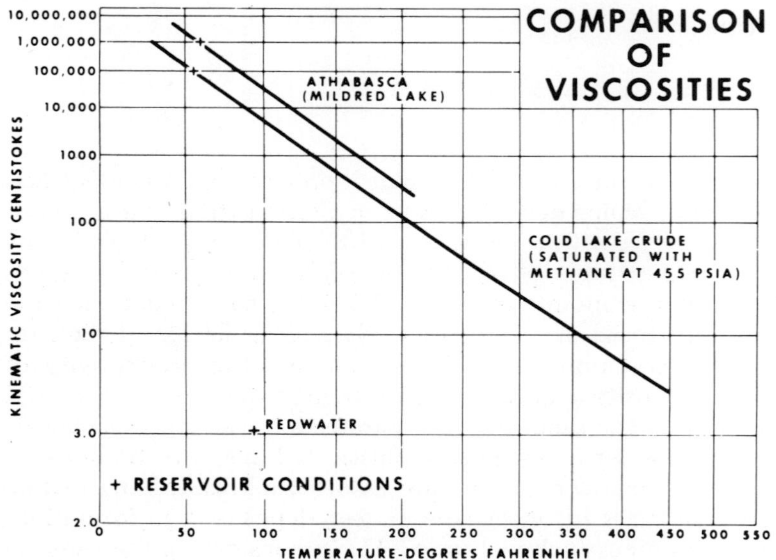

Fig. 8-1. Viscosities of Athabasca and Cold Lake crudes compared with Redwater crude. Kinematic viscosities versus temperature curves. (After Winestock [2].)

Properties of major tar-sand deposits

The Alberta oil-sand deposits are unconsolidated sands with a porosity ranging from 30 to 40%, having a good intrinsic premeability. The oil density ranges from 6 to 12° API. Viscosities are very high: approximately 1,000,000 cP in reservoir conditions for Athabasca bitumen and 100,000 cP for Cold Lake bitumen (Fig. 8-1). The sulfur content exceeds 3.5%.

The Orinoco tar belt (or "oil belt") consists of numerous permeable sands having a porosity between 12 and 38% (average 25%). The oil gravity ranges from 8 to 15° API. The lowest API-gravity values are found toward the southern edge, whereas API gravity increases toward the north. The general API-gravity average for tar-belt deposits is 9.9° [3]. Because of the lower density and the relatively greater depth of the tar-belt deposits (Chapter 2), oil viscosity at reservoir conditions is generally lower than that in the Alberta tar sands. Hence, some primary recovery is possible in several areas of the tar-belt deposits, which strictly speaking do not qualify as tar sands. It should be noted that sulfur content of the oil is close to 4% and that the oil contains high amounts of vanadium and nickel, as emphasized in Chapter 9.

The major tar-sand deposits of the United States are located in Utah. They occur in various rocks, ranging from low-porosity and low-permeability consolidated rocks to permeable unconsolidated sands. The range of bitumen properties is also very wide. Four of these deposits, Asphalt Ridge, Hill Creek, P.R. Spring and Sunnyside — in the Uinta Basin — contain over 10 billion bbl of oil containing less than 0.5% sulfur [4].

General review of in-situ recovery techniques

The first factor to be considered prior to developing an in-situ recovery technique for tar sands is the low injectivity of most of these deposits, which often makes it necessary to inject at a pressure above parting (fracturing) pressure. Fracturing would drastically change the fluid-flow characteristics of the reservoir, because a limited fracture system is induced through which fluids could flow, Consequently, sweep efficiency would tend to be very low. Furthermore, fracture orientation depends on relative magnitude of horizontal and vertical stresses in the formation rock which are difficult to control. Controllable communication between wells may be achieved when horizontal fractures are created, whereas in the case of vertical fractures prospects are much less favorable. Vertical fractures are induced at greater depths *, whereas in shallow wells there is great probability of inducing horizontal fractures. In cases where a low sweep efficiency is achieved utilizing only fractures or high-permeability natural streaks, it is essential to assist recovery by some appropriate method [see 5].

The second factor is the high viscosity of bitumen, which is the commonest characteristic of tar sands. The consequence is flow problems in the reservoir and in the wells.

Heating is a very efficient method for decreasing viscosity of heavy oil. As shown in Fig. 8-1, the Athabasca and Cold Lake crudes have a viscosity of only 10—20 cP at 350°F and, therefore, thermal methods appear to be very attractive in the case of tar-sand deposits.

A number of thermal methods have been developed: wellbore heating, electrical heating, hot-water injection, steam stimulation, steam flooding, forward in-situ combustion, forward in-situ combustion combined with water injection, and reverse combustion. Some of these processes, however, are not adequate for tar-sand deposits. For instance, wellbore heating reduces oil viscosity only around a well and is ineffective in reservoirs with low or nonexistent internal-drive energy.

Two main variations of electrical heating between two wells have been proposed for tar sands and oil shales [6]:

* The critical depth beyond which vertical fravtures are obtained can be estimated (see [40]) (editorial comment).

(1) On the imposition of sufficiently high voltage, electrical breakthrough occurs which may create a permeable path by carbonization and electro-volatilization of the hydrocarbon material.

(2) The electrothermic process relies upon an acceptable conductivity of the reservoir to carry electric current between electrodes in two wells. Upon water injection, some convective heating of the formation might be obtained. These methods, if feasible, have to be associated with more conventional techniques.

Cyclic hot-water injection has been occasionally tested in areas of the Orinoco oil belt where the crude is not entirely immobile (Morichal field) [7,8]. Hot water, however, is generally insufficient for mobilizing heavy tars.

On the other hand, steam injection has been widely used, either alone or in association with gas, emulsifiers or solvents, either in a cyclic or a drive process. If formation fracturing is necessary, however, the fluid-flow and heat-transfer characteristics of the process are significantly different from those that prevail in homogeneous formations. Furthermore, if additives are used, the risk of their chemical alteration and deposition [5] should be studied. Cyclic steam stimulation, or "huff and puff" technique, involves steam injection into a well for a limited period of time, followed by a shut-in period (soak time) and by production of the same well until a new injection phase is initiated. Steam drive, or steam flooding, involves a pattern distribution of separate injection and production wells [9]. Some steam projects carried out in tar sands have given encouraging results, even in the very viscous tar-sand deposits of Alberta.

In-situ combustion involves air injection into the reservoir. After ignition, a heat wave propagates within the formation due to the combustion of part of the in-place oil with oxygen of the injected air [10]. If ignition occurs at the injection well, the combustion front propagates to the producing wells in the same direction as the injected air; this process is referred to as forward combustion. The efficiency of the process may be significantly improved when water is injected along with air (fire—water flooding). The injected water vaporizes and transports heat ahead of the combustion front, thus increasing heat efficiency. In addition, a better sweep is obtained ahead of the combustion zone, and the amount of fuel burned decreases [11]. If ignition occurs at the production wells, air and the combustion zone travel in opposite directions; this process is called reverse combustion *. In the case of forward combustion, a cold oil bank occurs ahead of the combustion zone, which generally cannot be displaced without fracturing and eventually preheating the formation; an alternate is the use of forward combustion as a stimulation technique similar to the steam "huff and puff" method. The problem of a cold oil bank does not exist for reverse combustion, but even in that case it may be necessary to fracture the formation to attain an adequate

* Applying combustion in tar sands often implies the basic process to be adapted.

air injectivity. Furthermore, reverse combustion is subject to inherent limitations such as the risk of spontaneous ignition near the injection well. Presently, fire—water flooding seems to be the most promising variation of the in-situ combustion techniques.

Underground nuclear detonation is a combined fracturing and heating method which has been proposed as a means of releasing hydrocarbons from tar sands [6]. The process would include two steps: (1) creation of a rubble-filled, heated cavity by the nuclear device, and (2) subsequent gravity drainage of the heated bitumen to production wells. Conceivably, depletion during phase (2) could be accelerated and improved by injecting either air for supporting a combustion process in the cavity, or steam. This technique, however, is subject to a number of limitations. It can be applied only in deep, thick and continuous deposits with no possible communication with adjacent permeable formations, e.g., for a 10-kiloton device, thickness should be more than 150 ft and depth more than 1000 ft. Another problem concerns the extent of fracturing that would be achieved in plastic and deformable sands. The amount of heat usefully stored in the cavity at 200°F or more, however, might correspond to 50% of the energy released by the detonation [12] and might permit a dramatic improvement of bitumen mobility. *[1] Two nuclear projects have been proposed for Athabasca tar sands: one by Richfield Oil Company in 1959 [6] and a recent one by Phoenix Oil Company [13]. The proposed devices in both cases are in the 10-kiloton range. The problems associated with the development of such nuclear devices are likely to hamper on-site experimentation of this technique. *[2]

An alternate solution for decreasing bitumen viscosity is to blend it with a suitable solvent [14]. Viscosities of mixtures of Athabasca bitumen with two different solvents at various temperatures are shown in Figs. 8-2 and 8-3. The viscosity decreases rapidly with an increase in solvent concentration; however, large percentages of expensive solvents are needed to lower the viscosity to a value of 100 cP, which would be reached by heating the original bitumen to approximately 250°F (Fig. 8-1). Furthermore, viscous fingering and gravity overlay would impair the efficiency of bitumen displacement. For these reasons, solvents have been proposed either in conjunction with steam [15], or to create high-permeability paths in a formation [14]. Solvents are also being used for diluting heavy oils not pumpable at prevailing temperatures [7], or for formation clean up around the wellbore [14].

Another method of lowering oil viscosity is to emulsify the in-place oil,

---

*[1] This conclusion is doubtful (editorial comment).

*[2] The editors would like to point out that recent studies indicated that the heat released by nuclear devices of permissible yields is uneconomic compared to heat released by combustion of fossil fuels.

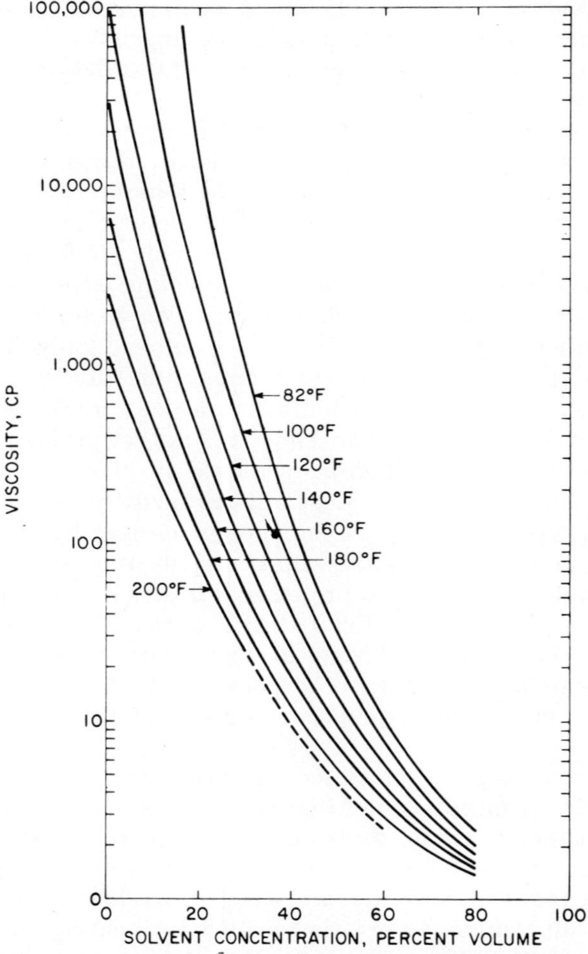

Fig. 8-2. Viscosities of mixtures of Athabasca bitumen with naphtha (specific gravity = 0.717, viscosity = 0.4482 cP). Viscosity versus solvent concentration curves. (After Farouq Ali and Abad [15].)

provided that the aqueous phase is the external phase. The emulsion viscosity, therefore, would be close to the viscosity of the aqueous phase. In order to obtain an acceptable sweep efficiency, however, it has been recommended to use this process in conjunction with steam [12].

Techniques based on alteration of bitumen by chemicals or bacteria have also been proposed for the recovery of bitumens from tar sands.

In-situ hydrogenation has been suggested to produce more fluid oil. It has also been claimed that bacteria could be capable of breaking down the

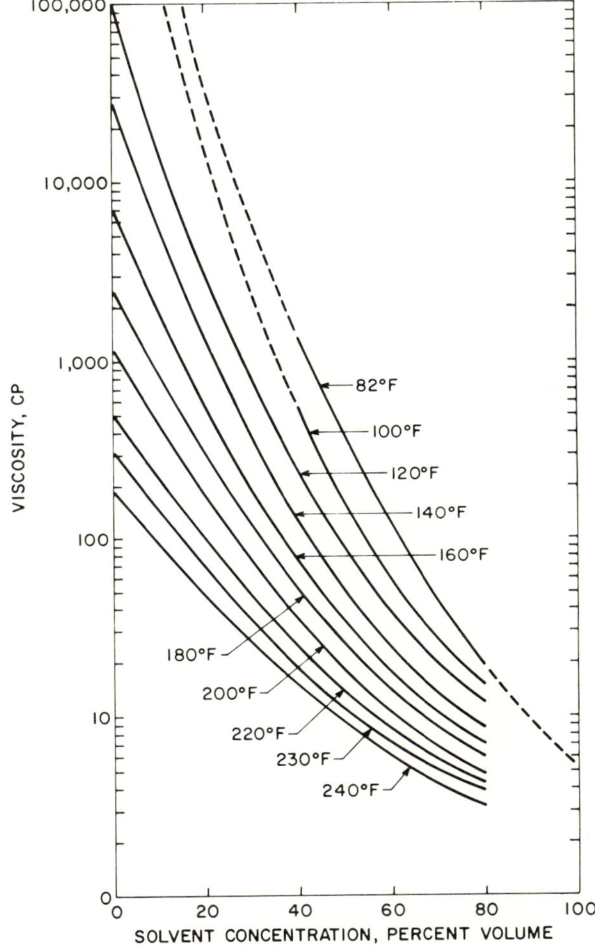

Fig. 8-3. Viscosities of mixtures of Athabasca bitumen with G.C.O.S. synthetic crude (specific gravity = 0.829, viscosity = 4.56 cP). Viscosity versus solvent concentration curves. (After Farouq Ali and Abad [15].)

hydrocarbon molecules; however, no adequate enzyme has been found at this stage for very heavy oils [6]. If such approaches were viable for upgrading in-place oil, however, they would imply the solution to one of the main problems faced by the in-situ processes, i.e., the distribution of the introduced agents throughout the formation.

It may be emphasized that a method combining mining, drilling from underground sites, and subsurface steam injection has been used successfully in Yarega (U.S.S.R.) since 1972. The normal viscosity of Yarega oil is 15,000—20,000 cP [16].

Cyclic steam injection

Cyclic steam stimulation is generally considered as not suitable for reservoirs with little or no primary energy [9]. In view of the lack of communication between wells in tar-sand formations, however, this technique may be selected, at least until a permeable path between wells would permit a steam-drive process. Maximum benefit of the supplemental energy resulting from pressure buildup near the well during injection can be reached by bringing the well to production as soon as possible, i.e., after a minimum soak period [9].

A number of "huff and puff" projects are currently on stream or in project in the Cold Lake area of the Alberta oil sands (Fig. 8-4).

Imperial Oil has been one of the most active companies in the Cold Lake area [2,17]. After an exploration program, carried out in the 1960's, it was decided to concentrate the work on the Clearwater Formation at a depth of about 1500 ft. The oil gravity here is 10—12° API, with a viscosity of 100,000 cP at the reservoir temperature of 55° F (Fig. 8-1). The average thickness is 140 ft, porosity is 37%, oil saturation is 55%, absolute permeability may be as high as 3000 mD, and reservoir pressure is 440 psi [2,18].

Three preliminary pilots (Ethel pilots) were operated from 1964 to 1970, in order to determine the injectivity of the sand unit and to evaluate performance of both steam stimulation and steam flooding. In 1972, the May pilot project was constructed with 23 new wells using 5-acre spacing (Fig. 8-5). Two previously operated wells were also reactivated, whereas two other Ethel wells were recompleted for steam tests in an overlying sand unit [17].

The original program indicated that essentially no steam would enter the formation at less than fracturing pressure. It was found that pressures of 1500—1600 psi were required for injecting at the envisioned rates of 1000 bbl/day. The experimental work involved the evaluation of additives to augment the benefits of steaming. The best results were obtained when a small volume of natural gas was introduced into the reservoir either before or after the steam. Additional work needs to be done to optimize this technique. Well-completion technique was evaluated by extensive work and various approaches were attempted. The currently favored technique uses casing of sufficient strength, fully cemented in place. Experience showed absence of sand problems and, consequently, gravel packs have been eliminated.

It was determined that preferred northeast—southwest directional permeability was present in the area. Eventually, with continuing cyclic steam injection, most wells communicated with one another in the "on-trend" direction. Furthermore, it was found that about half of the wells in the May pilot had a water zone in good vertical communication with the oil zone and, as expected, steam moved preferentially into this water. During the early cycles, these wells produced mainly water but, as the cyclic process con-

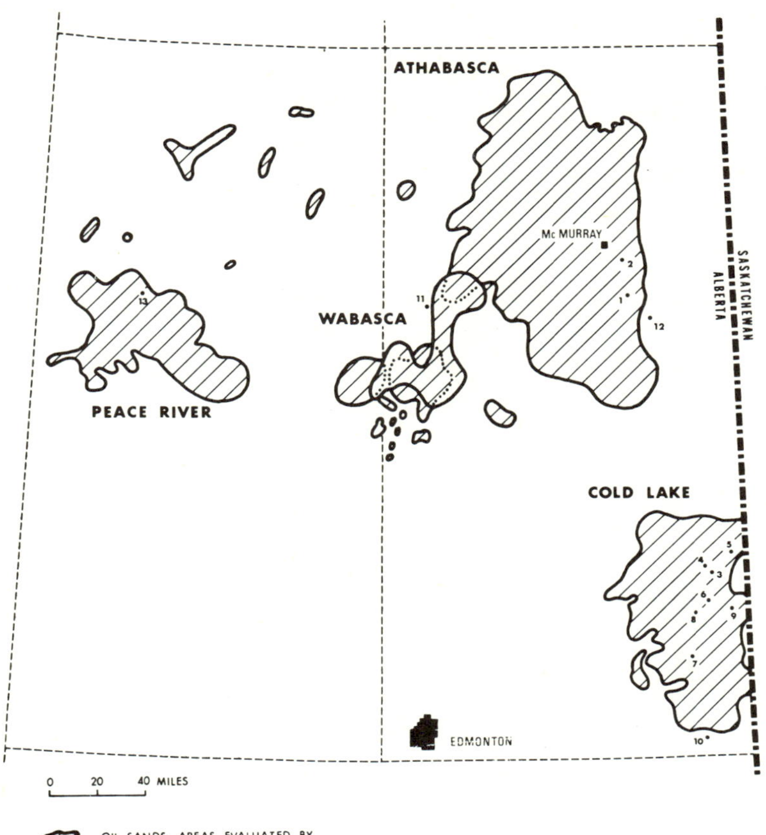

OIL SANDS AREAS EVALUATED BY ALBERTA CONSERVATION BOARD

Fig. 8-4. Recent in-situ projects in Alberta oil sands (Sept., 1976): *1* = Amoco combustion test — Athabasca; *2* = Texaco operations — Athabasca; *3* = Imperial Ethel-May pilots — Cold Lake; *4* = Imperial Leming pilot — Cold Lake; *5* = Norcen steam project — Cold Lake; *6* = Union Texas steam project — Cold Lake; *7* = Murphy 7-spot steam pilot — Cold Lake; *8* = Weco steam test — Cold Lake; *9* = Chevron steam test — Cold Lake (terminated); *10* = Murphy steam test — Cold Lake area (abandoned); and *11* = Gulf steam project — Wabasca. Planned in-situ projects in Alberta oil sands: *12* = Numac exploratory program on fracturing — Athabasca; *13* = Shell steam project — Peace River; and *14* = BP combined steam and combustion project — Cold Lake.

tinued, performance improved, more oil being heated directly or indirectly by the injected steam. The May pilot has been producing oil up to 1500 bbl/day.

It is felt that cyclic stimulation is only a prelude to flooding between wells. An approach will have to be found to achieve efficient oil displacement and reasonable recovery in spite of the directional-permeability trend

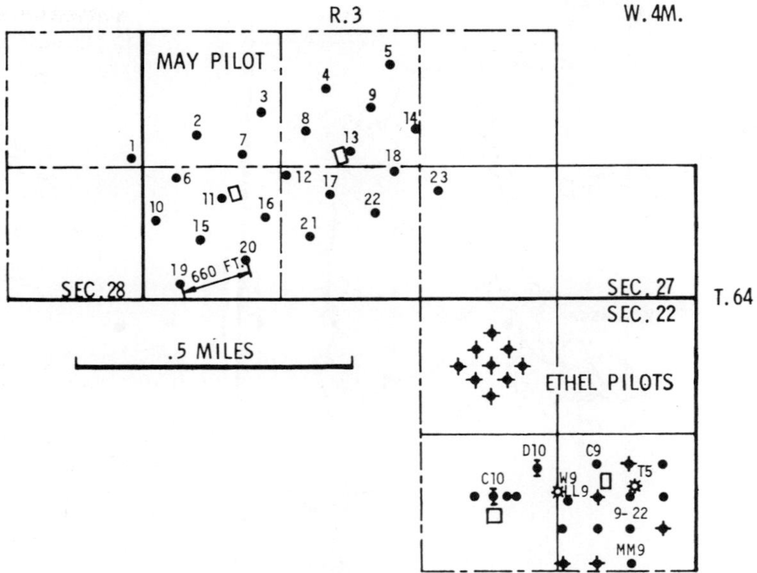

Fig. 8-5. Well configuration of Ethel and May pilots. (After Mungen and Nicholls [17].)

of the reservoir and of the presence of bottom water. The detrimental effect of bottomhole water is shown in Fig. 8-6.

A new pilot, the 56-well Leming pilot, was constructed by Imperial Oil on the same lease during 1974 and early 1975 (Fig. 8-7) in an area of Cold Lake, where the Clearwater Formation comprises about 160 ft of water-free

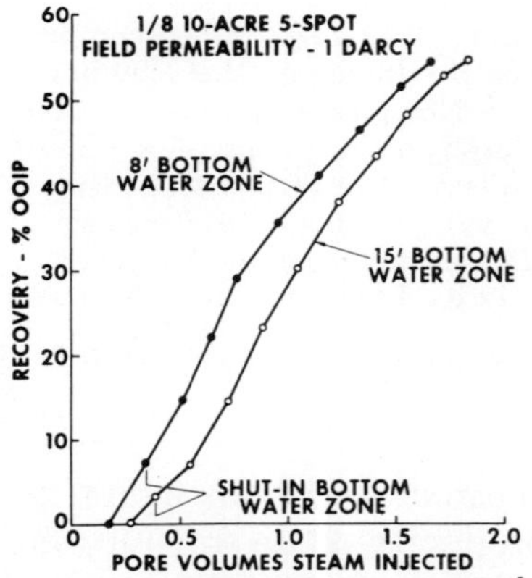

Fig. 8-6. Effect of water-zone thickness on steam-drive recovery. Laboratory tests simulating the conditions in a 1/8 of 10-acre, 5-spot pattern with a formation permeability of 1 D. Filled, black circles = 8-ft thick simulated bottom-water zone; open circles = 15-ft thick simulated bottom-water zone. (After Pursley [18].)

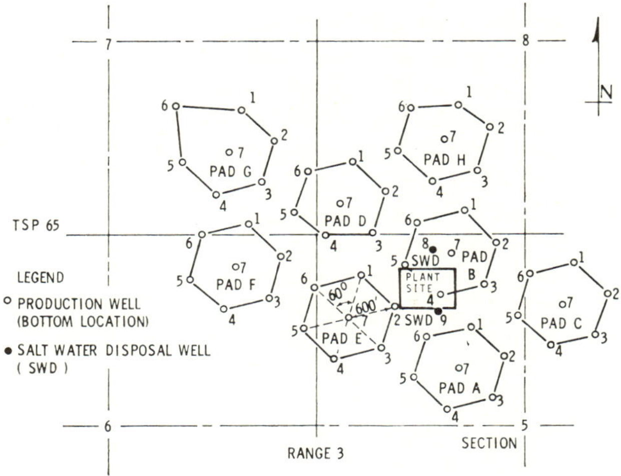

Fig. 8-7. Well configuration of Leming pilot. (After Mungen and Nicholls [17].)

oil sand [17,19]. The operating plans are to continue optimizing steam injection and production processes. One significant difference with the May pilot involves the change of spacing to a 21.5-acre, 7-spot pattern or 7.15-acre spacing with wells 600 ft apart (Fig. 8-7). The seven wells of a pattern have been drilled from the same pad, using the cluster-drilling concept. This procedure has a favorable impact on environment and on surface layout. Imperial Oil Company started steaming its Leming pilot in 1975 and added a cluster pad of 14 wells drilled directionally on 2.8-acre spacing in 1976.

Imperial Oil Company management expected to produce 5000 bbl of oil per day in 1976 from its May and Leming operations in Cold Lake [19].

In the same formation, Great Plains Company (presently Norcen Energy Resources) carried out an early single-well stimulation test in 1965, at a short distance from the Ethel pilot operations [20]. A sand thickness of 100 ft was present in the project area. An amount of 8400 bbl of water as 40% quality steam was injected down a gravel-packed well in 6 days, at a pressure of about 900 psi, with no apparent fracturing of the formation. An initial fluid-production rate of 180 bbl/day was achieved at a wellhead temperature of 180°F. A total of 578 bbl of oil and 2087 bbl of water were produced with no sand during the 47-day production period. Norcen, associated with a Japanese consortium, is currently starting a new steam project northeast of Cold Lake (Fig. 8-4). A total of 12 wells have been drilled and startup of steam-stimulated production of oil was targeted for mid-1976 at about 500 bbl/day [21].

Other steam-stimulation projects are being developed in the Cold Lake area [21,22] (Fig. 8-4). Murphy Company started a 7-spot pattern project in 1974 and Union Texas Company started a 15-well steam-injection project that went on production in September, 1975. Another project, activated by Weco Development Company in 1975 on a 1-well basis, is being expanded to a 5-spot pattern. So far, the results of the projects have been encouraging: for instance the first well of the Weco test produced at a relatively stable rate of 70 bbl/day of 12.5° API-gravity oil during the first months of 1976, following injection of 20,000 bbl of steam in November of 1975 [21].

A 2-well steam-stimulation test was carried out in a tar-sand deposit in California [23]. A thick reservoir (300—400 ft) occurs at a depth of 1870 ft, with formation porosity of 34% and average absolute permeability of the order of 5 darcies. The 5° API tar shows a rapid viscosity reduction at elevated temperatures. The results of four successive cycles were largely unsuccessful in the first well which received 2000—5000 bbl of water as steam per cycle, at a rate in the order of 300 bbl/day. Five cycles were conducted in the second well, which received 7500—23,000 bbl of water as steam per cycle at a rate of about 1300 bbl/day, and a maximum wellhead pressure of 1750 psi was achieved during Cycle 2. The response of this well was good: the oil/steam ratio was 0.23, 0.89, 1.53 and 1.23 bbl/bbl for the four complete cycles at the time of the publication [23].

Cyclic steam stimulation is being applied in some fields north of the Orinoco tar belt (Miga, Jobo, Melones, Morichal fields) [7,24,28], with encouraging results.

For instance, steam stimulation is being successfully applied using eleven wells of the Jobo field, in two reservoirs consisting of several unconsolidated sandstone lenses at a depth of 3600—4000 ft [28]. One of the formations has a net pay thickness of 60 ft and contains 13.5° API-gravity oil; the primary recovery factor has been estimated at 7.7%. The other reservoir, with a net pay thickness of 210 ft, contains 9° API-gravity oil; the estimated primary recovery factor is 2.1%.

The amount of steam injected per cycle corresponds to between 150 and 200 MMBtu/ft of pay. The soak period varies from one week to one month and production increases range from two- to fivefold.

Prestressed casings are used in the case of steam injection. During injection, each well is equipped with a bottomhole thermal packer, an expansion joint, and a gas lift mandrel for circulating into the tubing—casing annulus an insulating fluid having a commercial fatty acid compound as a base. This results in very low heat loss in the wells.

A special production equipment is used for the reservoir containing the 9° API-gravity oil. In this case fluids are lifted through the tubing—casing annulus with the 3.5-inch tubing kept full of water.

Future plans for the Jobo field are to stimulate all existing producers while carrying out an infill-drilling program. Furthermore, a steam-drive project has been proposed.

Combination of heating by steam and emulsification (Shell process)

In an early field work carried out by Shell Canada between 1956 and 1960, it was shown that combining caustic emulsification and steam injection was a viable approach for the in-situ recovery of Athabasca tar sands.

The Shell process, tested in 1960—1962, involves the following steps [12]:

(1) Achieving initial communication between wells, drilled in some suitable pattern, by horizontal hydraulic fracturing using dilute aqueous-alkaline solutions. The wells are completed in such a way that communication takes place only at or near the bottom of the tar-sand interval.

(2) Depleting sands adjacent to initial fracture by hot caustic emulsification and displacement of the emulsified tar. Hence, an horizontal flow path capable of conducting large volumes of fluid is developed.

(3) Injecting steam into a well and producing oil in aqueous emulsion from production wells. Injection is controlled to achieve optimum capture of heat within the oil-saturated zones and displacement of the heated crude.

Once it was proved that horizontal fractures could be induced in the Athabasca McMurray Formation, the Shell process was tested on a 3.6-acre, 5-spot pattern [25]. The 150—219-ft deep formation had a porosity of 38—42% and an original oil saturation averaging 55%. Step (3) of the process involved steam and aqueous-alkaline fluid injection. For approximately nine months, a total amount of 26,996 bbl of steam (in terms of condensate) and of 111,131 bbl of aqueous-alkaline fluids were injected into the center well; 52,604 bbl of fluids, including 2457 bbl of oil, were produced by the corner wells. An estimated 62% of the injected fluids, however, had been displaced beyond the 5-spot area. It was estimated that about 7600 bbl of oil had leaked away from the pattern. Finally, 22.9% of oil originally in place had been displaced or produced during the period. The final stage of the project involved the injection of 27,370 bbl of steam into the corner wells and production of 13,684 bbl, including 3795 bbl of oil from the central well, for a four-month period. Slugs of alkaline solutions were injected into the producing well, when necessary, to maintain or restore productivity. A total of 31.5% of the oil in place had been displaced beyond the pattern or recovered during the test. Furthermore, production was continuing at a steady water/oil ratio when the test was discontinued.

It was anticipated that a recovery efficiency above 50% was possible with the process. An expansion of the project was planned but abandoned for economic reasons.

Steam injection combined with various additives

Variations of the above-described process involve combined injection of steam and of an alkaline solution containing an interfacial tension reducer,

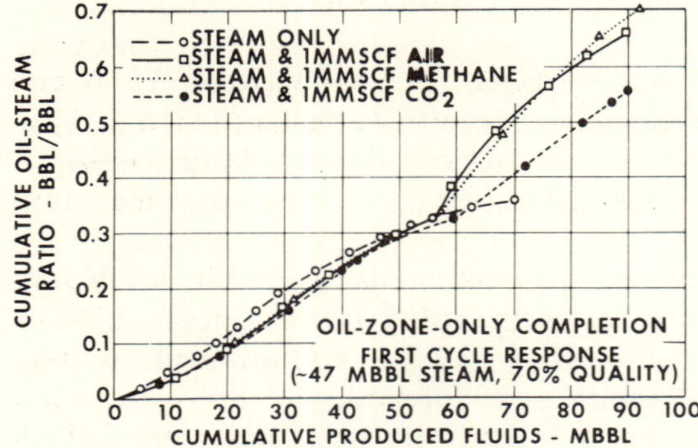

Fig. 8-8. Effect of gas injection on steam-stimulation performance. Laboratory tests simulating injection of 47,000 bbl of 70% quality steam plus 1 MMscf gas. First cycle response; gas injection during production phase of the cycle. (After Pursley [18].)

such as quinoline or pyridine [26,27]. Formation of an emulsion is claimed to improve the vertical conformance of a displacement by steam.

Injection of a solvent in conjunction with steam has also been proposed. Suitable solvents must be miscible with bitumen and must not cause asphaltene precipitation; these conditions are fulfilled for instance by aromatic solvents [14]. Laboratory tests carried out on samples from Athabasca tar sands have shown that the synthetic crude produced by Great Canadian Oil Sands was a suitable solvent to inject in conjunction with steam [15].

Several additives to steam were investigated in laboratory models under conditions existing in the Cold Lake area [18]. These included, beside solvents, water thickeners and various gases (air, methane and carbon dioxide). Water thickeners were found to increase oil recovery in steam drive, whereas addition of gas (particularly of methane or air) increased the oil/steam ratio achieved in steam stimulation significantly (Fig. 8-8). Some of these additives were tested in the Cold Lake area [17].

Field application of steam injection with introduction of caustic and condensate is planned in Wabasca by Gulf Oil [21]. Furthermore, combination of steam and additives may have been tested by Texaco in Athabasca (Fig. 8-4); their project, operated at a depth of 250 ft, has provided encouraging results and is currently being expanded [22].

Steam injection in the bottom-water zone

In some areas, a zone having a high water saturation underlies the formation. Shell had proposed to take advantage of this characteristic for heating the oil zone by conduction from the steam-flooded water zone. This approach implies, however, that a thin water zone underlies a relatively thick oil zone and that a procedure is found for achieving both conductive and convective heating of the oil zone. The process would involve the following steps [29]:

(1) Injection of high-quality steam at a high rate (1000—2000 bbl/day) until steam breakthrough. Back pressure at the producers is maintained at low values during this period to promote rapid expansion of the steam zone and enhance oil production. Only limited amounts of oil, however, are produced during this steam drive through the bottom-water zone where the oil content is low. Wells are stimulated by steam if necessary.

(2) Formation pressurization by increasing the back pressure at the producers, while maintaining steam injection at highest rates without exceeding fracturing pressure.

(3) Maintaining the pressure level in the steam zone for a period of time, by controlling steam-injection rate as well as production rate.

(4) Reservoir depressurization by increasing production rates, while continuing steam injection at low rates to provide additional pressure drive.

(5) Steps (2), (3), and (4) may be repeated to produce the remaining tar.

Shell had selected a site in Peace River for testing the process using seven 7-acre, 7-spot patterns at a depth of about 1800 ft. A 9-ft thick water zone with 54% oil saturation underlies the 81-ft thick oil zone with 79% oil saturation. Permeabilities to brine average 220 mD in the upper 74 ft of sand and increase to 1650 mD in the basal water zone. The 9° API-gravity tar has a viscosity of 200,000 cP at reservoir conditions. The Alberta Oil Sands Technology and Research Authority is expected to contribute in financing the project.

Stimulation by combustion

Stimulation by combustion consists in a limited combustion around an injection well followed by production from the same well. This technique may be attractive in deep reservoirs: (1) temperature is high enough for spontaneous ignition to occur in a relatively short time [30], and (2) contrary to steam injection, no heat loss occurs in the well. The duration of the combustion phase may range from less than one month, if the technique is used for increasing well productivity, to about three months if some depletion around wellbore is desired. Coking of oil during backflow through the hot zone must be limited by using appropriate methods.

A small number of wells in the Orinoco oil belt have been stimulated by combustion [7].

Forward combustion

Conventional dry forward combustion has been tested in reservoirs of the Orinoco tar belt, where because of burial depth, oil has some mobility at reservoir conditions. For instance, two forward-combustion pilot tests were carried out in the Morichal field (3000—4000 ft deep) [8]. The combustion process itself was satisfactory in both cases. But the first test, in a 10-ft sand containing 8° API oil, resulted in low oil production due to liner plugging by formation fines. The second test, in a sand containing 12° API oil with 100-cP viscosity at reservoir conditions, exhibited an excellent performance with the exception that it was impossible to pump some wells at satisfactory rates due to poor gas separation downhole.

Favorable performance was achieved in another fire flood operated in a 4000-ft deep reservoir of the Miga field, at the northern fringe of the Orinoco belt [31]. The 13° API-gravity crude had a 300—400-cP viscosity at reservoir temperature.

In the more general case when tar has no mobility, it is necessary to utilize fracturing techniques prior to initiating forward combustion. A pilot test carried out in 1959—1960 in a shallow tar sand of Kentucky has been recently described [32]. The process is schematically shown in Fig. 8-9. Mechanically propped horizontal fractures are created at the production wells and a pneumatic fracture is maintained at the injection well by the air injection. Combustion is initiated at the injection well and expands vertically from the fracture wherever oxygen contacts tar; flow capacity of the burned-out zone increases. The products of combustion and the displaced fluids move through the fracture system to the producing wells. The test has been carried out in a 100-ft deep and 20-ft thick sandstone of moderate consolidation; formation porosity was 22%, oil saturation, 64% and specific permeability, 2000 mD. The 10.6° API-gravity tar had a viscosity of 150,000 cP at a reservoir temperature of 56° F. The test pattern was an inverted 5-spot with a distance of 107 ft on a side. Producing wells were fractured near

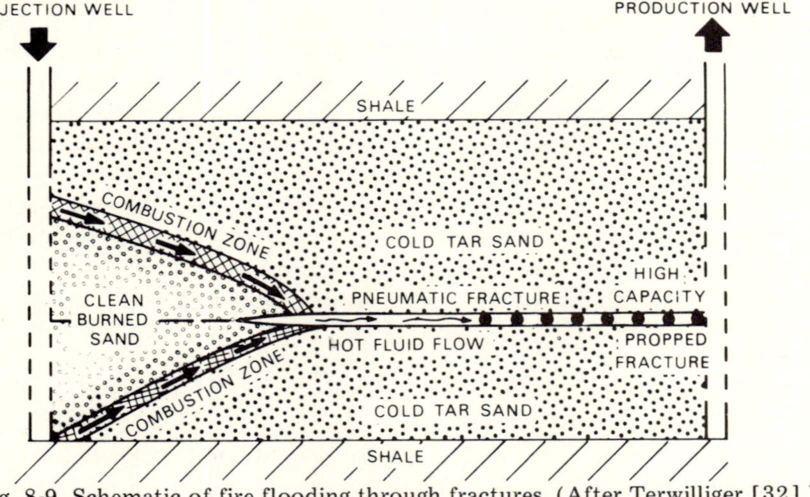

Fig. 8-9. Schematic of fire flooding through fractures. (After Terwilliger [32].)

the bottom of the sand from fracture wells drilled at a distance of 5 ft from producers. Horizontal fractures were initiated by air injection. They were extended and propped to a calculated radius of 25 ft by a water-base gel carrying 4—8-mesh sand. This procedure would not have been suitable in unconsolidated formations.

A downhole burner was used to initiate the combustion process. Production response was achieved ten days after the test started and the average oil production was 24 bbl/day for the four months required to complete the test. The total oil production constituted 54% of the oil originally in place in the pattern area; the injected-air/produced-oil ratio was 42,000 cu ft/bbl. The produced oil was upgraded to 14.5° API. The advance of the combustion zone occurred preferentially in the northeast—southwest direction and reverse combustion was used to correct the asymmetrical advance of the burning front. Oxygen utilization was not complete and loss of fluids to the outcrop was experienced.

Heating by dry combustion followed by fire—water flooding (Amoco process)

An original approach has been developed by Amoco for applying fire flooding to Athabasca tar sands [17] (Fig. 8-4). The process involves three steps:

(1) Heat-up phase: the formation is ignited and air is injected at, or slightly above, fracturing pressure to support in-situ combustion which produces heat to reduce the viscosity of bitumen. Production is controlled to maximize the volumetric heat-up of the formation.

(2) Blow-down phase: air injection is suspended or reduced and production at producing wells lowers the formation pressure.

(3) Displacement phase: fire—water flooding, with an air/water ratio in the injected fluids ranging from 500 to 3000 scf/bbl (COFCAW process) [33], supplies heat and driving fluids for the production of oil.

The first multiwell test, using a 0.5-acre, 5-spot pattern, was initiated in 1966 in a 1000-ft deep, 120-ft thick tar-sand formation [17]. The center well was fractured with water and coarse sand to permit starting the combustion front in a thin high-permeability section radiating 25—50 ft from the well. The well was ignited and forward combustion initiated; small amounts of water were injected to control the temperature in the injection well. Production-well response indicated a northeast—southwest permeability trend. The heating period was continued for about eight months; the blow-down period was four months, after which the fire—water-flooding operation was continued for about six months.

Of the 90,000 bbl of tar originally present in the tested portion of the formation, 29,000 bbl of upgraded oil were produced, i.e., 32% of the oil in place. This production was essentially achieved with two wells, owing to the permeability trend. The overall air/oil ratio (including the air for the heat-up phase) was 14,000 scf/bbl. A little over 7% of the oil in place was burned during the test, whereas about 60% was heated to 150° F or higher. It is estimated that a recovery of 50% is feasible with this process. Additional work has been carried out on a 10-acre, 9-spot pattern in order to study the influence of pattern size and to develop adequate completion techniques, which present a major problem in the Athabasca tar-sand area. Amoco is presently considering developing an even larger pilot of approximately 35 acres in 12 patterns of 2.5 and 4.3 acres. The orientation of the wells is expected to be guided by the directional permeabilities found in earlier tests. This project has been selected for funding by Alberta Oil Sands Technology and Research Authority.

Combination of steam and combustion

British Petroleum Company of Canada is planning to carry out a field research project in the Cold Lake deposit, to test a combination of steam injection and combustion as a recovery mechanism. This project is to receive financing from Alberta Oil Sands Technology and Research Authority.

Reverse combustion

In an early stage of its in-situ combustion work in Athabasca, Amoco performed two reverse-combustion 2-well tests, with spacings of 100 and 250 ft. The process was operative in the first test, but it was not possible to achieve communication between wells in the second case and the test was abandoned [17]. Later evaluation showed that formation fracturing had inadvertently occurred during the first test, which explained the satisfactory air-injection levels.

Oil production by reverse combustion was tried by Phillips Petroleum Co. in a deep reservoir of the Morichal field (Orinoco tar belt) containing 8° API oil, but the test failed due to spontaneous ignition at the injection well [8]. As a matter of fact, the oxidation rate at reservoir conditions is likely to be high in such a deep (~4000 ft) and hot formation [30,34].

These two unsuccessful attempts emphasize the problems related to applying reverse combustion in tar-sand deposits. The characteristics of a reverse-combustion front strongly depend on operating conditions, in contrast to forward combustion. Owing to the influence of air flux [10,35], the process is difficult to control in the field, especially when formation fracturing is likely to occur. Furthermore, the process is not adapted for deep reservoirs, due to the decrease in oil recovery when pressure increases [10] and to the risk of spontaneous ignition [30,34].

So far, only one reverse-combustion pilot test exhibited a certain degree of success [36]. It was carried out during the period 1955—1958 in Missouri tar sands, near Bellamy. The test zone was 12 ft thick and 50—60 ft deep. Porosity was 25%, tar saturation was 40—50%, and the effective air permeability of the saturated sand was 100—250 mD. The viscosity of the 10° API tar was 500,000 cP at reservoir temperature. Seven different small-scale patterns were used, including conventional 5-spot, 7-spot, 10-well radial pattern, and a 15-well line drive.

Several ignition techniques were tested. The best method consisted of packing the pay interval of the well with charcoal briquettes saturated with diesel oil, igniting the charge, and burning it by injecting air premixed with 1% propane into the injection well. The production performance of the line-drive test was satisfactory. A recovery factor of 67% of the tar originally in place was achieved, with a vertical-sweep efficiency of essentially 100%; the air/oil ratio was about 40,000 scf/bbl. The gravity and viscosity of the oil produced in the tests were 26° API and 10 cP, respectively. The oil contained a reduced concentration of nickel, vanadium, sulfur and nitrogen, a high percentage of olefins (21% by weight), and more oxygen than the original tar, due to the presence of partial-oxidation products such as carboxylic acids. The laboratory evaluation of the tests showed that the propagation velocity fell sharply towards zero as the air flux approached 19 scf/sq ft/hr; as a consequence, the air/oil ratio should approach infinity when air flux tends to reach this limit. A minimum air/oil ratio in the vicinity of 40,000 scf/bbl was attained for air flux between 30 and 50 scf/sq ft/hr. Reversal of the direction of propagation occurred when the value of air flux dropped below the critical.

Recently, the U.S. Energy Research and Development Administration conducted a reverse-combustion pilot test in the Asphalt Ridge tar-sand deposit near Vernal, Utah [4,37]. Two rows of three injection wells 120 ft apart delimited a rectangular pattern containing 2200 bbl of 10° API-gravity tar in a 10-ft thick interval at a depth of 295 ft; a row of three production wells was in the center of the rectangle. Injectivity problems were experienced and injection was made possible only above fracturing pressure in order to maintain the desired air flux. Only the center well was ignited, whereas air was injected into the two lateral wells of the center row and in four external wells of the pattern. The project was discontinued after 23 days of operation; total oil production was only 65 bbl of oil.

Solvent or diluent injection

Several processes involving formation fracturing followed by solvent injection have been developed. For instance, the U.S. Bureau of Mines proposed a technique involving fracturing by explosives [38].

A method patented by Texaco involves the following steps [39]: (1) creation of a permeable path by hydraulic fracturing or utilizing high-permeability streaks, and (2) injection at a pressure up to the overburden pressure of a suitable solvent saturated with a gas or containing appreciable quantities of dissolved gas.

The use of diluent is the principal production technique of the Orinoco tar belt [7]. In the major reservoir of Morichal field, 39° API gas oil is pumped to the bottom of the tubing to mix with the 8° API oil, as shown in Fig. 8-10. The 12° API blend can be desalted, dehydrated, and transported by pipeline. Oil from the Jobo field, in the same area, is treated similarly

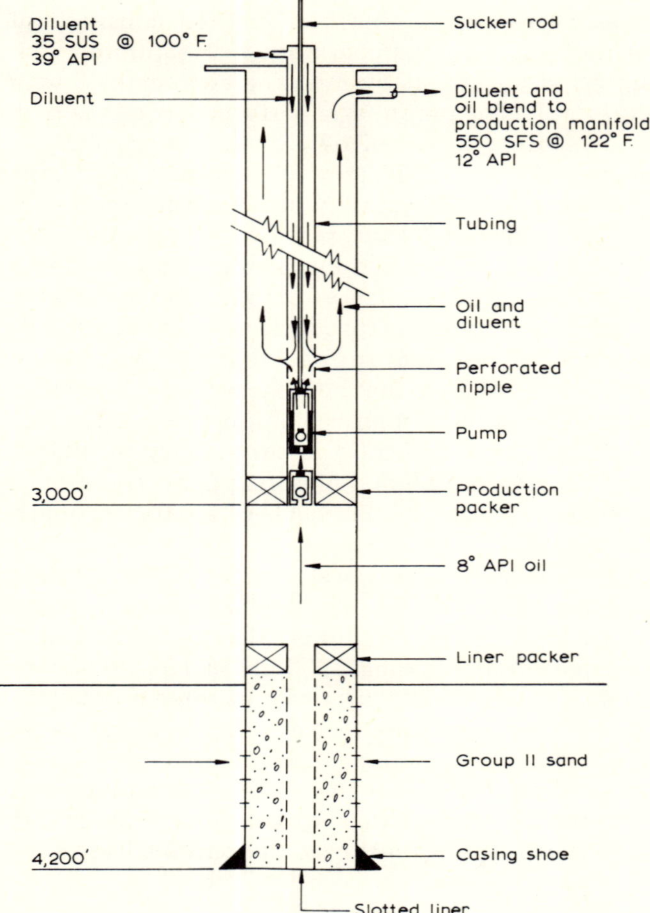

Fig. 8-10. Crude blending technique in Morichal field. (After Finken and Meldau [7].)

to that at Morichal. The total capacity of the treating and transport system is above 80,000 bbl/day.

Conclusions

A number of techniques have been proposed for in-situ recovery of oil from tar sands. It seems that the most promising techniques involve the application of heat, using some refined procedure compared with the basic processes. To date, several methods have been tested on pilot field operations, which unfortunately cannot be extrapolated to a commercial-scale operation. Larger field tests, therefore, are presently required to evaluate commercial feasibility of the various processes.

Technological problems are still to be solved and economic uncertainties have to be overcome for large-scale in-situ operations to be developed in tar sands. It is generally agreed that commercial in-situ projects are unlikely to be on stream prior to 1985. The development of field tests in tar sands of Canada and the United States, however, will benefit the recent decision of government agencies to contribute in the expenditures involved in these types of experiments. Furthermore, the Venezuelan government expects to develop large pilots in the Orinoco area in the near future.

References

1 G.W. Govier, "Alberta's Oil Sands in Energy Supply Picture", *AIME 104th Annu. Meet.*, New York, N.Y. (Feb. 18, 1975).

2 A.G. Winestock, "Developing a Steam Recovery Technology", in: *Oil Sands, Fuel of The Future, Can. Soc. Pet. Geol. Mem.*, *3*, 190—198 (1974).

3 J.A. Galavis and H.M. Velarde, "Geological Study and Preliminary Evaluation of Potential Reserves of Heavy Oil of the Orinoco Tar Belt, Eastern Venezuelan Basin", *Proc. 7th World Pet. Congr.*, *3*, 229—234 (1967).

4 L.C. Marchant, C.S. Land and C.Q. Cupps, "Experimental Approach to In-Situ Recovery from Tar Sands, *Proc. 15th Annu. ASME Resource Recovery Symp.*, Albuquerque, Ala. pp. 147—158 (March 6—7, 1975).

5 D.A. Redford and P.T. Cotsworth, "In-situ Production Problems", in: *The Athabasca Oil Sands, Proc. First Regional Conference, Western Region, Eng. Inst. Can.*, Edmonton, Alta., pp. 359—369 (April 17—19, 1974).

6 D.L. Flock and J. Tharin, "Unconventional Methods of Recovery of Bitumen and Related Research Areas Particular to the Oil Sands of Alberta", *J. Can. Pet. Technol.*, *14(3)*, 17—27 (July—Sept., 1975).

7 R.E. Finken and R.F. Meldau, "Phillips Solves Venezuelan Tar-Belt Producing Problems", *Oil Gas J.*, *70(29)*, 108—114 (July 17, 1972).

8 R.F. Meldau and W.B. Lumpkin, "Phillips Tests Methods to Improve Drawdown and Producing Rates in Venezuela Fire Flood", *Oil Gas J.*, *72(32)*, 127—134 (Aug. 12th, 1974). [Presented at *Symposium on Heavy Oil Recovery*, Maracaibo (July 1—3, 1974).]

9 G.J. Harmsen, "Oil Recovery by Hot-Water and Steam Injection", *Proc. 8th World Pet. Congr.*, *3*, 243—251 (1971).

10 L.A. Wilson, R.L. Reed, D.W. Reed, R.R. Clay and N.H. Harrison, "Some Effects of Pressure on Forward and Reverse Combustion", *Soc. Pet. Eng. J.*, *3*, 127—137 (1963).

11 J.G. Burger and B.C. Sahuquet, "Laboratory Research on Wet Combustion", *J. Pet. Technol.*, *25*, 1137—1146 (1973).

12 T.M. Doscher, "Technical Problems in In-Situ Methods for Recovery of Bitumen from Tar Sands", *Proc. 7th World Pet. Congr.*, *3*, 625—632 (1967).

13 S.D. Moore, "Nuclear Energy as a Subsurface Heavy Oil Recovery Technique", in: *The Athabasca Oil Sands, Proc. First Regional Conference, Western Region, Eng. Inst. Can.*, Edmonton, Alta., pp. 411—429 (April 17—19, 1974).

14 S.M. Farouq Ali, "Application of In-Situ Methods of Oil Recovery to Tar Sands", in: *Oil Sands, Fuel of the Future, Can. Soc. Pet. Geol. Mem.*, *3*, 199—211 (1974).

15 S.M. Farouq Ali and B. Abad, "Bitumen Recovery from Tar Sands, Using GCOS Synthetic Crude as Solvent in Conjunction with Steam", *26th Annu. Meet. Pet. Soc. C.I.M.*, Banff (June, 1975).

16 "Mining of Viscous Crude Claimed Viable", *Oil Gas J.*, *74(1)*, 46—47 (Jan. 5, 1976).

17 R. Mungen and J.H. Nicholls, "Recovery of Oil from Athabasca Oil Sands and from

Heavy Oil Deposits of Northern Alberta by In-Situ Methods", *Proc. 9th World Pet. Congr.*, *5*, 29—39 (1975).
18 S.A. Pursley, "Experimental Study of Thermal Recovery Processes", in: *Symposium on Heavy Oil Recovery*, Maracaibo (July 1—3, 1974).
19 "Recent Imperial Oil Review Highlights Operations at Cold Lake Related to Heavy Oil Development", *Daily Oil Bull.*, Calgary, Alta. (Aug. 14, 1975).
20 L.L. Samoil, "A Field Experiment in Recovery of Heavy Oil, Cold Lake, Alberta", *17th Annu. Meet. Pet. Soc. C.I.M.*, Edmonton, Alta. (May, 1966).
21 "Oil Sands and Mining Report. Economics Key to Development", *Oilweek, 27(5)* (March 15, 1976).
22 "Canadian Enhanced-Recovery Activity is Moderate, Centers on Thermal Projects", *Oil Gas J.*, *74(14)*, 128—129 (April 5, 1976).
23 R.C. Bott, "Cyclic Steam Project in a Virgin Tar Reservoir", *J. Pet. Technol.*, *19*, 585—591 (1967).
24 A. Franco, "Thermal Work Humming in Venezuela", *Oil Gas J.*, *74(14)*, 132—138 (April 5, 1976).
25 T.M. Doscher, R.W. Labelle, L.H. Sawatsky and R.W. Zwicky, "Steam Drive Successful in Canada's Oil Sands", *Pet. Eng.*, *36(1)*, 71—78 (Jan., 1964).
26 A. Brown, J.T. Carlin, M.F. Fontaine and S. Haynes Jr. (Texaco), "Method for Recovery of Hydrocarbons Utilizing Steam Injection", *U.S. Patent 3,732,926* (May 15, 1973).
27 A. Brown, M. Chichakli and M.F. Fontaine (Texaco), "Method of Improving Vertical Conformance of a Steam Drive", *Can. Patent 963,803* (March 4, 1975).
28 J.R. Ballard, E.E. Lanfranchi and P.A. Vanags, "Thermal Recovery in the Venezuelan Heavy Oil Belt", *27th Annu. Meet. Pet. Soc. C.I.M.*, Calgary, Alta. (June, 1976).
29 J.A. Dillabough and M. Prats, "Recovering Bitumen from Peace River Deposits", *Oil Gas J.*, *72(45)*, 186—198 (Nov. 11, 1974), [Original communication presented at *Symposium on Heavy Oil Recovery*, Maracaibo (July 1—3, 1974).]
30 J.G. Burger, "Spontaneous Ignition in Oil Reservoirs", *Soc. Pet. Engr. J.*, *16*, 73—81 (1976).
31 P.L. Terwilliger, R.R. Clay, L.A. Wilson and E. Gonzalez-Gerth, "Fireflood of the P_{2-3} Sand Reservoir in the Miga Field of Eastern Venezuela", *J. Pet. Technol.*, *27*, 9—16 (1975).
32 P.L. Terwilliger, "Fireflooding Shallow Tar Sands. A Case History", *50th Annu. Fall Meet. Soc. Pet. Eng. A.I.M.E.*, Paper 5568 (Oct., 1975).
33 F.F. Craig Jr., K.L. Hujsak and D.R. Parrish (Amoco), "Method of Forward In-Situ Combustion with Water Injection", *U.S. Patent 3,196,945* (July 27, 1965).
34 D.N. Dietz and J. Weijdema, "Reverse Combustion Seldom Feasible", *Prod. Mon.*, *32(5)*, p. 10 (May, 1968).
35 B. Sahuquet, "Mécanismes Réactionnels de la Combustion In-Situ. 1re Partie, Essais de Combustion à Contre-Courant", *C. R. 3me Colloque A.R.T.F.P.*, Editions Technip, Paris, pp. 709—720 (1969).
36 J.S. Trantham and J.W. Marx, "Bellamy Field Tests: Oil from Tar by Counterflow Underground Burning", *J. Pet. Technol.*, *18*, 109—115 (1966).
37 C.S. Land, C.Q. Cupps, L.C. Marchant and F.M. Carlson, "Field Test of Reverse Combustion Oil Recovery from a Utah Tar Sand", *27th Annu. Meet. Pet. Soc. C.I.M.*, Calgary, Alta. (June, 1976).
38 L.J. Heath, F.S. Johnson and J.S. Miller, "Solvents and Explosives to Recover Heavy Oil", *U.S. Bur. Mines Tech. Prog. Rep.*, *60* (Sept., 1972).
39 J.C. Allen and D.A. Redford (Texaco), "Method for Establishing Communication Path in Viscous-Petroleum Containing Formations", *U.S. Patent 3,913,672* (Oct. 21, 1975).
40 B.C. Craft, W.R. Holden and E.D. Graves, *Well Design (Drilling and Production)*, Prentice-Hall, Englewood Cliffs, N.J., 571 pp. (1962).

Chapter 9

VANADIUM: KEY TO VENEZUELAN FOSSIL HYDROCARBONS

GEORGE KAPO

Introduction

An unusual characteristic of Venezuelan coal and petroleum is their extraordinarily high vanadium content; several hundred parts per million are normal and, in isolated instances, over 1000 ppm have been detected. These values are not only one to two orders of magnitude higher than the vanadium contents found in most non-Venezuelan coals and crudes, but are also larger than the total content of all trace metals usually found in a typical fossil-hydrocarbon accumulation. Whereas in isolated instances one finds a fossil-hydrocarbon deposit outside Venezuela which has a comparable vanadium content, nowhere in the world is there any accumulation with a vanadium content and extension that approach those found in Venezuela.

Besides vanadium, high contents of nickel are also present in Venezuelan fossil hydrocarbons and these values again are usually larger than the nickel contents found in the crudes of other countries. Also, in samples of at least one petroleum deposit very high molybdenum concentrations have been reported, and in many coals quite large copper and molybdenum contents have been detected.

This author estimates that the amount of vanadium that has accumulated in the Venezuelan petroleum is equivalent to the total world reserves of this metal found in commercial ore deposits. If all of the vanadium was recovered from the present Venezuelan petroleum production, enough vanadium would be produced to almost satisfy the present world demand for this metal [1].

Whereas recovery of vanadium is important, the high content of this metal in Venezuelan petroleum, along with its high sulfur content, raises serious problems related to the refining and combustion of these crudes. Because of the fact that Venezuela's economy is almost totally dependent upon petroleum, and Venezuela's forthright role in the Third World and in the OPEC, solution to this unique problem has strong political overtones [2].

Geology and geochemistry

Apart from their high vanadium content, Venezuelan fossil hydrocarbons have other unusual properties, some of which can be correlated with the vanadium content. This author believes that the relation between these unusual characteristics and the presence of vanadium is not causal but, rather, that vanadium was responsible for at least some of them and is geologically associated with others. These relationships can be summarized as follows:

(1) For the coals and petroleums found in different sedimentary basins, there is a good correlation among their sulfur, nickel and vanadium contents. Approximate correlations can be expressed as follows:

$$\text{ppm vanadium} = (6 \pm 3) \times (\text{ppm nickel}) = (150 \pm 50) \times (\%\ \text{sulfur})$$

It has not been possible to establish trace-metal correlations from one basin to another for U.S.A. oils [24].

(2) The gravity (°API) of the crude increases with increasing vanadium content; this correlation takes the form of:

$$\text{ppm vanadium} = \text{antilog}\ [(4 \pm 1) - {}^{\circ}\text{API}/13]$$

(3) Whereas the sulfur contents of the coals are in the range of 1—2% and those of petroleums in the range of 2—5%, both associated and nonassociated natural gases have negligible sulfur contents.

(4) Although not as well documented as in the cases of vanadium and nickel, it appears that both coals and petroleums may also have unusually high contents of molybdenum.

(5) Except for the very oldest of the petroleum deposits, the connate waters of petroliferous formations in Venezuela have extraordinarily low salt concentrations. Typically, the sum of calcium- and magnesium-ion contents is less that 30 ppm and the total salt content is less than 3000 ppm [43].

(6) Some of the Venezuelan coals are sapropelic and many others contain significant amounts of allochthonous material. The properties of these coals differ from those of Tertiary coals found elsewhere in the world.

The total amount of vanadium accumulated in the Venezuelan fossil hydrocarbons is estimated to be at least 10^7 tons [1]. This is equivalent to the world reserves of vanadium in commercial ores [3]. Thus, an important question arises: "How did these large amounts of vanadium concentrate in the Venezuelan coals and petroleums?"

Whereas the presence of small amounts of many trace metals usually found in fossil hydrocarbons can be explained on the basis of a biological origin [4], the mega-amounts of vanadium found in the Venezuelan fossil hydrocarbons must have an inorganic origin. Fig. 9-1 is a map of Venezuela indicating the sedimentary basins and locations of coal and oil deposits as well as mineral deposits which contain either vanadium, nickel, or copper. The following explanation for the vanadium concentration in the fossil hydrocarbons has been offered by the writer [1].

In the center of Venezuela there is a mountain range where a large vanadium deposit has been found recently and where a large nickel deposit

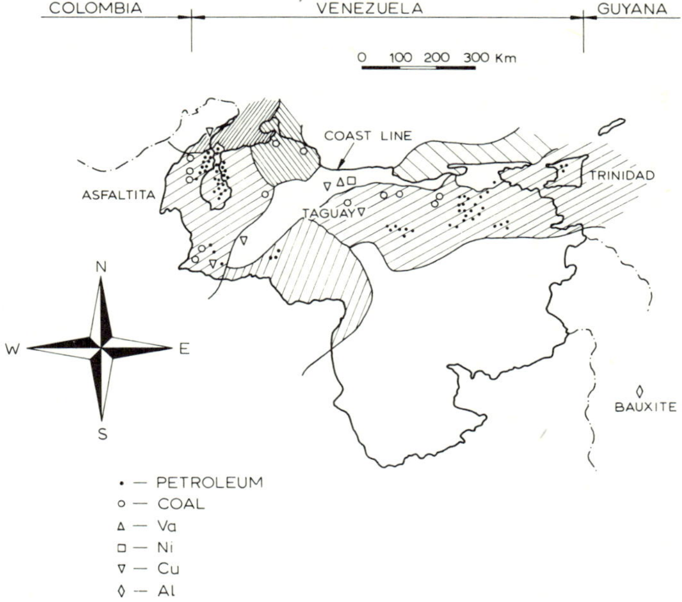

Fig. 9-1. Location of mineral and hydrocarbon deposits in Venezuela. Sedimentary basins are hatched.

was discovered some time ago. Before the last uplift of the Andes, this vanadium deposit had the highest elevation in this region. Rain water and thermal waters leached out and transported vanadium to accumulating biomasses in the surrounding fresh-water sedimentary basins, which were isolated from the sea. These metal-impregnated biomasses were later transformed into coal and petroleum. The latter then migrated to its present-day location, whereas the former, being a solid, was immobile. The location of coal deposits is probably near the original oil-generating biomasses.

The nickel, being close to the vanadium, would be expected to accumulate in the fossil hydrocarbons in a similar manner. Consequently, the vanadium/nickel ratio in the Venezuelan fossil hydrocarbons is more or less constant. This ratio of trace metals usually varies between 3 and 10 for both coals and petroleums in Venezuela. Some individual wellhead samples of the famous Boscan crude exhibited high values for the Va/Ni ratio, ranging up to 27, and the Taguay coals had values ranging down to slightly less than unity. In the rest of the world, vanadium/nickel ratios are quite variable and usually are less than unity. Some authors have suggested correlations between the vanadium/nickel ratio and the age of the petroleum or its chemical composition [24]. This is not true in the case of Venezuelan petroleums.

If the trace metals in fossil hydrocarbons originated from the organic sediments, one would not expect large and erratic variations in their contents in a single petroleum deposit or within a single seam of coal. Fig. 9-2 shows the variation in trace-metal content with depth for the two shallowest seams of the Taguay coal. As can be seen, there are large, erratic variations in the trace-metal contents. Fig. 9-3 is a nomogram of 27 analyses of wellhead samples of the Boscan crude; some of the samples are from individual producing strata, whereas others are mixtures of crudes from several closely related producing strata. Once again, there are large and erratic variations in the trace-metal contents. These large fluctuations in metal contents probably reflect changes in the pattern and amount of water drainage due to climatic and topographic variations which were occurring in the geologic epoch when the biomasses were being impregnated with trace metals.

In Colombia, some 1000 km to the west of the vanadium and nickel deposits of central Venezuela, there is an asphaltite deposit which contains 3000 ppm of vanadium and 2.8% sulfur and has a vanadium/nickel ratio of 5. Some 800 km to the east, the Trinidad asphalt contains 200 ppm of vanadium and 6% sulfur. About 400 km farther east, in Guyana, the bauxites are rich in vanadium content [1]. All of this could imply geological correlations of significant extent.

The Taguay coal is of special interest as it is the nearest fossil-fuel deposit to the vanadium and nickel accumulations, which are believed to be the source of the metals found in the coals and petroleums present in the

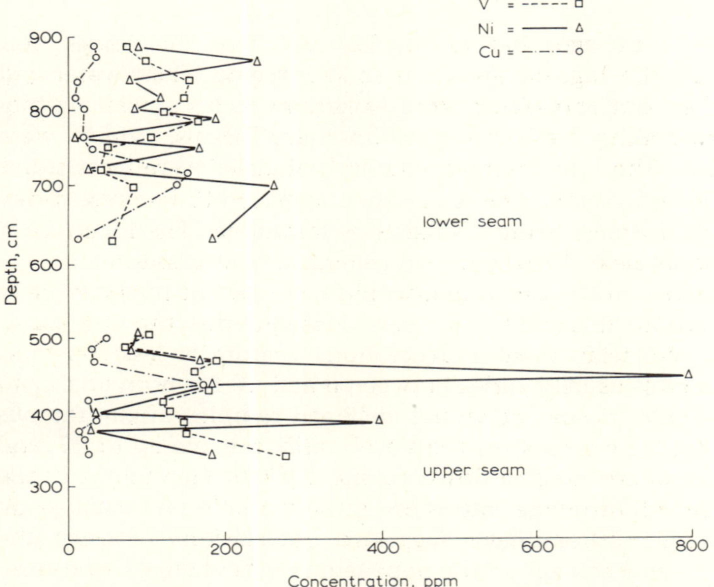

Fig. 9-2. Variation in V, Ni and Cu contents of Taguay coal with depth. (After Salcedo and Amez [15].)

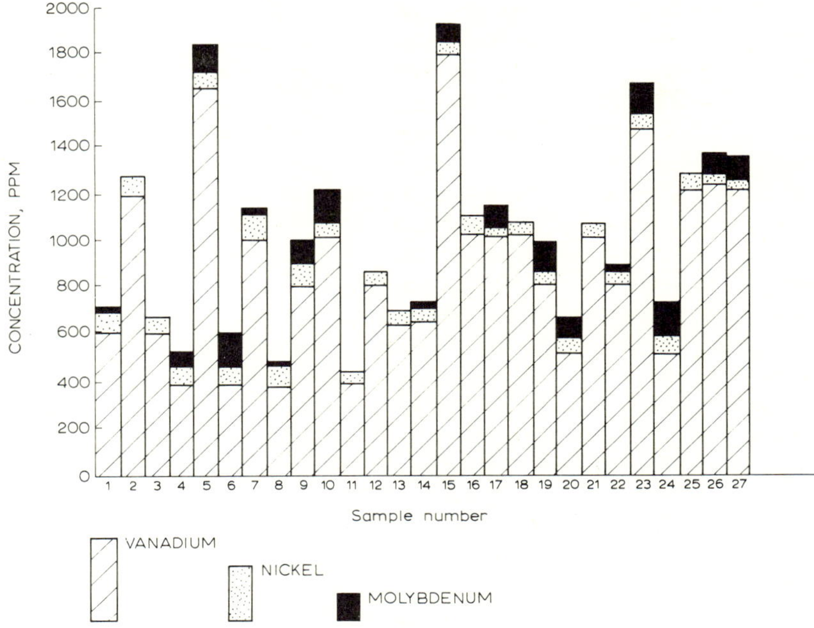

Fig. 9-3. Vanadium, nickel, and molybdenum contents of Boscan crude-oil samples. (After Ulrich and Roman [14].)

surrounding area of some thousand kilometers in extent. As expected, this coal has a higher vanadium content than other coals and petroleums farther away, and one sample exhibited over 3000 ppm of this metal. Its sulfur content is high; however, the vanadium/nickel ratio is somewhat lower than that found in other fossil fuels in Venezuela. The distance of this coal field from the vanadium and nickel deposits is about the same as the distance between the latter two metal deposits and, consequently, the drainage waters possibly did not have time to mix well. Thus, the Taguay coal received relatively more nickel than fossil hydrocarbons located at a greater distance from the source deposits.

The molybdenum concentrations as reported in Fig. 9-3 for individual samples of the Boscan crude oil are either undetectable or comparable to (more or less) that of the nickel. For the first time it was observed that molybdenum could be the second most important trace metal in the Venezuelan petroleum, instead of nickel. The average molybdenum concentration as shown in Fig. 9-3 is 60 ppm; Chevron Petroleum Company reported an average molybdenum concentration of 50 ppm [29] for Boscan crude oil. The low values of 4.6 [26] and 8 [28] reported in the literature for Boscan crude oil are probably due to the fact that the samples analyzed were not representative of the whole deposit. According to Fig. 9-3, if only one wellhead sample is taken for analysis, 9 out of 27 times the molybdenum concentration will be less than 10 ppm.

The molybdenum concentrations in Venezuelan coal ashes show the same large values and fluctuations similar to those for the Boscan petroleum (Fig. 9-3). The Colombian asphaltite, mentioned earlier, has a molybdenum concentration of about 1000 ppm. In the Athabasca tar sands, the molybdenum concentration is around 10 ppm and gradually decreases with depth. This gradual decrease is used as evidence that the origin of this metal in Athabasca bitumens is biological [26]. Based on the large and widely fluctuating metal concentrations reported for Venezuelan fossil hydrocarbons, it would seem that, just as for vanadium and nickel, a mineral origin for the molybdenum would be more likely. Whereas no large molybdenum deposit has yet been found in Venezuela, the writer would not be surprised if one were discovered in the future. The best place to explore for molybdenum deposits would be the same mountain range where the vanadium and nickel deposits were found. The fluctuations in molybdenum concentrations are much grater than those of vanadium, which in turn are greater than those of nickel. No explanation, however, can be given at this time and, unfortunately, no molybdenum determinations have been made on the Taguay coal.

There are several known copper deposits in Venezuela and these are located near the sedimentary basins where coal is found. These deposits were undoubtedly the source of this metal now found in the coal. The coals have copper concentrations 100 times higher than the petroleums. Apparently, the coals contacted the copper-laden surface waters after the petroleum migration.

The close association of the petroleum and coal in Venezuela could mean (1) that the coal deposits were also oil-generating beds, or (2) that at least the biomasses, which were converted into coal, contained large amounts of allochthonous material. The first possibility had been suggested many years ago [40]. Recent petrographic analyses of many Venezuelan coals have confirmed the latter theory [42]. The presence of 10—20% bituminous material in a low-rank coal can dramatically influence its technological properties. The bituminous material would displace the water in the pores and reduce the moisture content to very low values. Also, the bituminous material, functioning in the same way as the binder material for coke briquets, would impart coking properties to a young coal which, based on analysis (80—90% nonbituminous, humic material), would not be coking.

Thus, the geological sequence of events which explains some of the unusual properties of Venezuelan fossil hydrocarbons can be summarized as follows. The biomasses, which as a result of biogenesis, diagenesis, and metamorphism produced coal and petroleum, were deposited in several fresh-water sedimentary basins. Drainage into these basins was dominated by a mountain range having deposits of vanadium, nickel, and, perhaps, molybdenum. After the petroleum migration, the coals continued to adsorb trace metals, principally copper, from waters draining newly exposed deposits. The allochthonous material deposited formed some sapropelic coals and greatly modified the properties of coals of high vitrinite content. Swamp forests existed near a few islands which protruded up through the sedimentary basins, and classic humic coals were formed. These coals contain only small amounts of vanadium and nickel.

The inorganic theory, which explains accumulation of mega-amounts of metals in Venezuelan fossil hydrocarbons, may be applicable in some other areas of the world. A cursory look at the world map reveals the presence of vanadium-mineral deposits near fossil-hydrocarbon accumulations which are rich in vanadium (Peru, Argentina, Mexico, Canada, the U.S.A., Australia, Finland, and the U.S.S.R.) [1]. The writer also attempted to formulate hypotheses to explain (1) correlation between the vanadium and the sulfur contents in these hydrocarbons, (2) correlation between the density of the petroleums and their vanadium content, and (3) the absence of sulfur in the natural gases.

When vanadium-laden water is placed in contact with hydrogen sulfide, vanadium is precipitated as vanadium sulfide. This was probably the mechanism which initially immobilized the vanadium in the biomasses when the vanadium-laden drainage waters contacted them [4]. After precipitation as sulfide, the vanadium was slowly chelated into the organic structure of the fossil hydrocarbons being formed from the biomass. The large amount of accumulating vanadium used up large volumes of hydrogen sulfide. This could explain why the natural gases in Venezuela are not sour.

The precipitation of vanadium sulfide, however, does not explain the correlation between the vanadium and sulfur contents in coals and petroleums. The ratio of sulfur to vanadium in these hydrocarbons is about 50 times larger than the stoichiometric ratio of sulfur to vanadium in vanadium sulfide. Several authors have considered vanadium as a catalyst for geo-organic reactions. It has been suggested that the vanadium could have catalyzed the incorporation of sulfur into fossil hydrocarbons and also could have caused the polymerization of light petroleum into heavy asphalt [1]. The possible precipitation of vanadium and the sulfide, in addition to the suggested catalytic properties of vanadium, and the proposed inorganic origin of the vanadium and other trace metals, constitute a working hypothesis for explaining most of the unusual properties of the Venezuelan fossil hydrocarbons.

Ecology

It is too early to assess the magnitude of the ecological and environmental damage that will result from the liberation into the atmosphere of large amounts of vanadium and nickel due to the combustion of petroleum; however, one cannot be optimistic. It has been estimated that in the U.S.A. alone, some 100,000 tons of nickel and some 50,000 tons of vanadium are transferred from the petroleum to the environment each year. While the consumption of petroleum in Venezuela is small by comparison to that in the U.S.A., the fact that the crudes are so high in trace metals means that significant levels of contamination can be reached, especially in urban areas. Both vanadium and nickel are toxic when present in sufficient quantities, and dermatitis, respiratory disorder, cardiovascular disease, hypertension, and carcinogenesis are associated with lesser amounts of these metals [25].

Apart from the direct effect of these metals, secondary reactions can lead to an additional deterioration of the environment. Vanadium and nickel are catalysts for many reactions [19], so that in combustion and refinery processes there may occur chemical reactions which produce undesirable compounds. The best example is probably combustion where vanadium catalyzes the oxidation of sulfur dioxide to sulfur trioxide. Also, in a combustion reaction, vanadium and nickel could catalyze the formation of nitrogen oxides [34].

Whereas much more could be said about the metal-contamination problem, this short section will be terminated with the following thought: "In less than a century, the human race will release into the environment a mega-amount of trace metals which took thousands of centuries to accumulate in fossil hydrocarbons. Inasmuch as these trace metals have both biological activity and toxic characteristics, what dramatic consequences this will have, no one can prophesy."

Petroleum economics

Before the 1973 energy crisis, Venezuela was selling its high-sulfur, high-vanadium residual fuel oil at a price of $ 2.50/bbl. On the eve of the energy crisis, the environmentalists were having a field day, and strict sulfur-pollution legislation drove the price of low-sulfur fuel oil up to $ 4.50/bbl. At this time, Venezuela had its best technical minds at work trying to figure out how to desulfurize all of its residual fuel oil, so that its export value could almost be doubled. The crux of the problem was that vanadium and nickel present in the petroleum rapidly poisoned hydrodesulfurization (HDS) catalysts.

After the energy crisis, the differential of $ 2.00 between low- and high-sulfur fuel oils remained the same, but the price of the high-sulfur fuel oil and petroleum, in general, quintupled. With dirty fuel oil now selling at $ 12.50/bbl, it was of secondary importance to Venezuela if, by cleaning it up, it could be sold at $ 14.50/bbl. In fact, to minimize the financial damage of the energy crisis, ecological considerations in the First World were brushed aside and any available lower-priced fuel was being burned. There were few consumers interested in clean fuel oil at $ 14.50/bbl. Venezuela nationalized its petroleum industry in the beginning of 1976. As a result of the energy crisis, the price of sulfur was quadrupled and the problem of excessive production of this element by the local petroleum refineries disappeared.

The period of 1974—1977 must be looked upon as a technological recession for the Venezuelan petroleum industry. Both crude production and refinery throughput dropped considerably and the newly installed HDS unit was never started up. Petroleum revenues, however, were higher than ever, and any inefficiency in the recently nationalized industry could easily be tolerated. Except for the fact that a large quantity of refinery products are consumed locally, Venezuela could have closed its refineries and directly sold the crude oil and still made more money than ever in its history. This technological recession will continue until about 1980 when the Venezuelan petroleum industry will enter a new phase.

Traditionally, Venezuela never consumed internally more than 10% of its petroleum production (the other 90% being exported). Because of the gradual substitution of natural gas by fuel oil in thermoelectric plants and in heavy industries, the increased demand for gas by steel and petrochemical industries, the rapid increase in the internal demand for gasoline and other refinery products, and the reduction in the production of light petroleums. Venezuela can have its own small energy crisis in the 1980's unless new technology is introduced into the petroleum industry. Specifically, if by the 1980's the high-metal, high-sulfur, heavy crudes are not upgraded both to low-sulfur fuel oil as well as to gasoline, Venezuela will have to import light petroleums at international prices in order to satisfy domestic consumption. Already, for example, some thought has been given to the importation of low-sulfur crudes for the manufacture of petroleum coke needed for a proposed electrode plant.

Apart from the present bonanza that the energy crisis has meant for Venezuela, the most important impact that the quintupled price of oil had is that it converted the exploitation of the vast Orinoco heavy-oil field ($\sim 10^{12}$ bbl of oil) into an economic possibility. It was estimated that it will cost \$ 1.50/bbl to produce this oil and another \$ 2.50/bbl to refine it into a saleable product [41]. Clearly, before the energy crisis, this was an uneconomic possibility. The development of technology for the Orinoco heavy-oil field will dominate the Venezuelan petroleum industry for the last decade of this century and its production will be the dominating factor during the next century.

Whereas the technological problems associated with vanadium are considerable, the monetary value of the vanadium vis-a-vis petroleum is very small. A typical barrel of the Venezuelan residual fuel oil contains some 500 ppm of this metal. Upon combustion, about 0.2 pounds of vanadium pentoxide would be formed. Before the energy crisis, this vanadium in the form of a high-grade ash could be sold for around \$ 0.10 and more recently, for \$ 0.20. Consequently, before 1974, as a byproduct of a combustion operation, vanadium had a maximum value of only 6% of the fuel oil, whereas in 1975, this percentage dropped to 2%. Some of the vanadium fly ash will escape collection and that part which is collected will be of a low-grade variety. Hence, the value of the vanadium collected can be less than 1% of the total operating costs of a refinery or a thermoelectric plant. The largest vanadium fly-ash operation in the world is that directed by John O'Neal at the Long Island Lighting Company [9]. Whereas about 7% of the U.S.A.'s needs for this metal is derived from this operation, the vanadium revenues are a mere 0.4% of the total operating costs of this thermoelectric company [35].

Obviously, at the level of 0.4%, no thermoelectric plant nor petroleum refinery will make a serious effort to recover vanadium in the hope of improving its profit margin. A conservational attitude, coupled with the desire (or the obligation) not to contaminate the environment with trace metals, however, could result in a corporate policy in favor of vanadium recovery. For strategic reasons, taking into account the large amount of vanadium that could be recovered, the Venezuelan Government may take decisive steps in this area [2].

First- and second-generation petroleum refining

In Venezuela, the planning and development of the petroleum-refining industry has been and will be dominated by the following facts:

(1) Conventional HDS and fluid catalytic-cracking processes are inoperable with petroleums of high metal contents.

(2) The metals in the Venezuelan petroleums are associated with the non-volatile portion of the crude oil, so that practically all of the metals are found in the residuum fraction and the distillate fractions are extraordinarily free of metals (Table 9-I).

(3) The sulfur, on the other hand, is found in both the residuum and the distillate fractions, so that some sulfur can be removed by applying conventional HDS process to distillate fuel oils.

(4) The production and reserves of light and medium petroleums is declining rapidly. By the early 1980's, it is estimated that there will not be enough distillate fractions to satisfy either the installed refining capacity or the local demand for refinery products [29].

(5) The reserves of high-metal, high-sulfur, heavy crudes are great, whereas production and refining capacities are very limited.

As discussed in items (1) and (4) above, the possibility of Venezuela participating in the lucrative market of low-sulfur fuel which existed prior to 1974 was limited. Most of the strategy studies and research carried out then were aimed at resolving this problem. Since the energy crisis of 1974, the need to solve the low-sulfur fuel-oil problem disappeared and most of the related studies are now of only historical interest. The Venezuelan petroleum

TABLE 9-IA

Sulfur and metal contents in crude oil * of a commercial refinery (Mobil's El Palito refinery, Moron, Venezuela) (After González [12])

	Yield (vol%)	S (wt%)	Cu (ppm)	Fe (ppm)	Pb (ppm)	Cr (ppm)	Ni (ppm)	Na (ppm)	Al (ppm)	V (ppm)
Crude		1.81	0.44	1.63	3.09	3.98	29.59	3.62	0.86	194
FRM	14.1	0.0097	<0.01	<0.1	<0.01	<0.01	<0.1	<0.1	<0.1	<0.1
Kero	6.4	0.15	0.02	<0.1	0.22	<0.01	<0.1	0.08	<0.1	<0.1
LGO	10.9	0.59	0.02	<0.1	0.27	<0.01	<0.1	0.22	<0.1	<0.1
HGO	6.5	1.00	0.03	<0.1	0.40	<0.01	<0.1	0.38	0.1	0.1
Residuum crude	61.4	2.32	1.25	2.50	10.00	5.00	47.52	7.13	1.96	301

* Crude composition, % by volume: Barinas Blend, 8%; Tia Juana Medium, 86%; Mesa Blend, 6%; Corridor Block, 0%.

TABLE 9-IB

Sulfur and metal contents in crude oil * of a commercial refinery (Mobil's El Palito refinery, Moron, Venezuela) (After González [12])

	Yield (vol%)	S (wt%)	Cu (ppm)	Fe (ppm)	Pb (ppm)	Cr (ppm)	Ni (ppm)	Na (ppm)	Al (ppm)	V (ppm)
Crude		1.34	0.48	1.58	3.73	4.04	60.5	1.45	1.04	187
FRM	13.3	0.0030	0.002	<0.01	0.004	<0.01	<0.1	<0.1	0.06	<0.1
Kero	10.5	0.091	0.003	<0.01	<0.001	<0.01	<0.1	<0.1	<0.01	<0.1
LGO	9.1	0.49	0.007	<0.01	0.020	<0.01	<0.1	<0.1	<0.1	<0.1
Residuum crude	58.4	1.88	0.57	2.49	4.04	5.46	98.46	3.90	2.38	267

* Crude composition, % by volume: Barinas Blend, 81%; Tia Juana Medium, 1%; Mesa Blend, 3%; Corridor Block, 15%.

TABLE 9-IC

Metals and sulfur contents of Boscan crude, its residuum, and its gas oil (After Edler [18])

	Yield (vol%)	S (wt%)	V (ppm)	Ni (ppm)	Na (ppm)	Fe (ppm)	Zn (ppm)	Cu (ppm)	Pb (ppm)	Total metals content (ppm)
Whole crude		5.5	1080	100	46	4.9	1.1	0.4	0.1	1233
Residuum crude	75	6.0	1530	138	47	6.4	4.0	0.8	0.1	1726
Gas oil (HGO)	9	4.7	0.6	0.005	0.9	0.2	0.05	0.1	0.1	2

industry does not have to face any major technological challenge until the 1980's.

The problems that will have to be faced in the 1980's will be more complex than the desulfurization problem of the early 1970's. In the first place, when and if new pollution regulations are put into effect, they will include not only limitations on sulfur content, but also on nitrogen and metals contents. As discussed in an earlier section, only recently has the toxicity of vanadium and nickel been appreciated. Hence, total cleaning and not simply desulfurization will be required. In this respect, for example, a process such as Exxon Residfining [31], with its RT-2 and RT-3 catalysts, which selectively removes sulfur while leaving the metals behind, would not be a satisfactory process in the future. The other complex problem to be faced in the 1980's is the massive amounts of heavy oils which will have to be catalytically cracked (CC) to produce gasoline and other distillate products. In general, fluid catalytic cracking (FCC) will require a lower metal content in the feedstock than HDS. The feedstock for FCC process should have a nickel equivalent of less than 0.3 ppm; if the vanadium/nickel ratio in the Venezuelan petroleum is considered to have an average value of 5, the feedstock for FCC should not have more than 1 ppm of vanadium. The fixed-bed HDS units installed in Venezuela in the past can handle a maximum of 2 ppm of vanadium. Newer HDS processes with catalysts and using fluidized beds handle feedstocks with up to 100 ppm of metals. It is generally considered, however, that when metal content is above 100 ppm, direct HDS process is uneconomical [32].

There are large number of options available to the Venezuelan petroleum refineries; in one study alone, twelve different options were considered [30]. Fig. 9-4 shows the main options and designates four generations of development. Vacuum-distillation units, which have been installed for almost a decade in Venezuela, constitute the first generation. Compared to the atmospheric distillation, these units provide additional low-metal feedstock for the HDS units. Distillation is extraordinarily effective in eliminating metals in Venezuelan crudes. Typically, with vanadium content reaching 1000 ppm in the residuum, the heaviest distillate fuel oil will contain less than 2 ppm of this metal (Table 9-I). As the refinery feedstock grows heavier and distillate yields drop, however, a point will be reached where the refiner will have to install a second-generation process, which undoubtedly will be a coking unit.

Coking processes have been operating on a large scale for many years outside Venezuela and coking units could be installed relatively rapidly without any technological complications. The problem with a simple coker in Venezuela is that it would produce a coke with a large amount of sulfur (more than 5%), which would be difficult to use. Whereas there is a large demand for coke by both the steel and the aluminum industries in Venezuela, a high-sulfur coke would be unacceptable. In addition, the vanadium content

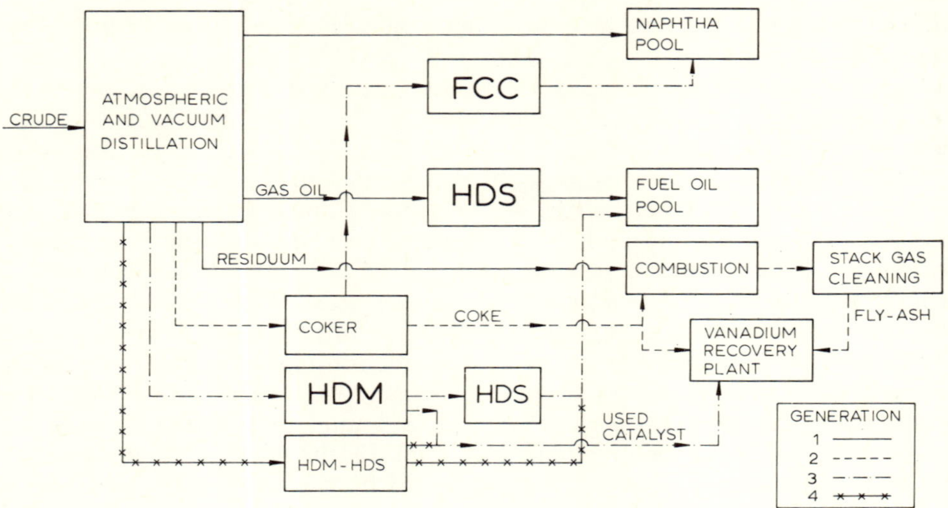

Fig. 9-4. Main options and four generations of development of the Venezuelan refineries.

in the petroleum coke would be less than 1%, making the coke worthless as a raw material for a vanadium plant. The thermoelectric plants in Venezuela are just beginning to phase out gas in favor of the more abundant residual fuel oil. None of the existing thermoelectric plants nor a large new one now under construction are equipped to handle a solid fuel. Perhaps within ten years a solid-fuel thermoelectric plant will be functioning; however, by then a high-sulfur petroleum coke would have to compete with low-sulfur coals (mines are now under development). Some coke could be consumed in the refinery for the production of hydrogen [41].

The main advantage of a coker is that, besides providing additional low-metal gas oil for HDS, it converts part of the residuum into the much needed naphtha and other distillates, thus postponing the necessity of installing FCC units. With some changes in the combustion system, the coker gas can be used as a refinery fuel, thus liberating natural gas for other, nobler, uses.

The undesirable characteristics of the petroleum coke produced by using a simple-coking process have been overcome by the Exxon's Flexicoking Process. In this process, by judicious selection of operating and design parameters, one can consume all of the low-BTU gas produced in the refinery and, at the same time, obtain a low-sulfur (less than 2%), high-vanadium (more than 2%) coke [31]. A coke of this composition would have considerable value as a raw material for a vanadium plant [13]. Even more interesting would be the use of this coke in an electric-furnace process for the direct production of ferrovanadium; this possibility will be discussed in a later section (metallurgy) of this chapter. At least one Flexicoking unit will probably be installed in Venezuela by 1980.

Deasphalting is sometimes considered to be a viable second-genera process for Venezuela. The writer, however, does not agree with this soluti for the following reasons: In the first place, deasphalting is not effectiv enough when applied to a residuum with a very high metal content. The metal removal in a deasphalting process refers to a 90% reduction of metals in the refined product, whereas a 99.9% reduction is achieved in atmospheric-, vacuum-, and destructive-distillation units (Table 9-I). If a residuum has a vanadium + nickel content of 10 or 100 ppm, a 90% reduction in metals will result in deasphalted oil with 1 or 10 ppm of metals, respectively. This oil would be suitable for some petroleum-refinery catalytic processes. With a typical Venezuelan residuum, the metal content of which is of the order of 1000 ppm, however, deasphalting will only bring the content down to 100 ppm, a value still too high for most refinery catalytic processes. A very low-metal oil could be produced by deasphalting a Venezuelan residuum only by using propane as the solvent and accepting a yield of less than 50% of refined oil. The large amount of asphaltenes byproduct produced in such operation raises the next objection to the deasphalting process [18].

Asphaltenes are semisolids and in some respects their material handling that the small-diameter pore systems are useless no matter how high their high sulfur content, in addition to the handling problems, make them an undesirable fuel for the refinery or for a thermoelectric plant. Although the vanadium content could be over 2000 ppm, this value is still too low to consider asphaltenes as a feedstock to a vanadium-recovery plant. As far as their use in highway paving and roofing is concerned, asphaltenes would have to compete with asphalt, which is already being produced in large quantities. Some Venezuelan scientists have accepted the challenge and are trying to convert asphaltenes into useful products. One should remember, however, that it is unlikely for deasphalting units to be installed in Venezuela.

Third-generation petroleum refining

To increase the flexibility in the refineries, FCC units will be installed in the 1980's. The low-metal gas oil produced by distillation and coking will be diverted from the old HDS units to the FCC units. New HDS units, which can handle crude oils with up to 100 ppm of metals, will be built. They will receive their feedstock from newly installed hydrodemetallization (HDM) units.

The HDM process involves the deposition of metals, in the presence of hydrogen, on catalysts [32], porous minerals [16], or manganese nodules [36]. The efficiency for metal removal in an HDM unit (around 90%) is of the same magnitude as that in a deasphalting process. Some of the shortcomings of the deasphalting process, therefore, also hold true for the HDM process. The HDM process, however, has an enormous advantage over deasphalting in that no undesirable residuum is produced and the vanadium concentration in rejected solids can range up to 40% [32]. This makes the rejected solids an extraordinarily valuable raw material for a vanadium-production facility. Also, in principle, by using several HDM units, one could reduce the metal content to as low a value as desired, with a corresponding increase in cost of the operation.

Fourth-generation HDM—HDS—CC units

The HDM process discussed in the last section is quite unlike any other refining process ever invented. For example, the old clay-refining techniques did not involve any chemical reactions and whole molecules were adsorbed, rather than just the undesirable atoms. Also, the basic principle of all heterogeneous catalytic processes has been that all of the reaction products should be continually desorbed from the solid catalytic surface. In the HDS process, for example, the undesirable atoms, sulfur and nitrogen, are catalytically converted to hydrogen sulfide and ammonia, which are then desorbed. Any material remaining on the catalyst's surface is usually considered to be highly detrimental to the catalyst's activity for the following reasons: (1) coverage of active sites by carbon and other inert material; (2) blockage of pore mouths by any deposited material; and (3) poisoning of the catalyst by small amounts of metals or other elements.

To increase catalyst activity, supports were developed which provided large pore surface areas. In order to achieve large internal areas, many small pores are needed, but with the exception of HDS, these small pore diameters were usually large enough to permit entrance of the reacting molecules.

The HDM process of a residual oil differs dramatically from the classical concepts of heterogeneous catalysts in three different ways:

(1) The reaction occurs with the largest molecules in the petroleums, so that the small-diameter pore systems are useless no matter how high their internal area may be.

(2) There is a massive deposition of material on the catalysts.

(3) The metals which are being deposited are catalytic and, consequently, an in-situ formation of a new catalyst could occur.

Based on the large size of residuum molecules, it is doubtful if a support can be prepared the available surface area of which in HDM reaction could exceed 100 m^2/g. Published results indicate that pore-effectiveness factors are low and the metals tend to accumulate on the exterior surface of the support [17].

Massive deposition, mentioned in item (2) above, means that much more material is deposited than that required for a monolayer adsorption. An adsorbed vanadium atom, in the +4 valence state, would be associated with four other atoms, e.g., oxygen, nitrogen, or sulfur. In addition, for every five atoms of vanadium, there should be one of nickel with its associated bonding atoms. Finally, there will be some deposition of carbon and other miscellaneous material. The writer estimates that for each vanadium atom adsorbed, at least 100 $Å^2$ of surface area will be covered. Using a value of 100 m^2/g as available surface area, one can calculate that the catalyst surface will be completely covered with a monolayer when it contains more than 1% vanadium, based on the original weight of the catalyst.

This simple calculation means that in the HDM process, where the metal content on the solid support can reach 40%, there is a multilayer adsorption, i.e., not just two or three layers, but ten, twenty or maybe even 100 layers.

The metals, mostly vanadium and nickel, which are being deposited, are known catalysts for many reactions, including HDM, HDS and CC [16]. This metal deposition constitutes an in-situ formation of a hydrogenation catalyst. The evaluation of the activity of this new catalyst depends on many factors. For example, if the original adsorbing solid is a high-activity cobalt—molybdenum catalyst, then one would expect some "poisoning" to occur as the surface composition changes from cobalt—molybdenum to vanadium—nickel—carbon and any other associated bonding atoms. If the original substrate is a noncatalytic, porous mineral, then autocatalytic behavior should be observed. The activity of a moderately active, catalytic mineral, such as manganese nodules, may not change much. Obviously, this in-situ formation of a catalyst will be more apparent if the metal/carbon ratio in the deposited layer is high. Venezuelan residuums, which have normal values of Conradson carbon (10—15%), but very high metal contents, are the most favorable candidates for this in-situ process. Many other residuums, which have been studied in the laboratories of international oil companies, had metal contents one or two orders of magnitude less than those of the Venezuelan residuums. In such cases, the beneficial effect of the deposited metals could be partially or completely cancelled by the simultaneous carbon deposition. Of course, the type of metals being deposited is of paramount importance, and the Venezuelan petroleum has an advantage because vanadium and nickel constitute 90% of its metals. Perhaps, if the metals were cobalt and molybdenum, the situation could be even better, whereas if the deposited metals were iron, copper, zinc, etc., the situation would be worse. If the lead, arsenic, sodium, etc., are the metals being deposited, then no in-situ catalyst formation would occur.

Parallel to the demetallization reaction, one would expect other hydrogenation reactions, such as desulfurization, denitrification and hydrocracking, to occur. The ratio of the rates of the last three reactions to the rate of demetallization can be considered as the selectivity of the process. Vanadium and nickel separately and together, have been used as HDS and CC catalysts [18], so that with the Venezuelan residuums, the HDM process would be expected to have a low selectivity. As discussed later, this may be a very useful phenomenon.

Carbon deposition has two negative effects. In the first place, carbon will dilute the metal concentration in the deposited layer. Consequently, carbon deposition can be considered as having reduced the specific activity of the surface without reducing the available surface area for reaction.

Inasmuch as every newly deposited layer would be expected to have the same composition as the previous layer, this type of carbon accumulation should not exhibit any evolutionary decrease in the activity of the solid substrate. Because of the low pore-effectiveness factor in the HDM reaction [17], one would expect a second type of carbon accumulation, i.e., a selective buildup of carbon at the pore mouths. This, of course, would cause a very rapid decrease in overall substrate performance, specially as the pore mouths become completely plugged. Here, the effect of the carbon deposition is a reduction in available pore surface area for the reaction and not in the specific activity of the available surface.

In a commercial process, by periodic combustion or gasification, in order to remove the carbon plugs, the solid substrate can be regenerated in situ and the HDM process continued in a cyclic manner. Eventually, however, the pores will become completely filled with metals. At this point, the overall activity will drop to almost zero and the solid substrate would be rejected from the HDM reactor. The solid substrate could contain at least about 10% vanadium and, as such, would constitute a valuable metallurgical raw material.

In 1970, based on the above concepts, a group of chemists and chemical engineers at the University Central de Venezuela (UCV), started to develop an HDM process for the Venezuelan heavy crudes [16]. The following guidelines were used by the UCV group:

(1) The solid substrate should consist of a mineral, the deposits of which are located near the Orinoco heavy-oil belt, with sufficient reserves to demetallize all of this heavy-oil deposit. In addition, in its natural state or after a simple calcination or acid treatement, this mineral must have a reasonably high surface area associated with large-diameter pores.

(2) The amount of vanadium deposited on the mineral before the latter permanently loses its activity should be at least 5%.

(3) The hydrogenation reactor would then operate under conditions enabling HDM, HDS, and CC reactions to occur simultaneously. Thus, an attempt should be made to solve the entire problem of upgrading the heavy crude in one reactor.

The latest results obtained by scientists and engineers are shown in Fig. 9-5. The following conclusions can be reached on examining this graph: The initial increase in HDM efficiency is due to the formation of the first catalytic layers of deposited metal. The subsequent decrease in activity is due to plugging of pore mouths. The removal of carbon by combustion reopens the pore mouths and, by exposing metals previously deposited, increases specific activity of the available surface. Perhaps, the oxidation also increases the specific activity of the surface by changing the valence state of the adsorbed metals. In addition to removing metals, the process also lowers the sulfur content of the feedstock and raises its gravity.

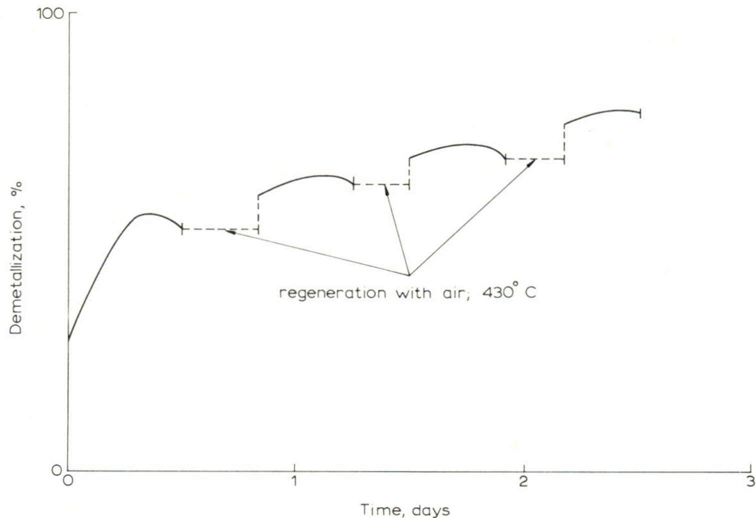

Fig. 9-5. Evolution of mineral activity for demetallization of Morichal crude, Venezuela.

Combustion, gasification, and stack desulfurization

With all of the difficulties associated with the Venezuelan residual oil, one may ask the following question: "Why not simply burn this residual oil and employ stack desulfurization and fly-ash removal techniques to meet ecological requirements?" In 1972, the writer proposed precisely such a strategy [34]. Both U.O.P. [33] and I.F.P. [5] agreed with the advantages of employing stack-desulfurization technologies to remove sulfur from the Venezuelan residual oil. As shown in Fig. 9-4, the combustion of residual oil and coke was included in refining strategy.

Many of the stack-desulfurization technologies developed for coal-fired thermoelectric plants can be used in an oil-fired one, but with one big difference: the fly ash and bottom ash collected from the operation, instead of being worthless, has a very high metallurgical value as a source of vanadium and nickel [6,34].

Combustion of the Venezuelan residual oil, however, is not without its difficulties. The vanadium—nickel fly ash produced can be extraordinarily corrosive and there is concern about its possible toxicity. Also, fly ash can catalyze the formation of sulfur trioxide [7] adding still another dimension to corrosion and contamination problems. The corrosion due to the fly ash itself occurs in the hot portion of boilers, where fly ash can be fluid; whereas the corrosion due to the sulfur trioxide formation occurs in cooler parts of an economizer. If the dew point is reached in the latter, condensation of sulfuric acid would occur. If the sulfur trioxide is not condensed or otherwise removed in the thermoelectric plant, the rain water in the region may in effect be a dilute sulfuric acid.

The principal cause for high-temperature corrosion is sodium vanadyl-vanadate. This compound has a melting point of 625° C and is easily formed in a combustion chamber if the residual oil contains sulfur, vanadium, and sodium (the latter can be in the form of an emulsified brine) and if excess oxygen is supplied to the flame [8]. The corrosion mechanism of sodium vanadate is related to its ability to dissolve iron.

In a reducing atmosphere, corrosive nickel sulfide can be produced. This can cause intergranular sulfide attack and decarbonization of boiler tubes built from nickel alloys [10].

There are three different ways of combating vanadium corrosion in combustion systems:

(1) The technology of using magnesium oxide or other fuel additives has been well developed by John O'Neal [9] at the Long Island Lighting Company, which produces some 7% of the U.S.A.'s vanadium consumption.

(2) A second method usually preferred by petroleum refineries, is the use of special alloy steels.

(3) By using a stoichiometric amount of air and attaining perfect mixing in the combustion process, vanadium pentoxide will not be formed at any temperature [8]. The latter technique has been perfected by Eisenklam in a low-pressure, two-stage combustion pilot plant with interstage heat removal [10]. By maintaining reducing conditions and intense mixing in the first stage, only the lower vanadium oxides are formed and these are not corrosive. By completing the combustion in the second stage with more than the stoichiometric amount of air, the formation of vanadium pentoxide is again avoided. Nickel sulfide, however, is formed, probably due to the reducing conditions in the first stage.

By avoiding the presence of excess oxygen in the two-stage combustion chamber, sulfur trioxide formation was considerably reduced and, by a combination of deficit oxygen and lowering of the peak-flame temperature, nitrogen oxide formation was also considerably suppressed. The partial combustion in the first stage, however, produced coke and soot which could only be partly eliminated during the second stage [37].

Eisenklam's [10] work is of great historical and academic interest. During the petroleum boycott of the Suez crisis, the British Navy had to switch immediately from the low-vanadium Arab fuel oil to the high-vanadium Venezuelan fuel oil. After the explosion of a few ship boilers, the British Navy began research programs to determine methods of how to handle the high-vanadium fuel oil. One of these programs was the two-stage combustion pilot plant of Eisenklam. During World War II, the Canadian Navy had a similar problem, which was resolved by using basic metal oxides as fuel-oil additives. The Venezuelan Navy also controls its vanadium corrosion problem with fuel-oil additives. Because there are no pollution laws on the high sea, low-cost fuel oils, which have high sulfur and metal contents, are preferred by naval and merchant-marine organizations.

Two-stage combustion with stoichiometric air eliminates vanadium corrosion and the formation of nitrogen oxides and sulfur trioxide. It does not, however, suppress emission of either sulfur dioxide or vanadium—nickel fly ash. In addition, unless a large second-stage combustion chamber is used, there will be carbon emissions in the form of coke and soot. Hence, from an ecological point of view, the two-stage combustion alone is unsatisfactory. In order to be acceptable, it would have to be equipped with a stack-cleaning device.

Two-stage combustion is a radical departure from classical fuel technology and, while complicating equipment design, it only partly solves the difficulties of burning the Venezuelan residuums. Hence, it is doubtful if Eisenklam's prototype will ever leave the pilot-plant stage. The basic idea of breaking up the combustion into two stages, however, is a good one; this leads discussion into the subject of gasification.

The use of a combined cycle (gas and steam turbines) in a thermoelectric plant, backed by a heavy-oil gasifier and hot-gas cleaning, offers an extraordinary opportunity to handle high-sulfur, high-metal residuums. Squires [11] estimated that gas turbines, which can support temperatures of up to 1330° C, will be on the market by 1978. Based on this operating temperature, he estimated an efficiency of 41% for a combined-cycle facility. Furthermore, by employing hot-gas cleaning, he claims that efficiencies of up to 45.6% could be achieved in a combined-cycle generation plant. With either cold- or hot-gas cleaning, the sulfur and vanadium would be byproducts and there would be no air pollution. In the partial combustion gasifier, no corrosive vanadium compounds would be expected to occur.

One of the salient characteristics of gasification processes is that they operate at a pressure of some 20 atm. This high pressure reduces proportionally, by a factor of twenty, the size of the gas cleaning equipment when compared with equipment employed in a conventional thermoelectric plant operating at 1 atm. Whereas the cost of stack desulfurization and fly-ash removal in thermoelectric plants is very high, it would be quite low in a combined-cycle plant. High pressures, of course, are necessary if one wants to take advantage of gas turbine.

One novel application of vanadium in flame technology is as follows: To achieve high-efficiency electrostatic precipitation in a coal-fired thermoelectric plant, the fly ash should have a very low resistivity, corresponding to a conductivity on the order of 10^{-10} Ω^{-1} cm^{-1}. One of the dominating factors which influences the resistivity of a fly ash is the sulfur trioxide content of the flue gas. The higher the trioxide concentration, the lower is the resistivity of the ash. The probable cause of this relation is the chemisorption of sulfur trioxide and water vapor on the fly ash with the subsequent formation of highly conductive salts. At the operation temperature of electrostatic precipitators, the theoretical equilibrium conversion of sulfur dioxide to sulfur trioxide is above 95%. Because of the high rate of heat transfer in a thermoelectric plant and the bovinity of the homogeneous oxidation of sulfur dioxide, however, this reaction is effectively frozen at the lower conversion level of several percent. Nonetheless, this several-percent conversion is sufficient to impart a low resistivity to the fly ash, providing that the coal being burned contains a high (e.g., 3%) sulfur content. When low-sulfur coals are burned, however, not enough sulfur trioxide is formed and the electrostatic precipitator's efficiency drops. To overcome this problem of low-sulfur-trioxide formation when burning low-sulfur coals, some thermoelectric plants have adopted the unusual procedure of injecting sulfuric acid directly into the combustion chamber. As an alternative solution to this problem, it has been proposed to use the Venezuelan residuum as a supplementary fuel, with the idea that the vanadium fly ash would catalyze the oxidation of sulfur dioxide to sulfur trioxide [23]. Thus, in combustion, as well as in the refining processes, advantage could be taken of the in-situ formation of a vanadium catalyst.

A novel stack-desulfurization process for the Venezuelan residuum-fired thermoelectric plants has been proposed [7,34], which takes advantage of this in-situ formation of an oxidation catalyst. This process, which is called VANOX, is similar to the Monsanto's CATOX process.in that it removes the sulfur from the flue gas by first converting all of the sulfur to sulfur trioxide and then absorbing it in a wet scrubber.

In the VANOX process, the sulfur trioxide is absorbed, along with the vanadium fly ash, in a floating-bed wet scrubber, and the sulfuric acid—vanadium fly-ash slurry is sent to an adjacent hydrometallurgical plant. As discussed earlier in the economics section of this chapter, the recovery of vanadium would not significantly improve the profit margin of a thermoelectric plant. Stack desulfurization and demetallization, however, would permit the direct combustion of the Venezuelan residuum without contamination of the environment.

Metallurgy

Whereas vanadium compounds find many uses such as catalysts, semiconductors, superconductors, metalloceramic materials, pigments, and protective coatings [19], these diverse applications represent only 10% of the world's vanadium consumption. The other 90% is consumed by the metallurgical industry. Vanadium is a powerful and versatile alloying agent for high-strength, low-alloy steels (HSLA) and other special steels. Also, vanadium-base alloys have been considered for nuclear-reactor construction [21]. Most of the consumption is for the HSLA steels, where the vanadium concentration in the finished product is in the range of 0.01—0.1%. In addition to improving strength characteristics, vanadium also imparts better temperature, welding, working, and surface characteristics to steels [20]. Vanadium also finds applications in nonferrous metallurgy; the principal use here is in the production of titanium—aluminum—vanadium alloys for the aerospace industry.

Useful and versatile as it is, the world consumption of vanadium is very small, i.e., less than 10^5 tons/year. World reserves of vanadium in economic vanadium-mineral deposits, on the other hand, are on the order of 10^8 tons, making vanadium, at its present rate of consumption, practically an inexhaustible resource [3]. Vanadium is a very fugacious * element and this is why it is not widely used in the industry. Rarely does a vanadium mineral assay more than 2% of V_2O_5; hence, normally vanadium is a by-product of a principal process which is recovering the metals in the other 98% of the mineral. Since the beginning of the century, the metallurgical industry has been served by the cornucopia of vanadium sources shown in Table 9-II. This cornucopia of sources generated numerous processes to produce vanadium oxides, vanadium metal, and other vanadium products. This wide variety of raw materials and processes, however, was not conducive to the creation of a stable vanadium industry on a worldwide scale.

Many of the sources presented in Table 9-II have been closed down or mined out, whereas others are of minor importance. In fact, present non-Soviet vanadium production is dominated by two countries: the U.S.A. and South Africa. These two countries have contrasting technologies and political situations.

In the U.S.A., where the vanadium is usually found as a low-assay mineral or already dissolved in the mother liquor of another hydrometallurgical process, the logical tendency has been to first produce the intermediate vanadium pentoxide and then to reduce it to a ferrovanadium, vanadium carbide, or vanadium metal. The principal proponent of this technology is the Union Carbide Company; one of their main sources of V_2O_5 has been the residue from the uranium-production operations.

In South Africa, iron slag which assays 10—30% vanadium is the main source of this metal. Whereas this slag can also be processed in hydrometallurgical circuits, it is more advantageous to convert it directly to ferrovanadium using pyrometallurgical techniques. In this way, the Foote Mineral Company, a U.S.A.-based firm, competes with domestic vanadium producers, even though this company imports its raw materials from South Africa.

Ashes produced from the Venezuelan petroleums have vanadium assays in the same range as the South African iron slags. The rejected catalysts from future refinery operations are expected to have also around the same assay. In addition, the major vanadium mineral in Venezuelan deposits, which were discussed earlier in the geology section of this paper, is similar to that in South Africa. Consequently, by using similar technology, the slags produced from the processing of this mineral would be almost identical to those of South Africa. Thus, the writer concludes that pyrometallurgical techniques,

* Does not tend to concentrate as ores; found in a dispersed form in sediments.

TABLE 9-II

Brief history of the world's vanadium production

Period	Location	Mineral	Principal product	Byproduct
1900—1950	Peru	patronite (asphalt)	vanadium	
1915—1922	Colorado and Utah, U.S.A.	carnotite	radium	vanadium
1922—1931	Colorado, U.S.A.		vanadium	
1925—present	S.W. Africa	descloizite	lead	vanadium
1932—1948	North Rhodesia	descloizite	lead	vanadium
1935—1940	Colorado, U.S.A.	carnotite	vanadium	
1939—1944	Germany	magnetite	iron	vanadium
1941—(?)	New York, U.S.A.	magnetite	titanium	vanadium
1950—present	Colorado, U.S.A.	carnotite	uranium	vanadium
1950—present	U.S.S.R.	iron ore	iron	vanadium
1953—present	Finland	magnetite	iron	vanadium
1958—present	South Africa	magnetite	vanadium and iron	vanadium
1941—present	Idaho	?	phosphate	vanadium
1960—1965 (?)	Argentina	?	vanadium (?)	
1960—1962	Zambia	?	vanadium (?)	
1960—1961	Sweden	ilmenite	titanium	vanadium
1965—1970	Canada	petroleum (Venezuela)	refined oil	vanadium
1964—1972	Canada	bauxite	aluminum	vanadium
1966—present	Norway	iron ore	iron	vanadium
1968—present	Arkansas, U.S.A.	secondary deposition	vanadium	
1969—(?)	Chile	iron ore	iron	vanadium
1969(?)—present	France	bauxite (origin)	aluminum	vanadium

and not hydrometallurgical processes, are best suited for the Venezuelan vanadium industry.

At the Universidad Metropolitana in Venezuela, by applying the aluminum thermite reaction on calcined fly ashes, Arturo Medina produced a ferrovanadium—nickel alloy, which was baptized FERVANIKA [38]. Also, by adjusting the composition of the charge, ferronickel alone could be produced. By refiring the slag from the first charge, ferrovanadium is recovered [39]. Some iron was present in the fly ash used and adjustments in the iron content of the alloy were made by supplementary additions of calcined iron ore to the charge.

Another attractive pyrometallurgical technique is that which could be used for the recovery of vanadium from a petroleum coke [13]. If a petroleum coke rich in vanadium is used for reducing iron ore, then the vanadium should concentrate in the slag, just as it does in the South African operations. In the latter case, however, the vanadium is in the iron ore and not the coke. The resulting slag could be reduced with aluminum scrap to produce vanadium metal. If the petroleum coke has a high sulfur/vanadium ratio, this process may not be economical because of the large consumption of limestone. Flexicoke may be the most promising raw material for this process.

Union Carbide Company, the world's largest vanadium producer, is not worried about an undersupply, but rather about an oversupply situation in the near future because of the large number of projected new vanadium-recovery operations [27]. It is important to consider the great potential of vanadium production from the Venezuelan petroleum [22]. The following facts should be considered in evaluating the impact of this production:

(1) The amount of vanadium that can be recovered is comparable to the present-day, non-Soviet, world production.

(2) Vanadium could be in the form of high-assay material, fly ash and used catalysts, to which pyrometallurgical processes can be applied advantageously.

(3) The production from fly ash and used catalysts will be a totally involuntary operation independent of the vanadium market situation.

(4) Electric energy is cheap in Venezuela and almost the entire Venezuelan metallurgical industry (iron ore and alumina reduction, ferroalloys, and foundries) is based on electric-oven processes. Consequently, equipment and technology for a pyrometallurgical vanadium industry already exist in Venezuela.

(5) The large vanadium-mineral deposits recently discovered in Venezuela, which probably acted as the source rock for the vanadium now found in the Venezuelan fossil hydrocarbons, could eventually constitute an additional future source of this metal.

Several years ago, due to the massive desulfurization of fossil hydrocarbons, an oversupply of sulfur developed which had a short-range disastrous effect on the voluntary sulfur producers. The oversupply situation, however, did stimulate many research programs for new uses of this versatile element, especially in the construction industry. Sulfur prices have since recovered and currently are much higher than those before the sulfur crisis of 1970—1972. The new uses for sulfur that have been developed now present a viable competition to the sulfuric acid plants in using the available sulfur. Historically, 90% of the world's sulfur production was consumed by sulfuric acid plants and this new competitive factor will stabilize sulfur prices. The writer believes that an analogous scenario can be painted for the vanadium industry upon initiation of vanadium recovery from the Venezuelan petroleum on a large scale.

The national steel company of Venezuela, Siderurgica del Orinoco (SIDOR), consumes some 100 tons of vanadium per year and is interested in increasing its consumption of this metal if there is a concomitant increase of vanadium recovered from petroleum [20]. Since the beginning, this company has been a 100% government-owned enterprise and the iron mines were nationalized one year before the petroleum nationalization. SIDOR has been a pioneer in iron and steel technology and has one of the largest electric-oven shops in the world. Whereas its initial effort in direct reduction using the coal-based Strategic-Udy process failed, a new installation using natural gas seems to be performing better, and others are in construction. The foundry industry is booming, and new ferroalloy and aluminum plants will be inaugurated soon.

Acknowledgment

The preparation of this manuscript was partially funded by a cooperative program of CORDIPLAN, CONICIT, and the OAS. The help extended by Drs. T.F. Yen and George V. Chilingarian is also greatly appreciated.

References

1 G. Kapo and V. Lopez, "Correlations between the Hetero-Atoms in Venezuelan Fossil Hydrocarbons", in: G. Kapo (Editor), *International Symposium on Vanadium and Other Metals in Petroleum*, Aug. 19—22, Universidad del Zulia, Maracaibo, Venezuela, Vol. II-J (1973).

2 A. Silva, "A Fiscal Policy for Vanadium"; A. Materan, "The Vanadium Problem and Technological Dependence of Venezuela", in: G. Kapo (Editor), *International Symposium on Vanadium and Other Metals in Petroleum*, Aug. 19—22, Universidad del Zulia, Maracaibo, Venezuela, Vol. I-B, I-C (1973).

3 Y. Jacquin, A. Deschamps, J.F. Le Page and A. Billon, "I.F.P. Stack Desulfurization Technology", in: G. Kapo (Editor), *International Symposium on Vanadium and Other Metals in Petroleum*, Aug. 19—22, Universidad del Zulia, Maracaibo, Venezuela, Vol. V-A (1973).

4 E. Edler, "Geochemistry of Inorganic Trace Elements in Petroleum", in: G. Kapo (Editor), *International Symposium on Vanadium and Other Metals in Petroleum,* Aug. 19—22, Universidad del Zulia, Maracaibo, Venezuela, Vol. II-I (1973).
5 Y. Jacquin, A. Deschamps, J.F. Le Page and A. Billon, "I.F.P. Stack Desulfurization Technology", in: G. Kapo (Editor), *International Symposium on Vanadium and Other Metals in Petroleum,* Aug. 19—22, Universidad del Zulia, Maracaibo, Venezuela, Vol. V-A (1973).
6 M. Minicozzi, "The Economics of Vanadium Recovery in a Thermoelectric Plant Equipped with a Stack Desulfurization Unit", in: G. Kapo (Editor), *International Symposium on Vanadium and Other Metals in Petroleum,* Aug. 19—22, Universidad del Zulia, Maracaibo, Venezuela, Vol. V-B (1973).
7 G. Kapo, K. Goméz, F. Peña, E. Torres, J. Bilbao and K. Mazeika, "The Vanox Process for Stack Desulfurization and Vanadium Fly Ash Recovery", in: G. Kapo (Editor), *International Symposium on Vanadium and Other Metals in Petroleum,* Aug. 19—22, Universidad del Zulia, Maracaibo, Venezuela, Vol. V-D (1973).
8 D. Rodriquez, "Flame Chemistry and Corrosive Properties of Vanadium Fly-Ashes", in: G. Kapo (Editor), *International Symposium on Vanadium and Other Metals in Petroleum,* Aug. 19—22, Universidad del Zulia, Maracaibo, Venezuela, Vol. V-E (1973).
9 J. O'Neal, "Vanadium Recovery from the Combustion of Venezuelan Fuel Oil", in: G. Kapo (Editor), *International Symposium on Vanadium and Other Metals in Petroleum,* Aug. 19—22, Universidad del Zulia, Maracaibo, Venezuela, Vol. V-C (1973).
10 P. Eisenklam, "Avoidance of High Temperature Vanadium Corrosion in Two Stage Spray Combustion Systems", in: G. Kapo (Editor), *International Symposium on Vanadium and Other Metals in Petroleum,* Aug. 19—22, Universidad del Zulia, Maracaibo, Venezuela, Vol. V-F (1973).
11 A. Squires, "Vanadium Recovery in the Production of Low-Btu gas by Partial Oxidation of Venezuelan Residuum", in: G. Kapo (Editor), *International Symposium of Vanadium and Other Metals in Petroleum,* Aug. 19—22, Universidad del Zulia, Maracaibo, Venezuela, Vol. V-H, 14 pp. (1973).
12 A.J. González, "Metal Distribution in a Crude Distillation Tower", in: G. Kapo (Editor), *International Symposium on Vanadium and Other Metals in Petroleum,* Aug. 19—22, Universidad del Zulia, Maracaibo, Venezuela, Vol. III-A (1973).
13 E. Edler, "Production of Vanadium from Blast Furnace Slag", in: G. Kapo (Editor), *International Symposium on Vanadium and Other Metals in Petroleum,* Aug. 19—22, Universidad del Zulia, Maracaibo, Venezuela, Vol. VI-D (1973).
14 R.B. Ulrich and W. Roman, *Analysis de Metales en Petroleo,* Trabajo Especial de Grado, Universidad del Zulia, Maracaibo, Venezuela (Oct., 1974).
15 Z. Salcedo and C. Amez, Metales Traces en el Carbon de Taguay, Trabajo Especial de Grado, Universidad Metropolitana, Caracas, Venezuela (Oct., 1976).
16 T. Martinez, W. Arvelo, A. Mata, G. Sepúlveda, L. Katan, P. Roa, N. Rosa-Brussin and G. Kapo, "Simultaneous Demetallization and Hydrodesulfurization with Minerals of High Vanadium Crudes", in: G. Kapo (Editor), *International Symposium on Vanadium and Other Metals in Petroleum,* Aug. 19—22, Universidad del Zulia, Maracaibo, Venezuela, Vol. III-G (1973).
17 Y. Jacquin, A. Deschamps, J.F. Le Page and A. Billon, "Problems of Pollution by Sulfur from Crude Oil in European Countries"; A.M. Squires, "Vanadium Recovery and the Production of Methane and Light Aromatics by Hydrogenation of Venezuelan Residuum in High-Velocity Fluidized Beds", in: G. Kapo (Editor) *International Symposium on Vanadium and Other Metals in Petroleum,* Aug. 19—22, Universidad del Zulia, Maracaibo, Venezuela, Vol. IV-E, IV-F (1973).

18 E. Edler, "Techniques for Removal of Metals from Petroleum", in: G. Kapo (Editor), *International Symposium on Vanadium and Other Metals in Petroleum*, Aug. 19—22, Universidad del Zulia, Maracaibo, Venezuela, Vol. III-C (1973).
19 S.M. Aharoni, "Vanadium Compounds as Stereospecific Catalysts in the Polymerization of Olefins and Dienes"; J.H. Ott, "Applications of the Superconductivity of Vanadium Materials"; S.M. Aharoni and J.H. Ott, "Novel Chemical and Physical Uses of Vanadium Oxides", in: G. Kapo (Editor), *International Symposium on Vanadium and Other Metals in Petroleum*, Aug. 19—22, Universidad del Zulia, Maracaibo, Venezuela, Vol. II-K, L, M (1973).
20 A. Mago and O. Gago, "Utilization of Vanadium in the Venezuelan Steel Industry", in: G. Kapo (Editor), *International Symposium on Vanadium and Other Metals in Petroleum*, Aug. 19—22, Universidad del Zulia, Maracaibo, Venezuela, Vol. VI-G (1973).
21 G.A. Whitlow, "Vanadium Alloys for Fuel Cladding in Nuclear Reactors", in: G. Kapo (Editor), *International Symposium on Vanadium and Other Metals in Petroleum*, Aug. 19—22, Universidad del Zulia, Maracaibo, Venezuela, Vol. VI-F (1973).
22 G.P. Castillo, "Venezuela and the World Vanadium Industry", in: G. Kapo (Editor), *International Symposium on Vanadium and Other Metals in Petroleum*, Aug. 19—22, Universidad del Zulia, Maracaibo, Venezuela, Vol. VI-I (1973).
23 R. Coughlin and G. Kapo, "The Utilization of Venezuelan Residual Fuel Oil to Increase the Efficiency of Electrostatic Precipitators in Coal Fired Thermoelectric Plants", in: G. Kapo (Editor), *International Symposium on Vanadium and Other Metals in Petroleum*, Aug. 19—22, Universidad del Zulia, Maracaibo, Venezuela, Vol. V-G (1973).
24 R.H. Filby, "The Nature of Metals in Petroleum", in: T.F. Yen (Editor), *The Role of Trace Metals in Petroleum*, Ann Arbor Science, Ann Arbor, Mich., ch. 2, pp. 31—58 (1975).
25 I.C. Smith, T.L. Ferguson and B.L. Carson, "Metals in New and Used Petroleum Products and By-Products — Quantities and Consequences", in: T.F. Yen (Editor), *The Role of Trace Metals in Petroleum*, Ann Arbor Science, Ann Arbor, Mich., Ch. 7, pp. 123—148 (1975).
26 W.K.T. Gleim, J.G. Gatsis and C.J. Perry, "The Occurrence of Molybdenum in Petroleum", in: T.F. Yen (Editor), *The Role of Trace Metals in Petroleum*, Ann Arbor Science, Ann Arbor, Mich., Ch. 9, pp. 161—166 (1975).
27 F.J. Shortsleeve, "Free World Supply and Demand for Vanadium from 1973 through 1980", in: T.F. Yen (Editor), *The Role of Trace Metals in Petroleum*, Ann Arbor Science, Ann Arbor, Mich., Ch. 14, pp. 207—214 (1975).
28 R.H. Filby and K.R. Shah, "Neutron Activation Methods for Trace Elements in Crude Oils", in: T.F. Yen (Editor), *The Role of Trace Metals in Petroleum*, Ann Arbor Science, Ann Arbor, Mich., ch. 5, pp. 89—110 (1975).
29 Primeras Jornadas Venezolanas de Refinación, Centro de Ingenieros, Punto Fijo, Venezuela, Tema 5, Chapters I and II (Nov., 1974).
30 Primeras Jornadas Venezolanas de Refinación, Centro de Ingenieros, Punto Fijo, Venezuela, Tema 5, Ch. IV.
31 Primeras Jornadas Venezolanas de Refinación, Centro de Ingenieros, Punto Fijo, Venezuela, Tema 5, Ch. V.
32 Primeras Jornadas Venezolanas de Refinación, Centro de Ingenieros, Punto Fijo, Venezuela, Tema 5, Ch. VI.
33 Primeras Jornadas Venezolanas de Refinación, Centro de Ingenieros, Punto Fijo, Venezuela, Tema 5, Ch. VII.
34 G. Kapo, "Stack Desulfurization and Vanadium Pentoxide Fly Ash Recovery from Flue Gas Produced from Venezuelan Residuums", *AIChE Symp. Ser., 68(126)*, 147—154 (1972).

35 J. O'Neal, "Vanadium Recovery from Combustion of Venezuelan Fuel Oil", in: *National Ash Association Symposium on Ash Utilization*, St. Louis, Mo. (March, 1976).

36 P.B. Weisz and A.J. Silvestri, "Demetallization of Hydrogen Charge Stocks with Manganese Nodule Catalyst", *U.S. Patent 3,813,331* (May 28, 1974).

37 D. Rodriguez Polanco, *Spray Combustion of Residual Fuel Oil: Particulate and Gaseous Emissions in a Two-Stage System*, Thesis, Imperial College of Science and Technology, London (Nov., 1975).

38 A. Medina, *Obtención de Aleación de Ferrovanadio a partir de Mineral Ferrico y de Cenizas Provenientes de la Combustion de Petroleo Venezolano*, Trabajo Especial de Grado, Universidad Metropolitana, Caracas (Oct., 1975).

39 J. Gutierrez and H. Casado, *Estudio de las Variables en la Producción de Ferrovanadio*, Trabajo Especial de Grado, Universidad Metropolitana, Caracas (Dec., 1976).

40 L.M. Banks, "Oil-Coal Association in Central Anzoátegui, Venezuela", *Bul. Am. Assoc. Pet. Geol.*, *43*, p. 1998 (Aug., 1959).

41 A.G. Reyes, *Rev. Soc. Venez. Ing. Pet.*, *10*, 27—39 (Dec., 1972).

42 M.K. Salsamendi, *Informe de Análisis de Carbón para la Corporación Venezolana de Fomento*, Universidad Metropolitana, Caracas (1976).

43 G. Kapo and E. Pérez, "Characterísticas de Aguas de Formación en Venzuela, III", *Jornadas Pet. Soc. Venez. Ing. Petroleo*, pp. 41—50 (Oct. 14—16, 1971).

Chapter 10

ROLE OF ASPHALTENE IN REFINING

W.K.T. GLEIM

Introduction

Liquid hydrocarbons of relatively low volatility, but which are still distillable (i.e., gasoline, jet fuel, diesel oil, etc.), represent the most concentrated, most convenient, and most versatile form of stored energy. They are shipped and stored easily, and can be kept for prolonged periods in even the most primitive container. These properties set them apart from other sources of stored energy such as liquefied gases, gases under pressure and electric batteries.

Liquid hydrocarbons are derived from crude petroleum. The first step in refining crude petroleum, after the removal of salt and water, consists of separating the volatile hydrocarbons from the nonvolatiles by distillation. To obtain the maximum amount of distillate, the higher-boiling hydrocarbons must undergo vacuum distillation to keep the temperature below 700° F, because the nonvolatile residues of the vacuum bottoms contain heat-sensitive materials which would crack to form coke, gas, and liquid if heated above 700° F. Uncracked vacuum bottoms can be utilized as a source of residual fuel, in asphalt production, and as raw material for coke production.

Coke formation by the vacuum bottoms would have an undesirable effect on the still pot. When heated above 700° F, the distilling crude would gradually deposit an ever-increasing layer of coke on the walls of the distillation vessel. Inasmuch as coke is an insulator, the increasing layers of coke would demand an increasing heat input, leading to even higher degrees of overheating and, consequently, faster coke formation. Thus, the still pot would be filled with coke in a short period of time.

The materials responsible for the effect described above are known as asphaltenes. Asphaltenes, the structure of which has been elucidated primarily by Yen [1], are a heterogeneous colloidal mixture of heterocyclic "sandwiches" with hydrocarbon side chains. Asphaltenes are soluble in benzene but insoluble in light hydrocarbons. Most organic-metal complexes occurring in crude petroleum are either enveloped or buried in asphaltenes.

Role of asphaltenes in refining

Until recently, no commercial attempts were made to upgrade asphaltenes to distillate for the following reasons:

(1) The presence of small amounts of asphaltenes would make subsequent refining operations difficult and hardly commensurate with the small increase in yield.

(2) There is a ready market for asphalt and residual fuel oil extracted from vacuum bottoms kept below 700° F.

Now the continuous increase in the demand for transportation fuel has revived attempts to upgrade asphaltenes to distillate.

Additional transportation fuels can be obtained by hydrocracking and hydrogenating residual fuels and coal. The addition of H_2 is a major cost factor in the upgrading of residual fuels. Transportation fuels contain between 12 and 13.5% H_2; vacuum bottoms, 10% H_2; and coal, 5% H_2. Thus, it would be advisable to first upgrade the vacuum bottoms because they need only 2—3.5% H_2. Yet, if all of the crude is upgraded, none would be left for H_2 production and for generation of power necessary to operate the refinery.

Because of the low overall H_2 content of 5%, only those portions of coal should be upgraded which are most readily liquefied and hydrogenated. The remainder of coal would be used (1) to produce H_2 for the upgrading of oil and liquefied coal; (2) for generation of power necessary to operate the mining of coal; and (3) for the refining of coal and oil.

At first, major efforts have been directed toward the conversion of crude oil to distillate. This was brought about by the continuous increase in the consumption of transportation fuels which, in turn, led to a higher production of vacuum bottoms. In the U.S.A., this increase in consumption was readily disposed of by the electric utilities, which doubled their power output every ten years.

When the demand for crude petroleum exceeded domestic production, importation of high-asphaltene, high-sulfur crude was initiated to meet this increased demand. The increased burning of these high-sulfur fuels increased the SO_2 concentration in the stack gases to such a degree that the health of the population in densely settled industrial areas was endangered. Consequently, efforts have been made to remove, or at least decrease, the sulfur in residual fuels.

Several desulfurization plants are now operating commercially, but only with great difficulties. The processes being currently used are only modifications of those basically designed for the desulfurization of heat-stable distillate charges, namely, gasoline, kerosene, gas oils, and vacuum gas oils. The application of these processes to heat-labile, nondistillable liquids, an operation for which they were never designed, results only in insufficient desulfurization.

These desulfurization processes involve treating the charge with H_2 under pressure in a fixed catalyst bed, which consists of alumina, having large surface area, or silica—alumina impregnated with cobalt or nickel oxides and molybdenum trioxide. These catalysts are activated by sulfiding. Inasmuch as all operations must occur at or above the temperature at which the asphaltenes begin to coke, a high H_2 pressure must be applied to repress the coking. The process is further complicated because the next two reactions, i.e., adsorption of the charge on the catalyst and the subsequent hydrogenation, are exothermic.

As a result of catalyzing the rate of hydrogenation, the rate of heat production increases. The greater the heat production per unit time, the more difficult it is to dissipate this heat fast enough to prevent local overheating and subsequent coke production. The use of a more active catalyst would only further aggravate this problem.

In the so-called tricklebed reactor, the charge is blown downwards (downflow) with H_2 gas as a froth over a fixed-bed catalyst. The purpose of this procedure is to spread the charge as a thin film on the large surface of the catalyst. This facilitates the access of both feedstock and H_2 to the active catalyst sites. The thinner the oil film, the faster is the diffusion of H_2 through the film to the active catalyst sites.

Theoretically, in the case of very active catalysts, the H_2 diffusion rate to the catalyst site controls the rate of hydrogenation. Use of a more active catalyst would not increase the hydrogenation rate. This consideration holds for the processing of distillate, i.e., gas oils and vacuum gas oils, where the heat of adsorption and hydrogenation of the charge is dissipated by the volatilization of the product. Very little catalyst deactivation occurs from carbon laydown caused by the inability of this product to dissipate the heat of adsorption and reaction.

If this were not the case, catalyst-site temperatures would rise due to the inability of the bulk of the catalyst skeleton to dissipate these local temperature rises fast enough, and carbonization of some of the product would occur. The resulting carbon formation would deactivate the catalyst, the support of which is not a good conductor of heat because it consists of alumina or silica—alumina.

When the charge to the hydrogenating desulfurization reactor is a non-volatile petroleum residue, i.e., a vacuum bottom, then the above-described phenomenon occurs. Some of the heat of adsorption and hydrogenation is used up by carbon—carbon bond breaking of the microcrystalline-wax components of the vacuum bottoms and by subsequent volatilization of the resulting smaller molecules.

The other components of the vacuum bottoms, the asphaltenes, carbonize the catalyst in two ways. In the first place, they are heat labile at the process temperature as mentioned in the beginning of this chapter and, secondly, they are large molecules which are adsorbed onto more than one catalyst site. This multiple adsorption restricts the thermal movement of the asphaltenes and prevents the dissipation of the heat of adsorption and the subsequent hydrogenation.

Local temperature increases will cause a shift to the right in the following equilibrium reaction:

Naphthenes + hydroaromatics ⇌ aromatics + free radicals + olefins + H_2

This reaction is completed in a very short period of time due to the presence of the catalyst. The mixture shown on the right-hand side polymerizes immediately and forms coke. This can be observed by monitoring the catalyst-bed temperature in a small continuous plant with a fresh catalyst. In the first five to ten hours, the bed temperature will rise 20°F above the applied temperature, after which it will decline to normal after a period of additional eight hours. In this short period of 18 hours, 12—18% coke will form on the catalyst causing the catalyst's performance to be reduced to a mediocre level. Hence, the activity level of the catalyst and not the rate of H_2 diffusion is the limiting factor. Furthermore, both surface activity and hydrogenating activity are greatly decreased, meaning that the rate of coke formation also is greatly diminished in conjunction with the degree of desulfurization.

If the flow in the reactor is reversed to an upflow with a concurrent H_2 flow, i.e., liquid-continuous, gas-discontinuous system, there should be poor H_2 diffusion, but better heat transfer than in the tricklebed due to the immersion of catalyst in the liquid charge. The temperature in the catalyst bed was not monitored during the first 24 hours under upflow conditions. The same degree of desulfurization and hydrogenation, however, resulted whether running upflow or downflow, provided it was done in the same plant under the same conditions with the same catalyst.

In general, the more active the catalyst, the more coke is formed. Inasmuch as most of the reaction occurs inside the pores of the catalyst, it makes little difference whether the flow of charge is in an upward or downward direction or whether the oil molecules enter the pores from a thin film or from a liquid bulk.

In addition to the coke, metals present in the asphaltenes will also fill up the pores of supported catalysts within a few weeks. In other words, use of active supported catalysts with a large surface area and porosity is counterproductive in the cleanup of asphaltene-containing residual fuels. High catalyst activity in a complex organic mixture implies indiscriminate non-selective activity. For the process being discussed, the opposite is needed, i.e., a very selective catalyst. Also, to minimize coking, the first reaction step should consist only of heat stabilization and depolymerization of the asphaltenes to oil-soluble materials.

To achieve this, an unsupported, mediocre catalyst suspended in the charge stock was employed, thus eliminating the heat of adsorption by the catalyst support. By operating the process in an upflow condition as a liquid-continuous, gas-discontinuous system, it was hoped to dissipate heat of reaction. In choosing colloidal MoS_2, which is not very active as a catalyst, it was expected that MoS_2 will be preferentially adsorbed on the asphaltenes. In general, colloidal molecules have a greater surface activity than molecules dissolved in the oil.

If these assumptions are correct, then a selective catalyst is brought together with its substrate. Preliminary experiments using autoclaves demonstrated the feasibility of such an approach.

There is another problem, however. If vanadium is considered as a catalyst poison, then MoS_2 would have to be continuously regenerated. An inorganic refinery besides an oil refinery, therefore, would be required.

If one considers the possibility that V is or can become catalytic, then possibly one can start out with V. The charge stock would continuously provide new V, while coked or otherwise deactivated V is withdrawn. Studies by the writer have shown that vanadium sulfides do indeed catalyze the conversion of asphaltenes to oil-soluble materials in the presence of H_2, but deactivate very quickly. This deactivation can be kept to a minimum by operating under a partial H_2S pressure. Another advantage is that on using vanadium, coke formation occurs at a much lower rate than in supported Mo or W catalysts.

Process for conversion of asphaltenes to oil-soluble materials

Using the above-described findings, Gatsis and Gleim [2] have developed a process which allows conversion of asphaltenes to oil-soluble materials while simultaneously removing metals. The process operates under pressure in an H_2—H_2S atmosphere, using the same materials as those being converted or removed (see Tables 10-I and 10-II). This process can be outlined as follows:

Step 1—catalyst formation: The H_2S converts the metals to sulfides. The resulting VS_x serves as a selective hydrogenation catalyst for the depolymerizing hydrogenation of asphaltenes and metal complexes.

Step 2—catalyst carrier: The asphaltenes act both as a catalyst carrier and substrate.

Step 3—operation: This process is carried out as a slurry operation where the catalyst sludge, i.e., the VS_x adsorbed on the asphaltenes, moves upward with the charge stock through an empty tube, which serves as the reactor at temperatures ranging from 750 to 800° F (399—426° C). The gas phase (H_2 +

TABLE 10-I

Process of converting asphaltenes to oil-soluble materials (After Gatsis and Gleim [2a, 2b] and Gleim et al. [2c])

Reaction conditions	
H_2 + 12—20 mol% H_2S	2000—3000 psi
LHSV *	0.5—2
Temperature	750—800° F (400—425°C)

*Liquid hourly space velocity.

TABLE 10-II

Process of converting asphaltenes to oil-soluble products, with removal of metals (After Gatsis and Gleim [2a,2b] and Gleim et al. [2c])

Gravity and composition	Charge	Product
Gravity (° API)	3—9	15—20
C_7 insolubles (%)	8—18	0.2—0.5
H(%)	10—10.5	11.3—11.9
S(%)	4.5—6	3—4.9
Me (ppm)	up to 1600	10

15% H_2S) also passes upwards through the liquid phase. This gives rise to a liquid-continuous, gas-discontinuous system.

Step 4—product separation: The reaction mixture is first degassed and then deasphalted by a standard deasphalting procedure, i.e., the product is separated from the unconverted asphaltenes and catalyst with C_4 or C_5 hydrocarbons.

Step 5—catalyst recycle: The resulting deasphalted bottoms are then recycled and mixed with the fresh charge stock.

Step 6—start-up: The process is initiated by adding about 10% VS_x to the charge.

Step 7—catalyst concentration: To keep the catalyst concentration constant, a dragstream is installed just after the deasphaltor.

Step 8—conversion per pass: Inasmuch as asphaltenes are necessary as catalyst carriers, total asphaltene conversion per pass must be avoided.

Step 9—product desulfurization: The process product, being asphaltene free, is readily desulfurized (sulfur content below 0.1%).

Step 10—charge stocks: Venezuelan vacuum bottoms, Boscan, de-ashed Athabasca tar-sand oil, and the Orinoco tar-sand oil have been treated successfully using the above process.

The UOP Company is commercializing this process under the name of Aurabon [3] and is currently building a demonstration plant together with the Venezuelan government.

As far as analytical work is concerned, the basic contribution to the improvement in refinery processing was made by Duffy [4]. His findings made it possible to fractionate C_7 insolubles according to polarity and size by a simple solvent extraction. On spreading C_7 insolubles on teflon, the extraction with successively more polar solvents will yield successively more polar extracts of higher molecular weights. Such a series of solvents would be n-C_6, cyclo-C_6, ether, MEK, benzene, and $CHCl_3$.

As found by this method, the asphaltenes are not what they were originally described to be, i.e., a bituminous material soluble in benzene and insoluble in heptane.

Actually the first 50% of extract are soluble in heptane and are of relatively low molecular weight (in the low thousands). Of the remainder, 25% are benzene soluble and 25% are benzene insoluble but $ChCl_3$ soluble. The molecular weight of the fraction soluble in chloroform is in the 20,000—40,000 range. The oxygen and nitrogen contents do not increase with increasing molecular weight, whereas the sulfur content does.

Duffy's method is simple and makes it easy to observe the effects of process conditions on any asphaltene fraction, as well as changes in heteroatom content and molecular weight. Heat stability can be measured through thermal analysis or, more accurately, with the differential scanning microcalorimeter. The microcalorimeter makes it possible to study coking in the pressure cell in the presence of H_2 at a pressure of 1000 psi with or without the presence of a catalyst. By connecting the cell to a mass spectrometer, one can analyze the volatiles produced under a given set of conditions and determine whether the reactions are exo- or endothermic.

A revived old method, "Photo-acoustic Spectroscopy", pioneered by Alan Rosencwaig of Bell Laboratories [5], holds great promise for asphaltene analysis. His method shows that ultraviolet and visible spectra can be directly recorded from solids such as leaves, blood smears, and TCL chromatograms. This method is thus ideal for asphaltene-fraction analysis.

As mentioned previously, if one converts all of the crude oil to distillate and gas, none will be left to provide the energy needed to operate the refinery. Hydrocarbons lighter than gasoline are too valuable as raw material for chemicals, and methane is too valuable as a household fuel to be utilized as a source of H_2 for the refinery. Consequently, one would have to resort to coal.

Coal can serve not only as a source of energy by combustion, but also as a source of liquid hydrocarbons. Coal refining would be greatly facilitated by the refining of coal and oil together. Time would be saved because an oil-soluble coal extract could enter refinery simultaneously with the crude-oil fractions. This change would require no new technology, but only an enlargement of the existing facilities. Additional personnel also would not be required.

Approximately 40—50% of the coal entering the refinery will be used to provide essential services for the refining of crude petroleum and coal. As an example, 100 g maf Illinois No. 6 coal contains: 81.0% C = 6.7 atoms; 5.2% H = 5.2 atoms; 10.9% O = 0.7 atoms; and 2.9% S = 0.09 atoms. For every atom of C, approximately 0.8 atoms of H are present.

In paraffins, $H_3C—(CH_2)_n—CH_3$, the H_2/C ratio is 2 : 1, because the terminal CH_3 groups do not affect this ratio very much. In reality, fuels which are mixtures of saturated paraffins and naphthenes like kerosene, do not reach this ratio of 2 : 1 due to the presence of small amounts of benzene rings and condensed naphthenes. A more realistic ratio for saturated fuel should be 1.8, corresponding to 13.03% H.

For example, in order to upgrade 4 of the 6.7 carbons in the 100 g maf coal by increasing the atomic H/C ratio from 0.8 to 1.8, an addition of 2 moles of H_2 to the 4 C atoms, producing 55.2 g of oil (4 C = 48 g and 4 × 1.8 H = 7.2 g) would be required. Approximately 0.5 mole of H_2 would also be necessary for both the removal of sulfur and some of the oxygen. At least another 0.5 mole of H_2 would be necessary for the operation of the oil refinery. Three moles of H_2 from the 100 g maf coal, therefore, must go towards producing the hydrocarbons and hydrotreating the coal and crude petroleum. Out of the 6.7 atoms of carbon, 4 are being hydrogenated to a hydrocarbon oil and 2.7 C are available to produce H_2 via the water—gas and the subsequent shift reaction.

The autothermic equation, coupling the maximum H_2 production with a minimum C consumption:

$$2.2\ C + 0.7\ O_2 + 3\ H_2O \rightarrow 3\ H_2 + 2.2\ CO_2$$

$$\begin{array}{ll} +207\ \text{kcal} & -204\ \text{kcal} \\ (2.2 \times 94) & (3 \times 68) \end{array}$$

demonstrates that an absolute minimum of 2.2 C atoms of the remaining 2.7 C atoms is necessary for H_2 production. Only 0.5 C, i.e., 6% of the maf coal, is available for the energy needed to run the oil refinery, which is probably barely sufficient.

The above crude approximation demonstrates that the production of 55 g oil from 100 g maf coal is the upper limit that one should expect from the liquefaction of coal, if coal has to provide all the energy for its own liquefaction, as well as for the refining of petroleum.

This 50—55% limit means that only about half of the coal has to be liquefied. The extraction becomes a simple process, therefore, and should result directly in a high-grade product, namely, heptane-soluble oil. This prevents all downstream refining difficulties caused by the presence of asphaltenes. By eliminating asphaltenes, coal extraction would be simplified to two steps: directly dissolving the oil-soluble, asphaltene-free portion and separating it from the insolubles by distillation.

Explicitly, the solvent should perform two functions: (1) dissolve 50—60% of the coal selectively, e.g., asphaltene free; and (2) thermohydrocrack using H_2 from the H_2 atmosphere to achieve a sufficiently low molecular weight of item (1) in order to permit its separation from the undissolved coal by distillation. Before proceeding with the discussion of the solvent, however, the conditions under which the solvent must operate must be established first, i.e., the temperature and H_2 pressure.

Bituminous coal acts like a giant potential asphaltene. The coal is heat labile and begins to soften at about 380° C. At 400° C it starts coking, which produces gas, volatile liquids, tar, and coke. If temperature is not permitted to exceed 400°C, up to 20% tar can be obtained. Coking can be repressed to some extent by the application of high H_2 pressure. Beyond 440°C, however, the thermodynamic equilibrium favors complete aromatization. Obviously, it would be advantageous to dissolve the coal below the coking temperature to prevent coke and asphaltene production.

Yet, the temperature should be sufficiently high to achieve a reasonable rate of decomposition. It was found by Pott and Broche [12] and later confirmed by Pier [6] that 400°C should not be exceeded. Increasing the temperature to 430°C only increases the asphaltene content of the extract. High H_2 pressure is also beneficial because it represses coking and stabilizes heat-labile materials. For practical operating reasons, however, the pressure should not exceed 200 atm.

Organic materials proposed as coal solvents include: (1) paraffins [7], (2) hydroaromatics, (3) aromatics, (4) coal-tar fractions, and (5) the coal extract itself.

The latter, a very economical solvent, was used by a subsidiary of the Gulf Oil Co., the Spencer Chemical Division, for the first commercial coal-liquefaction venture in the United States, i.e., the SRC (Solvent Refined Coal) process [8]. This process produced nearly a 100% coal extract having a melting point of 170°C and containing 40—50% benzene insolubles and 20—30% heptane insolubles. The API gravity of coal extract ranged from minus 20° to minus 30° and contained about 0.1% ash.

It has been suggested that this material should be upgraded to a distillate. This process, however, would not be economically feasible, because it would require about 3000 to 4000 psi H_2 atmosphere, with a slurry catalyst to convert this material just to an oil having API gravity of ±0°. Thus, the API gravity would be lower than that found in even some of the worse crude-oil vacuum bottoms. In addition, the catalyst would have to be continuously freed of ash by an expensive separation process. Consequently, the low-ash, high-Btu fuel oil derived from the SRC process would not make a suitable raw material for the production of distillate from coal.

Pott and Broche [12] proposed a coal solvent consisting of 80% tetralin and 20% coal-tar cresol which could dissolve up to 90—95% of coal. Tetralin, without the cresols, is a little more discriminating, and will dissolve only 80—85% coal. Such an extract will contain 25—35% heptane insolubles, but only a relatively small amount (~5%) of benzene insolubles.

Many investigators have shown that part of tetralin's ability to liquefy coal is due to the transfer of its H_2 to coal, forming naphthalene. Review of literature shows that no one has proved that tetralin can also transfer H_2 from the atmosphere to coal, i.e., carry out a catalyst-like reaction. This type of claim, however, would be very difficult to substantiate because even if the H_2 material balance did show an uptake of H_2, this uptake could always be attributed to the catalytic activity of the walls of the reactor and/or the catalytic activity of coal constituents. In addition, one would not expect a large effect, and statistical correlations drawn from very small differences in the analytical hydrogen data would raise doubts as to their validity.

There is another way of proving hydrogenation activity, namely, to demonstrate different degrees of hydrogenation with different solvents in the same reactor vessel with the same coal and under identical temperatures and H_2 pressure. In addition, a solvent should be employed which leaves no doubt as to its direct role in the promotion of hydrogenation. If such a solvent were found, one would expect it to be more selective, e.g., extract only a minimum of heptane insolubles and no benzene insolubles. In other words, one would expect that a greater hydrogenation activity would preferentially act on the asphaltenes, thereby decreasing their content in the extract.

Such a solvent was found in a partially hydrogenated refinery stream by Gleim and O'Hara [9] similar to the one used by Langer and co-workers [10] for the thermal hydrocracking of petroleum residues.

Vacuum bottoms were hydrocracked using a typical Ni—Mo on Si—Al_2O_3 desulfurization catalyst under rather severe conditions, i.e., 760° F and 3000 psi H_2 pressure. The product boiling below 840° F was removed by distillation. The results are shown in Table 10-III under Run 1. The oil obtained consisted of 50% hydroaromatics and 40% saturates, but, unfortunately, contained 10% polycyclic aromatics.

For comparison, the following solvents were used: (A) tetralin (Run 2), (B) catcracker slurry oil, which is practically all aromatic (Run 3); (C) hydrogenated catcracker slurry oil, producing hydroaromatics similar to tetralin (see Table 10-III) (Run 6); (D) raw, nonhydrogenated vacuum

TABLE 10-III

Thermal hydrocracking of petroleum residues (After Gleim and O'Hara [9a])

Product	Run*					
	1	2	3	4	5	6
Initial boiling point (° F)	840		545		475	
Gravity (°API)				8.2	18.8	11.2
C_7 insolubles (%)	1.8		1.3	4.7	1.1	
H(%)	11.6	9.1	7.2		12	9.1
P + N(%)	40					
Hydroaromatics(%)	50	100				
Condensed polycyclic aromatics(%)	10					

* *1* = Hydrocracked vacuum bottoms (V.B.'s); *2* = tetralin; *3* = catcracker slurry oil; *4* = V.B.'s; *5* = hydrocracked V.B.'s; and *6* = hydrogenated catcracker slurry oil.

bottoms (Run 4); unmodified hydrocracked vacuum bottoms (light part of the product was not removed by distillation as was the case in Run 1); this solvent had the highest H_2 content (12%), whereas the slurry oil had the lowest H_2 content (7.2%) (Run 5).

As shown above, even the hydrocracked vacuum bottoms still contain some C_7 insolubles; however, whether or not these figures are reliable is questionable. Very often, C_7 insolubles in hydrocracked vacuum bottoms are microcrystalline waxes soluble in warm heptane, but insoluble in C_7 at ambient temperatures [11], and not asphaltenes.

When 200 g Illinois No. 6 coal was extracted in autoclaves with 200 g solvent at 400°C under 200 atm H_2 pressure, the maf coal had the following composition: ash = 7.8%, H_2O = 7.2%, C = 81.0%, H = 5.2%, O = 10.9%, and S = 2.9%.

Overall, 170 g maf coal was extracted, and its composition is shown in Table 10-IV. The data confirms that the extraction of the hydrogenated vacuum bottoms is a selective process, as shown by Run 1 (48%) and Run 5 (56%). Tetralin (Run 2) and slurry oil (Run 3) yielded the same extraction result, namely, 83%. More total coal products, however, were extracted by the hydrocracked vacuum bottoms of Run 5 than by the heavier vacuum bottoms of Run 1. This result was totally unexpected, because it was originally thought that the lighter components would prevent, not enhance, solubility. In Run 4, only 10% liquid product were extracted using raw, unhydrogenated vacuum bottoms and, consequently, the products from this run were not investigated any further.

The percentages in the extract (Table 10-IV) were calculated from the percentages in the solvent and in the solution: % in extract = % in solution — % in solvent. In the case of H_2 contents, the H_2 in the unconverted coal solids was used to arrive at the H_2 material balance.

On examining the extract figures, one can note that the C_7 insolubles of Run 1 are disappointingly high (9.6%), whereas in Run 5 they were much lower (3.8%). But even the latter figure is still too high to achieve easy refining. The cause is most likely due to the presence of 10% polycyclic aromatics in the solvent of Run 1.

This effect is especially apparent in the case of Run 3, where the entire solvent, slurry oil, consists of polycyclic aromatic. The extract contained 43% C_7 insolubles and an additional 14% benzene insolubles, with 43% of the extract being soluble in C_7. Such a product could not be easily refined because it would have to be deasphalted, a very expensive process yielding only 43% C_7 solubles.

The hydroaromatics, tetralin, and hydrogenated slurry oil (Runs 2 and 6) extract a slightly lower amount of undesirable material: 34% C_7 insolubles and 6% benzene insolubles. For refining purposes, however, these hydroaromatics are not practical because half of the extract still consists of material which is not C_7 soluble.

TABLE 10-IV

Composition of extracted maf coal (After Gleim and O'Hara [9a])

	Run*					
	1	2	3	4	5	6
Liquid product						
Coal maf dissolved (%)	48	83	83	10	56	84
Gravity (°API)				15.8	18.6	4.3
C_7 insolubles of solution (%)	4	14	20	4	2	13
C_7 insolubles of extract (%)	9.6	34	43		3.8	
Benzene insolubles of solution (%)	0.01	2.5	5.7			
Benzene insolubles of extract (%)		6	14			
H_2 of solution (%)	11.5	8.4	7.1	10.6	11.4	8.2
H_2 of extract (%)	11.5	7.4	7.0		9.0	
Unconverted solids						
H (%)	4.5	2.8	1.7		3.8	
Moles H_2 added to 100 g maf coal	1.5	0.8	0.5		1.35	

* *1* = Hydrocracked vacuum bottoms (V.B.'s); *2* = tetralin; *3* = catcracker slurry oil; *4* = V.B.'s; *5* = hydrocracked V.B.'s; and *6* = hydrogenated catcracker slurry oil.

TABLE 10-V

Hydrocracking of vacuum bottoms (After Gleim and O'Hara [9b])

Hydrocracked VB's	Solvent		Product	
Run:	1	2	1	2
Initial boiling point (IBP) (°F)	850	475		
Gravity (°API)	20	18.8	20.1	20.5
C_7 insolubles (%)	1.0	1.1	1.1	1.2
P + N (%)	40	46		
Hydroaromatics (%)	50	49		
Condensed polycyclic aromatics (%)	10	4		
H in solids (%)			5.2	5.9
Extract of maf coal (%)			47.0	58.1
Benzene extract (%)			35.5	18.6

H_2 content and H_2 uptake

The values in the last line in Table 10-IV indicate the amounts of H_2 consumption for 100 g of maf coal. According to this table, the hydrogenated vacuum bottoms are not only the more selective solvents but are also much better H_2-transfer agents, i.e., catalysts. The absolute accuracy of these figures is questionable due to the increasing error in arriving at the smaller absolute H_2 values. Even though the values are not very accurate, however, they still show real differences, e.g., 1/2 or 1/3 as much. Whether there was more hydrogenation in Run 1 than in Run 5 is open to discussion. In other words, is a hydrocracked vacuum bottom minus its front end better than the total product as a H_2-transfer agent or not?

Experiments with two new solvents produced as before from domestic and mideast vacuum bottom [9] showed a greater selectivity (Table 10-V). The solvents contained 1% C_7 insolubles and the resulting coal solution had the same content of insolubles. Thus, the extract itself also contained 1% C_7 insolibles. Solvent 1 extracted only 47% of the coal, but this extract was totally distillable, demonstrating a simple solution to the problem of separating the coal extract from the insolubles. These solvents actually upgrade the undissolved solids.

The H_2 content of the maf coal is 5.2%. In evaluating the H_2 content of the residue, comparison tests must be made with the H_2 content of a moisture-free, but not ash-free coal, which is 4.8%. In the case of the hydrocracked vacuum bottom without front end as solvent (Run 1), Table 10-V shows 5.2% H_2 content in the undissolved solids. In the case of the total hydrocracked vacuum bottom as solvent (Run 2), the H_2 content is 5.9%, indicating an increase of 1.1%.

Furthermore, the fact that these residues were upgraded could be demonstrated by the additional material extracted from them by boiling with benzene at atmospheric pressure. For example, in Run 1, where 47% was extracted in the first stage, an additional 35.5% was extracted in the second stage. In Run 2, where the first stage already yielded 58%, the second stage yield was only 18.6%. In both cases, this benzene extract consisted of 50% C_7 insolubles and 50% solubles.

In a model refinery visualized by the writer for the conversion of crude, a unit such as the one previously described could be set up which would permit complete conversion of petroleum asphaltenes.

Repeated use of the first coal extract without removing the dissolved coal would increase the dissolved-coal content of the solvent and thereby increase its hydroaromaticity. Consequently, the solvent would dissolve more coal, thereby losing its selectivity, and thus the % C_7 insolubles would increase with the number of extractions.

Rehydrogenating the coal solution between extractions will not totally reconstitute the selectivity of the original coal solvent. The paraffins in the original solvent derived from petroleum contribute to the selectivity. These paraffins are diluted by the addition of dissolved coal. In order to maintain this selectivity, a continuous fresh stream of a new solvent from the petroleum refinery would be required. This is another reason for combining coal liquefaction with petroleum refining.

As shown in Table 10-V, 58% of the maf coal can be dissolved, i.e., 50% of the raw coal. Assuming that all of the refined crude boiling above 500° F could serve as an extract, approximately 70% of the total crude oil could be used as solvent using a coal/solvent ratio of 1 : 1.

How much additional distillate could such a scheme provide for the United States, for example? According to Table 10-VI [3], in 1975, $8.5 \cdot 10^6$ bbl/day were produced in the U.S.A.; in 5—6 years with the development of Prudhoe Bay deposits, production will be about $10 \cdot 10^6$ bbl/day. With $6.5 \cdot 10^6$ bbl/day ($\sim 1 \cdot 10^6$ tons/day), $1 \cdot 10^6$ tons/day of raw coal can

TABLE 10-VI

Oil industry four-week average of supply, demand, stocks, and drilling operations in the U.S.A. in 1975, and changes from one year ago (After O.G.J. Newsletter [3])

U.S. Industry Scoreboard 2/3	Latest week 1-24-75	Change from: Week before	Change from: Year ago	% change year ago
Demand (bbl/day) (4-week avg.)				
Motor gasoline	6,197,000	— 76,000	341,000	+ 5.8
Middle distillates	3,845,000	+ 147,000	— 445,000	—10.4
Jet fuel	1,014,000	— 28,000	+ 25,000	+ 2.5
Residual	3,343,000	+ 112,000	+ 330,000	+11.0
Other products, etc.	3,605,000	0	— 145,000	— 3.9
Total demand	18,004,000	+ 155,000	+ 106,000	+ 0.6
Supply (bbl/day) (4-week avg.)				
Crude production	8,595,000	— 32,000	— 556,000	— 6.1
Crude imports	3,768,000	+ 197,000	+ 1,385,000	+58.1
Product imports	2,878,000	— 35,000	+ 82,000	+ 2.9
Crude runs to stills	12,399,000	— 136,000	+ 761,000	+ 6.5
Stocks (bbl)				
Crude	270,555,000	+5,793,000	+39,040,000	+16.9
Motor gasoline	232,455,000	+3,448,000	+20,299,000	+ 9.6
Middle distillates	205,488,000	—5,741,000	+ 741,000	+ 0.4
Jet fuel	30,139,000	+ 348,000	+ 1,304,000	+ 4.5
Residual	56,739,000	—2,548,000	+ 9,458,000	+20.0
Drilling (4-week avg.)				
Rotary rigs	1,615	— 3	+ 243	+17.7

be extracted yielding 500,000 tons/day of coal extract, i.e., $3.25 \cdot 10^6$ bbl/day. At present, the U.S.A. uses approximately $11 \cdot 10^6$ bbl/day of distillate. In ten years, distillate requirements will probably be at least $13 \cdot 10^6$ bbl/day. The deficit could be balanced by $3.25 \cdot 10^6$ bbl/day oil taken from coal, in addition to the $10 \cdot 10^6$ bbl/day from petroleum production.

Thus the yearly requirement of coal will be $365 \cdot 10^6$ tons. Substituting coal for the $3.3 \cdot 10^6$ bbl of residual oils that the U.S.A. uses daily, would require another $0.5 \cdot 10^6$ tons/day, or $182 \cdot 10^6$ tons/year. A total of $550 \cdot 10^6$ tons of coal per year, therefore, will be needed to substitute for imported oil.

The present U.S.A. coal production is about $650 \cdot 10^6$ tons/year. Consequently, the U.S.A. coal industry would have to double its output in ten years, which is hard to achieve. Thus, the oil industry may have coal-refining installations ready long before the coal industry can supply sufficient coal.

The ideal conditions for the concept presented here exist in the Midwest (U.S.A.). Oil, coal, and refineries are situated close together, with sufficient H_2O available. One would phase-in this operation gradually, beginning with the conversion of oil, including asphaltenes, to distillate. For this, coal would serve as the refinery fuel and H_2 source. When this is achieved, the second step, namely, the partial extraction of coal could be undertaken, leaving the asphaltenes with the undissolved coal to serve as a source of energy to power the combined liquefaction and refining operations. Further details will appear in the next "Petroleum Science Series" book by T.F. Yen and G.V. Chilingarian (1979).

References

1 a. J.P. Dickie and T.F. Yen, "Macrostructures of the Asphaltic Fractions by Various Instrumental Methods", *Anal. Chem.*, *39*, 1847—1852 (1967).
b. J.P. Dickie, M.N. Haller and T.F. Yen, "Electron Microscopic Investigations of the Nature of Petroleum Asphaltics", *J. Colloid Interface Sci.*, *29*, 475—484 (1969).

2 a. J.G. Gatsis and W.K.T. Gleim, "Conversion of Asphaltene-Containing Hydrocarbonaceous Charge Stocks", *U.S. Patent 3,723,294* (March 27, 1973).
b. J.G. Gatsis and W.K.T. Gleim, "Conversion of Asphaltene-Containing Charge Stocks", *U.S. Patent 3,785,297* (March 27, 1973).
c. W.K.T. Gleim, J.G. Gatsis and M.J. O'Hara, "Desulfurization and Conversion of Black Oils", *U.S. Patent 3,785,958* (Jan. 15, 1974).

3 *O.G.J. Newsletter* (Feb. 3, 1975).

4 A.P.I. Research Project 60, Report No. 13, *Characterization of the Heavy Ends of Petroleum* (July 1, 1972—June 30, 1973).

5 Pittsburgh Anal. Conf., F. 1975, "Photoacoustic Effect with Solids: A Theoretical Treatment", *Science*, *190*, 556—557 (1975).

6 M. Pier, "Production of Hydrocarbons of High Boiling Point Range", *U.S. Patent 1,988,019* (Jan. 15, 1935).

7 W.G. Scharmann and F.T. Barr, "Method of Production of Motor Fuels", *U.S. Patent 2,436,938* (March 2, 1948).

8 Office of Coal Research, Dep. of the Interior, Washington, D.C., *Research and Development Report No. 9*, Contract No. 14-01-0001-275, Spencer Chemical Div., Gulf Oil Co., p. 254 (1963).

9 a. W.K.T. Gleim and M.J. O'Hara, "Coal Liquefaction Process", *U.S. Patent 3.849,287* (Nov. 19, 1974).
b. W.K.T. Gleim and M.J. O'Hara, "Coal Liquefaction Process", *U.S. Patent 3,867,275* (Feb. 18, 1975).

10 a. C.S. Carlson, A.W. Langer, J. Stewart and R.M. Hill, "Thermal Hydrogenation", *Ind. Eng. Chem., 50,* 1067—1070 (1958).
b. A.W. Langer, J.Stewart, C.E. Thompson, H.T. White and R.M. Hill, "Thermal Hydrogenation of Crude Residua", *Ind. Eng. Chem., 53,* 27—30 (1961).
c. A.W. Langer, J. Stewart, C.E. Thompson, H.T. White and R.M. Hill, "Hydrogen Donor Diluent Visbreaking of Residua", *Ind. Eng. Chem., Process Design and Development, 1,* 309—312 (1962).

11 J. Knotnerus and C.J. Krom, "The Composition of Wax From Bitumen", *Bitumen—Teere—Asphalte—Peche und verwandte Stoffe, 16,* 299—303 (1965).

12 A. Pott and H. Broche, "The Solution of Coal by Extraction under Pressure — The Hydrogenation of the Extract", *Fuel, 13,* 91—95, 125—128, 154—157 (1934).

Chapter 11

FORMATION EVALUATION OF TAR SANDS USING GEOPHYSICAL WELL-LOGGING TECHNIQUES

WALTER H. FERTL and GEORGE V. CHILINGARIAN

Introduction to geophysical well logging

Well logging denotes any operation wherein some characteristic data of the formation penetrated by a borehole are recorded in terms of depth. Such a record is called a log. The log of a well, for example, may simply be a chart on which abridged descriptions of cores are written opposite the depths from which cores were taken. A log may also be a graphic plot with respect to depth of various characteristics of these cores, including porosity, horizontal and vertical permeability, and water and hydrocarbon saturations.

In geophysical well logging, a probe is lowered in the well at the end of an insulated cable, and physical measurements are performed and recorded in graphical form as functions of depth. These records are called geophysical well logs, well logs, or simply logs. Often, when there is no ambiguity, geophysical well-logging operations are referred to shortly as well logging or logging.

Various types of measuring devices can be lowered on cables in the borehole for the sole purpose of measuring (logging) both borehole and in-situ formation properties. These logging tools, or logging sondes, contain sensors which measure the desired downhole properties, whether thermal, magnetic, electric, radioactive, or acoustic. Insulated conductive cables not only lower these sondes in the borehole but also pass power to the sondes and transmit recorded signals (data) to the surface, where the latter are recorded as a log. Hence, geophysical well-logging methods provide a detailed and economical evaluation of the entire length of drilled hole. Most of these logging operations are performed with the sonde moving uphole. Most well-logging methods available today have been initially designed and developed to assist in answering some major questions associated with exploration, evaluation and production of oil and gas [11,13,26,27,29,30,45].

Today, however, many of these logging sondes and associated interpretation concepts are applied with equal success in exploring and evaluating groundwater resources, geothermal steam reservoirs, metallic deposits [2,5,9,10,45], and nonmetallic minerals [4,38], and have also found application in mine design [23].

Well logging, as it is known today, was born on September 5, 1927, when the first resistivity log was run in well Diffenbach No. 2905, Pechelbronn oil field, Alsace, France. This discovery resulted from experiments intended to achieve another purpose, a feature which is not uncommon in scientific research.

Well logs provide one of the main keys to realistic formation evaluation. The latter is a blend of science and art used to determine the composition and physical or chemical rock properties in situ and the nature and amount of fluids contained in the pore space, thereby giving an indication of the economic worth of natural resources accumulated in the subsurface, such as oil, gas, coal, uranium, etc.

Despite major technical advances over the last several decades, it is practically impossible as yet to directly measure in-situ engineering and economically important properties of geologic formations penetrated by exploratory holes. Hence, the oil and mining industry has to resort to measurements of specific secondary rock properties (resistivity, radioactivity, etc.). These are then empirically related to the amount of oil and gas in place, the quality and size of coal deposits, the "richness" of particular ore concentrations in host rocks, oil yield per ton in oil-shale deposits, tar content of tar deposits, etc.

Geophysical well logs for tar-sand evaluation

Several geophysical well logs can be used to evaluate tar sand deposits, as far as lithology variations, thickness, and oil yield are concerned. Before starting a discussion of properly selected logging program and associated interpretation methods, however, geophysical well logs applicable for tar-sand evaluation are reviewed briefly here.

Caliper log

The caliper log basically measures and records the size (diameter) of the borehole. Basically, large flexible springs or arms, which ride against the wall of the borehole, are mounted on the body of a sonde. Owing to changes in hole size, the movement of these springs or arms generates voltage changes within the sonde which are then transmitted to the surface and recorded in terms of hole diameter. The caliper log works equally well in boreholes filled with fresh- or salt-water muds (drilling fluids), oil-base drilling fluids, or gas (e.g., air).

Spontaneous potential (SP)

The SP curve is a record of naturally occurring potential differences between a surface electrode and the movable electrode in the mud column [14,20,33]. The downhole variations are caused by differences of potential in the mud column due to currents flowing around the junction points of

(1) the permeable beds, (2) the adjacent shales, and (3) the mud column (emf's of electrochemical origin). Another phenomenon which causes an EMF is the movement of mud filtrate through the mud cake and into the formation (emf of electrofiltration).

Principal uses of the SP measurement include the determination of formation-water resistivity, detection and correlation of permeable zones, bed thickness, etc. The recorded SP curve loses detail in conductive muds where the salinity of the formation water approaches the salinity of the drilling-mud filtrate. Furthermore, no measurements are possible in empty or gas-filled boreholes, in oil-base muds, and in cased well bores.

Focused-resistivity logs

Several types of focused-resistivity logs are routinely used by the oil industry for the purpose of measuring the apparent resistivity of a thin segment of the formation perpendicular to the borehole. These logging sondes perform well in the salt-water muds and in detecting highly resistive and relatively thin zones, but are ineffective in gas-filled boreholes and in oil-base muds.

Generally speaking, the resistivity is measured by recording the voltage at a central electrode from which a constant current is so focused that it goes laterally a certain distance into the formation and then fans out vertically. This focusing is accomplished by a proper arrangement of several electrodes, situated above and below the measuring electrode. These electrodes are maintained at the same potential as the measuring electrode [12,15].

Gamma-ray log

The gamma-ray log is used as a lithology log and to determine shaliness and/or radioactive heavy minerals in zones investigated. It basically measures the intensity of the natural gamma rays of the formations penetrated by the borehole. The natural radiation intensity is measured by proper downhole detectors, amplified, and then transmitted to the surface. The gamma-ray log can be run in cased or uncased holes and is independent of the type of fluid in the borehole [21,24].

Scattered gamma-ray density logs

Formation density logs basically measure the electron density (i.e., number of electrons per cubic centimeter) of downhole formations. This electron density is directly related to the true bulk density which, in turn, is a function of the composition of rock-matrix material, the formation porosity, and the density of the fluids and/or gases filling the pore space.

A gamma-ray source emits intermediate-energy gamma rays into the formation where collision takes place with electrons in the rock. This collision causes a backscattering and a decrease in gamma-ray energy (Compton effect). These scattered gamma rays reaching the detector (on the logging sonde), which is situated at a fixed distance from the radioactive source, are then recorded as a measure of formation density. These values can then be used to estimate formation porosity. Density logs can be run equally well in all types of borehole fluids [1,35,39].

Acoustic logs

Acoustic logs record the time required for a compressional sound wave to traverse one foot of formation. This interval of transit time is the reciprocal of the velocity of the compressional sound wave and is a function of formation lithology, porosity, and types of fluids in the pore space [25,28,37,44]. Hence, the acoustic log (1) is a good porosity tool, provided the lithology is known; (2) assists in interpreting seismic data; and (3) indicates the quality of cement bonds in cased holes [18].

A transducer creates elastic-wave pulses which travel a given distance through the formation and are then picked up at receivers in logging sonde. Travel time, total acoustic-wave trains, signal amplitudes, etc., can be recorded.

Acoustic logs can be run in any type of borehole fluids but not in air-filled or gas-filled boreholes.

Neutron logs

The basic purpose of neutron logs is to measure the abundance of hydrogen atoms in downhole formations or the relative abundance of epithermal or thermal neutrons arriving at the detector.

High-energy, fast neutrons are continuously emitted from a radioactive source which is mounted in the sonde. Owing to collision with nuclei of the rock, the neutrons will be slowed down until finally captured by the nuclei of atoms, such as chlorine, hydrogen, silicon, etc. Thereby, the capturing nucleus emits a high-energy gamma ray of capture.

Depending on the type of neutron-logging sonde, either these capture gamma rays or neutrons themselves are counted by a detector in the sonde [34,36]. Usually, the neutron logs can be run equally well in uncased and cased boreholes and are independent of the type of fluid present in the borehole.

Concluding remarks on logging tools

For the purpose of logging shallow tar-sand deposits, both maximum pressure and temperature ratings of the appropriate logging sondes are of no interest, because most tools withstand approximately 20,000 psi pressure and 400° F temperature. In the above review of logging tools, however, attention has been given to the fact that some of these tools can be run in water-filled and empty (air- or gas-filled) boreholes, whereas others are limited only to boreholes filled with drilling fluid.

Basic considerations in formation evaluation of tar sands

Coring provides the only *direct* measurement of the tar content of tar sands. This approach, however, is time consuming and expensive, particularly if large and rather thick tar-sand deposits are to be sampled and evaluated by a multi-well program. *Geophysical well-logging techniques* provide *indirect* tar-content estimates which result in a faster, economical, and more attractive evaluation of tar-sand deposits. Nevertheless, core-analysis data, in at least the first test hole, is always desirable to allow an *empirical* correlation between a measured specific quality of the tar sand and the response of geophysical well-logging sondes.

Core-analysis procedures

The importance of reliable core analysis and the factors which may affect these results have to be considered in detail, because core data frequently provide the yardstick, i.e., reference base, in evaluating the reliability of well-log analysis results over the cored subject interval.

For example, *the tar content* of core samples can be determined by different methods, including the Dean-Stark, modified Dean-Stark, Retort, and Pressure Elusion-Fisher Titration techniques [8,16].

An interesting comparative study of four commercial and four company in-house laboratory techniques used in tar-sand analysis has been completed recently [16]. This comparison was carried out through a Round Robin test procedure, whereby tar-sand samples have been analyzed from common, specially mixed, batch samples. The general conclusions of this extensive investigation have been summarized by Eade [16] as follows:

"(1) The Round Robin study indicated that all laboratories examined, generally showed good agreement in analysis results with consistent low standard deviations. The average 95% confidence-level values for the percent bulk-weight tar sand and percent dry-weight tar categories were very close to or below the expected tolerance of 5%.

(2) Consistency of results indicate that no one laboratory or technique showed superiority over another within the accuracy demanded for the tar grades examined. The modified Dean-Stark technique, however, has the distinct advantage of a built-in material-balance check on the results. The level of consistency between the in-house and commercial laboratories was generally the same.

(3) Although the ~5% acceptability tolerance was a goal set out in our terms of reference, it was obvious that a reproducibility expectation as low as ~3% with a 95%

confidence level can be achieved for rich material. This level increases with decrease in tar grade and extrapolates to ~10—12% for 6% bulk-weight grade material.

(4) The water loss or gain errors incurred during the sampling process can be significant ones if extreme care is not taken. This is especially so in the lower tar-saturation ranges where water content is higher. The problem can be avoided if a dry-weight approach is used.

(5) Water losses or gains during the laboratory analytical procedure, while not quite as large in magnitude as experienced during sampling, can be as significant since they have direct effect on tar content. The problem cannot be avoided through the dry-weight approach.

(6) No evidence was apparent that fines escaping into the solvent phase during refluxing was a problem in the range of tar grades examined. This of course does not preclude the problem in the low-bitumen—high-fines material which was not available for our study.

(7) This study must be considered a quality-controlled one and, therefore, the study on the basis of routine work is lost. Quality control on routine work must be considered the responsibility of each company having work done."

Owing to the friable nature of the core material, a direct laboratory *porosity* measurement is usually not feasible. Core analysis is, therefore, restricted to the measurement of the weight of water and solids and, by difference, tar as a fraction of the bulk weight of the core (Dean-Stark method). If the pore space is completely occupied by tar and water, and the densities of the components are known, the data is sufficient to calculate the formation porosity [4]. Furthermore, inasmuch as the water loss due to evaporation is significant, the porosity values determined from the weight of water and tar, and their densities will represent a minimum value of the porosity that exists at the time the core is analyzed. The mathematical relationship between these sample properties and laboratory porosity can be expressed as follows:

$$\phi_{\mathrm{lab}} = (\rho_{\mathrm{ma}} \times GW_{\mathrm{f}})/(\rho_{\mathrm{ma}} \times GW_{\mathrm{f}} + \rho_{\mathrm{f}}) \tag{11-1}$$

$$\phi_{\mathrm{lab}} = (\rho_{\mathrm{ma}} \times BW_{\mathrm{f}})/[\rho_{\mathrm{ma}} \times BW_{\mathrm{f}} + \rho_{\mathrm{f}}(1 - BW_{\mathrm{f}})] \tag{11-2}$$

where: ϕ_{lab} = laboratory porosity, a minimum value
ρ_{ma} = matrix density of reservoir rock (g/cc)
ρ_{f} = density of fluid components in pore space (g/cc)
GW_{t} = tar weight as a fraction of weight of solids
GW_{w} = water weight as a fraction of weight of solids
$GW_{\mathrm{f}} = GW_{\mathrm{t}} + GW_{\mathrm{w}}$
BW_{t} = tar weight as a fraction of bulk weight
BW_{w} = water weight as a fraction of bulk weight
$BW_{\mathrm{f}} = BW_{\mathrm{t}} + BW_{\mathrm{w}}$

Figs. 11-1 and 11-2 show the graphical solution to eqs. 11-1 and 11-2. One has to keep in mind, however, that log-derived porosity values, which are determined in situ, are generally much lower than the laboratory minimum

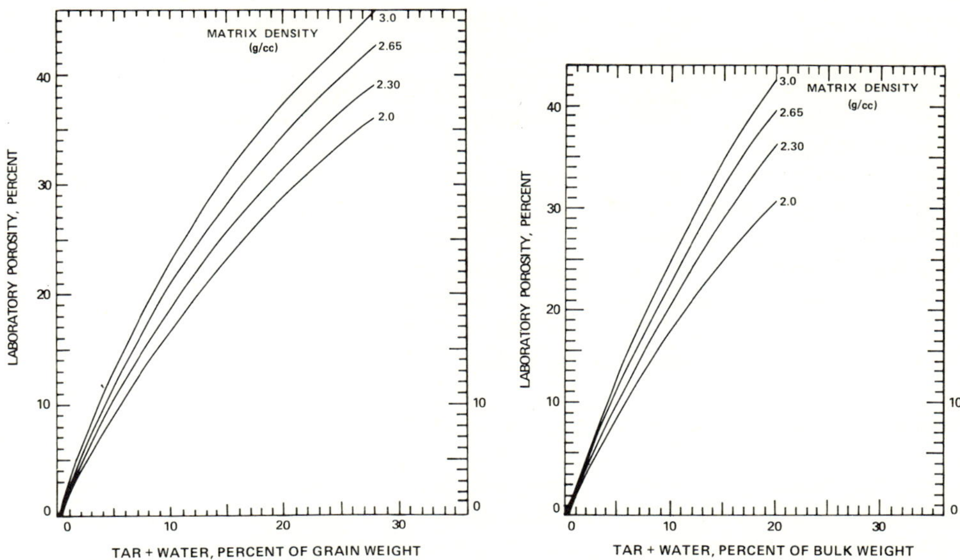

Fig. 11-1. Relationship between porosity determined in the laboratory and weight fraction of tar + water, expressed in percent of dry grain weight. (After Woodhouse [41], fig. 3, p. 11; courtesy of SPWLA.) Assumption: pore space is occupied by tar and water, having density of 1.0 g/cc.

Fig. 11-2. Relationship between porosity determined in the laboratory and bulk-weight fraction of tar + water. (After Woodhouse [41], fig. 4, p. 11; courtesy of SPWLA.) Assumption: pore space is occupied by tar and water, having density of 1.0 g/cc.

TABLE 11-I

Comparison of typical downhole and laboratory analyses showing the effects of 6% and 9% core expansion (After Woodhouse [41])
Porosity = 34%; oil saturation = 90%. matrix density = 2.65 g/cm^3; and fluid density (water + tar) = 1.0 g/cm^3

Core analysis parameter	Original downhole values	Laboratory conditions	
		6% expansion of core	19% expansion of core
Water, % bulk weight	1.6	4.6	9.8
Tar, % bulk weight	14.4	13.9	13.4
Apparent porosity, %	34	37.7	44.5
Water, % grain weight	1.9	5.4	12.8
Tar, % grain weight	17.5	17.5	17.5

porosity values determined from core samples.

Excess core recovery, i.e., greater core length than the actual length of the subsurface interval drilled, is frequently observed in the tar sands [3]. The reported data suggest a linear expansion of about 6% due to decompaction (depressuring) effects.

Table 11-I shows the difference of typical downhole and laboratory analysis due to effects of core expansion for a typical McMurray tar-sand sample as reported by Woodhouse [41] recently.

These effects have to be considered if the goal is a realistic comparison of core and log-derived properties of the tar sands.

Tar-sand evaluation using geophysical well logs

Basic physical properties of Athabasca tar sands important to well-log analysis

Several reservoir properties of importance to well-log evaluation exhibit a wide range of values rather than a simple typical value, which could be considered representative for the entire Athabasca tar-sand deposits.

Grain-size distribution. In general, three typical units having characteristic grain size distribution patterns are encountered, which are related to the stratigraphic location within the deposit. Unit A consists of rather poorly sorted, coarse-grained, quartzose sandstone and pebble conglomerate deposited in ancient river channels. Unit B consists of well-sorted, fine-grained sands, including the richest tar sands as found in the middle section of the McMurray Formation having thicknesses of 100 ft or more. These are overlain by micaceous sands and silt which represent Unit C. The latter constitute the typical overburden consisting of very lean, i.e., poorly impregnated with tar, parts of the deposit.

Interbedded at random within these stratigraphic units, one may encounter thin coal seams and lenses of clay and gravel, which are characteristic of the fluvial—deltaic depositional environment.

There is an apparent correlation between (1) increasing amount of silt- and clay-size material and (2) a decrease in the tar content of this geologic interval. This observation is based on core-sample analysis (Fig. 11-4) and has been also confirmed by detailed computerized interpretation of modern well-logging suites.

The range of sand-grain particle size versus cumulative sample weight as measured by Carrigy [69] is illustrated in Fig. 11-3.

Sample *fines* are defined as solids which pass through a 230-mesh Tyler sieve (63 μm), and basically consist of silt and clay. Sample *colloidal fines* are defined as solids which pass through a 325-mesh Tyler sieve (44 μm). The latter consist only of the clay components, which are troublesome during surface-extraction process [32].

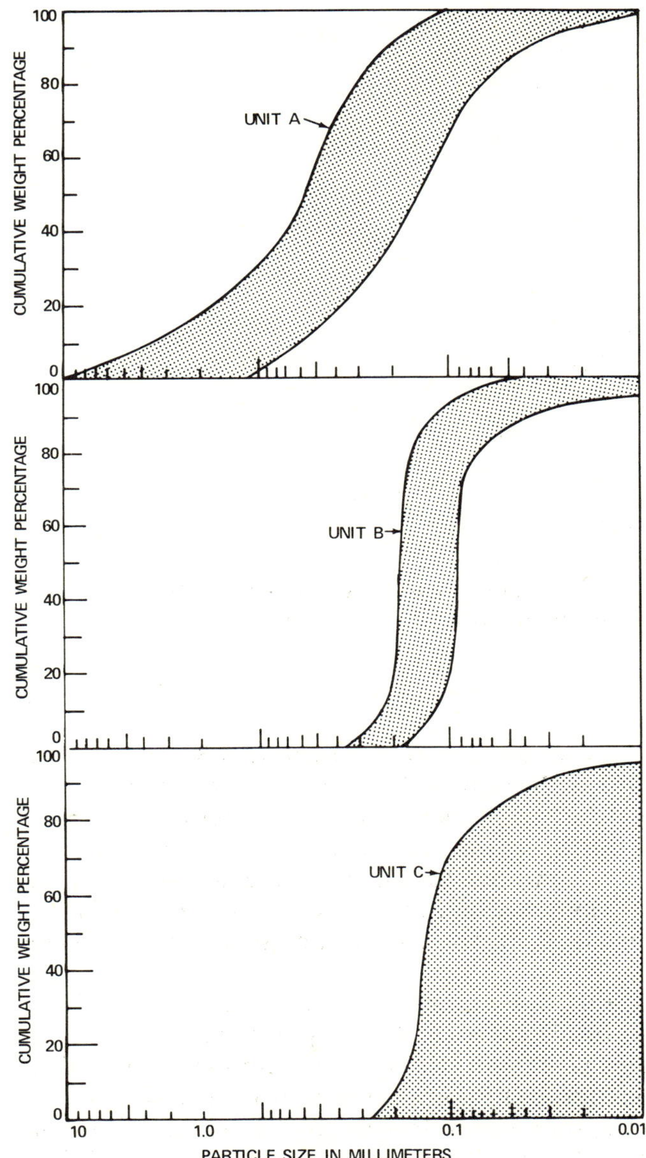

Fig. 11-3. Limiting envelopes of grain-size distributions of typical Athabasca tar sands: A = coarse-grained poorly sorted sands; B = fine-grained well-sorted sands; C = very fine-grained sands and silts. (After data by Carrigy [6].)

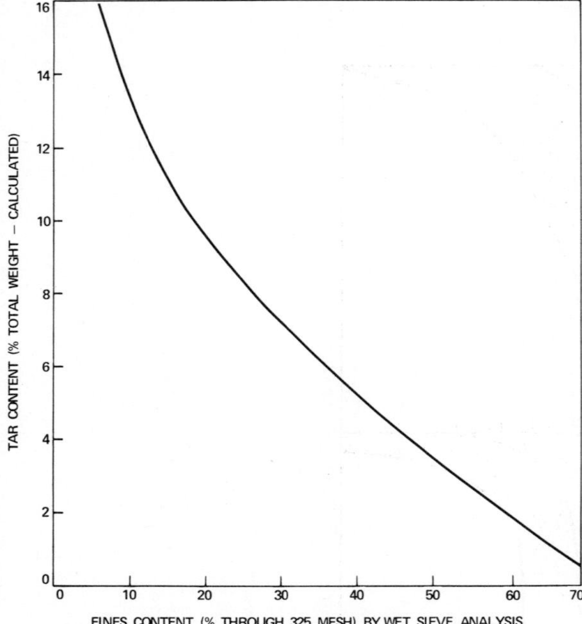

Fig. 11-4. Relationship between tar content and fines content (percentage passing through 325-mesh screen) of Athabasca tar sands, Alberta, Canada.

Fig. 11-5 compares the fines content (% passing through a 325-mesh sieve) by wet-sieve analysis with the deflection of the gamma-ray log over the zone of interest. This gamma-ray deflection can be used to determine the shale bulk volume of the sand by one of several mathematical relationships [17].

Standard log-interpretation concepts for shaly-sand analysis allows the volumetric determination of hydrocarbons, formation water, clay, silt, and quartz. The detailed mathematical treatise has been presented in the logging literature by Poupon et al. [31].

Figs. 11-6 and 11-7 show the comparison of core- and log-derived data for the percentage bulk volume of (1) silt and (2) clay-plus-silt in an Athabasca test well. Fig. 11-8 compares the amount of clay (as weight percentage of the matrix) calculated from well logs to the corresponding core-analysis data.

Special plotting techniques, such as the Holgate plot [22], are also available, which allow a graphical correlation between the amount of *fines* obtained from well logs and that shown by core analysis.

Reservoir permeability. Laboratory measurements on undisturbed tar-sand cores have shown that permeabilities generally are about 50 mD [7].

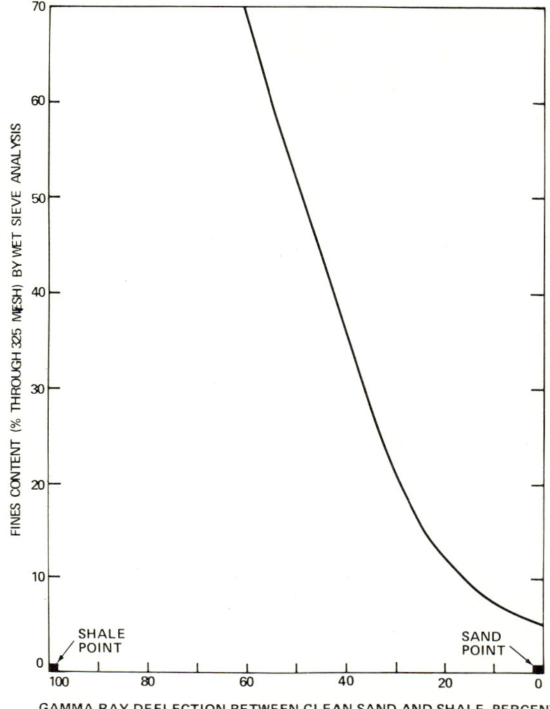

Fig. 11-5. Relationship between gamma-ray log response and fines content, Athabasca tar sands, Alberta, Canada.

Formation-water salinity. Fresh water is found in the upper layers of the Athabasca tar-sand deposits. Water becomes more saline with increasing depth. Total dissolved solids of formation water produced from the McMurray Formation may vary between 3,000 and 18,000 mg/l. For example, analysis of water samples from Athabasca tar sand in Lease No. 24 showed a total solids concentration of 15,000 mg/l to be characteristic of the bottom water associated with the main accumulation [41].

Depending on the formation-water salinities and the tar content of rich, medium, or lean "ore", the measured formation resistivities may be as high as 4,000 Ωm.

Tar properties. Tar exhibits properties which are quite different from those of conventional oils. Gravity is only 6° API at 60° F, and viscosity is extremely high and may range from 10^4 to 10^7 P at 40° F. Fig. 11-9 shows the viscosity as a function of temperature for Athabasca tar from several localities. Consequently, effects due to mud-filtrate invasion and displace-

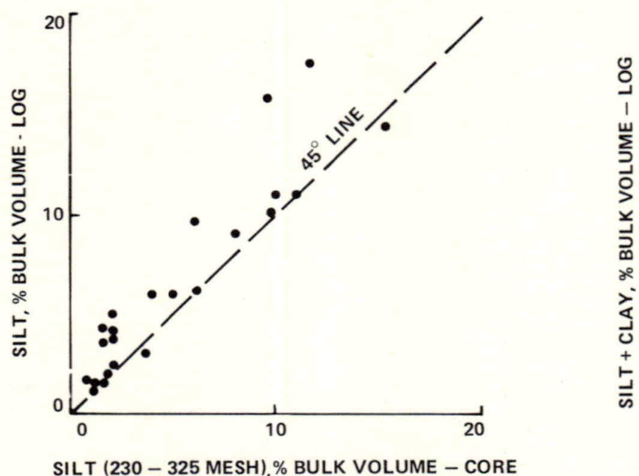

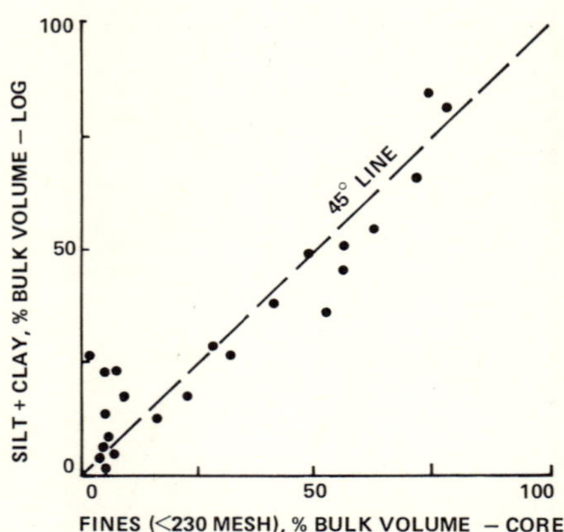

Fig. 11-6. Comparison of silt content (% bulk volume) determined from logs with silt content determined from core analysis (% of bulk volume of sediment passing through a 230-mesh sieve but retained by a 325-mesh sieve). (After Sah et al. [32], fig. 13b, p. 11; courtesy of SPE of AIME.)

Fig. 11-7. Relationship between the content of clay + silt (fines) of tar sand determined from logs and the content of sediment passing through a 230-mesh sieve, as determined from core analysis and expressed as percent of bulk volume. (After Sah et al. [32], fig. 13c, p. 11; courtesy of SPE of AIME.)

ment of tar from around the borehole are practically nonexistent and do not need to be specially considered in the log analysis of typical rich tar-sand deposits.

Size variations of drill holes. Hole-size variations may affect log interpretations, if the wellbore is severely washed out (enlarged). Experience in Canadian test holes shows that hole-size variations in tar sands often exhibit characteristics opposite to those of typical sand—shale sequences. Frequently, clean, rich tar sands tend to wash out the most, whereas shaly zones are drilled closer to bit size. Repeat runs of well logs, however, usually show satisfactory reproducibility of the measurements.

Reservoir porosity. Depending on grain-size distribution, a wide range of porosity can be present. Scattered gamma-ray density logs and neutron logs, properly analyzed, provide a realistic and continuous in-situ porosity estimate. As discussed earlier, however, porosities determined from cores are always higher (more optimistic) than the in-situ log-derived ones.

Water saturation. Rich bituminous sands usually have a very low water content ranging from 3 to 6% by weight. Standard shaly-sand water-saturation equations, particularly parabolic equations of the form $y = bS_w + cS_w^2$, can be

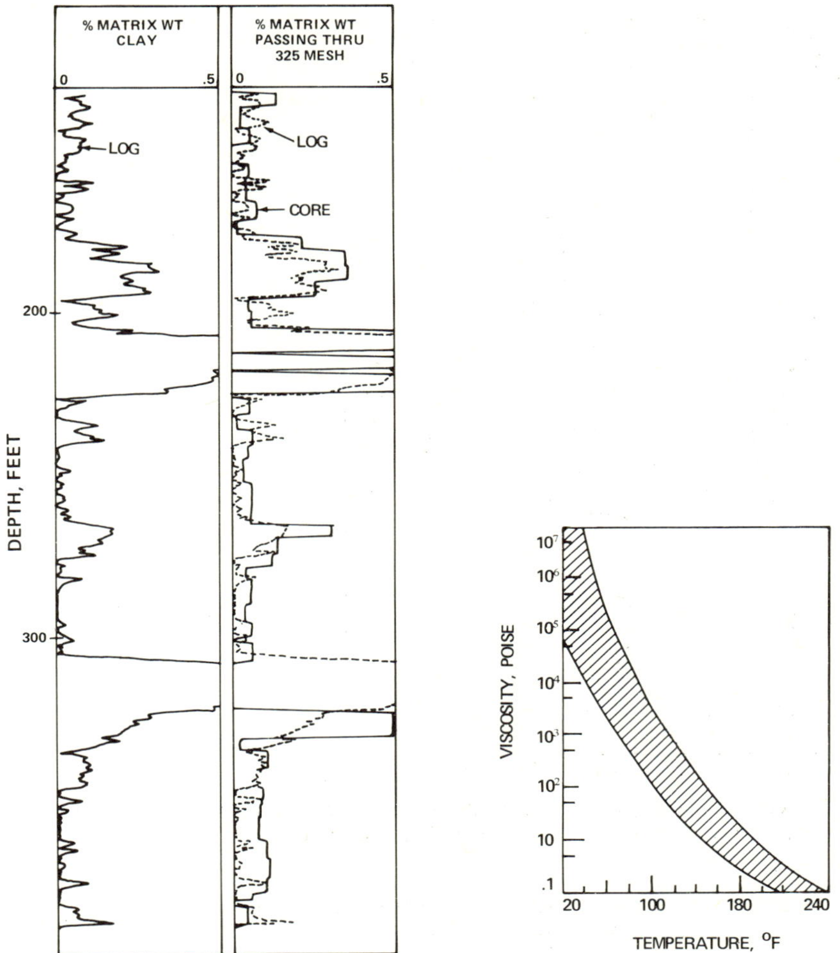

Fig. 11-8. Comparison between the clay content determined from logs and the amount of sediment passing through a 325-mesh sieve (core analysis), both expressed in percentages of the weight of matrix. (After Sah et al. [32], fig. 14a, p. 12; courtesy of SPE of AIME.)

Fig. 11-9. Relationship between viscosity and temperature of Athabasca bitumen from various localities. (After data of Ward and Clark [40].)

applied successfully [17,31], where S_w is the percentage of pore space occupied by water. A detailed formula can be presented as follows:

$$\frac{1}{R_t} = \frac{\phi^2 S_w^2}{aR_w} + \frac{V_{sh} S_w}{R_{sh}}$$

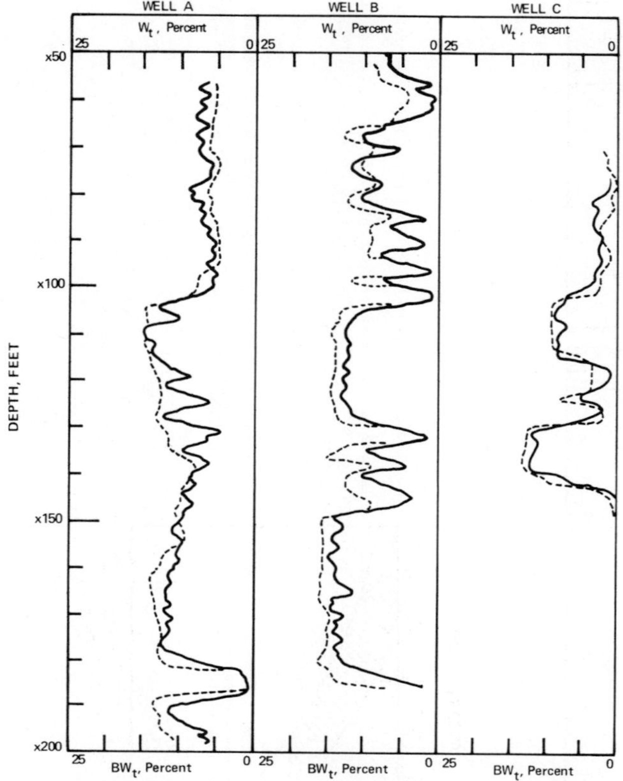

Fig. 11-10. Tar content (BW_t) expressed as the percentage of bulk weight based on core analysis (dashed curves) and determination from logs (solid curves).

where: R_t = true formation resistivity in zone of interest (shaly sand)
R_{sh} = resistivity of the adjacent shale beds
R_w = formation-water resistivity at formation temperature
V_{sh} = shale fraction in reservoir rock
ϕ = effective reservoir porosity
S_w = water saturation

Tar content. The tar content is equal to $1 - S_w$, and the tar weight as the percentage of the bulk weight (BW_t) can be calculated as follows:

$$BW_t = \phi \times S_t \times \rho_t \tag{11-3}$$

where: ϕ = effective reservoir-porosity fraction
S_t = tar saturation, which is equal to $(1 - S_w)$ in %
ρ_t = bulk density of the tar

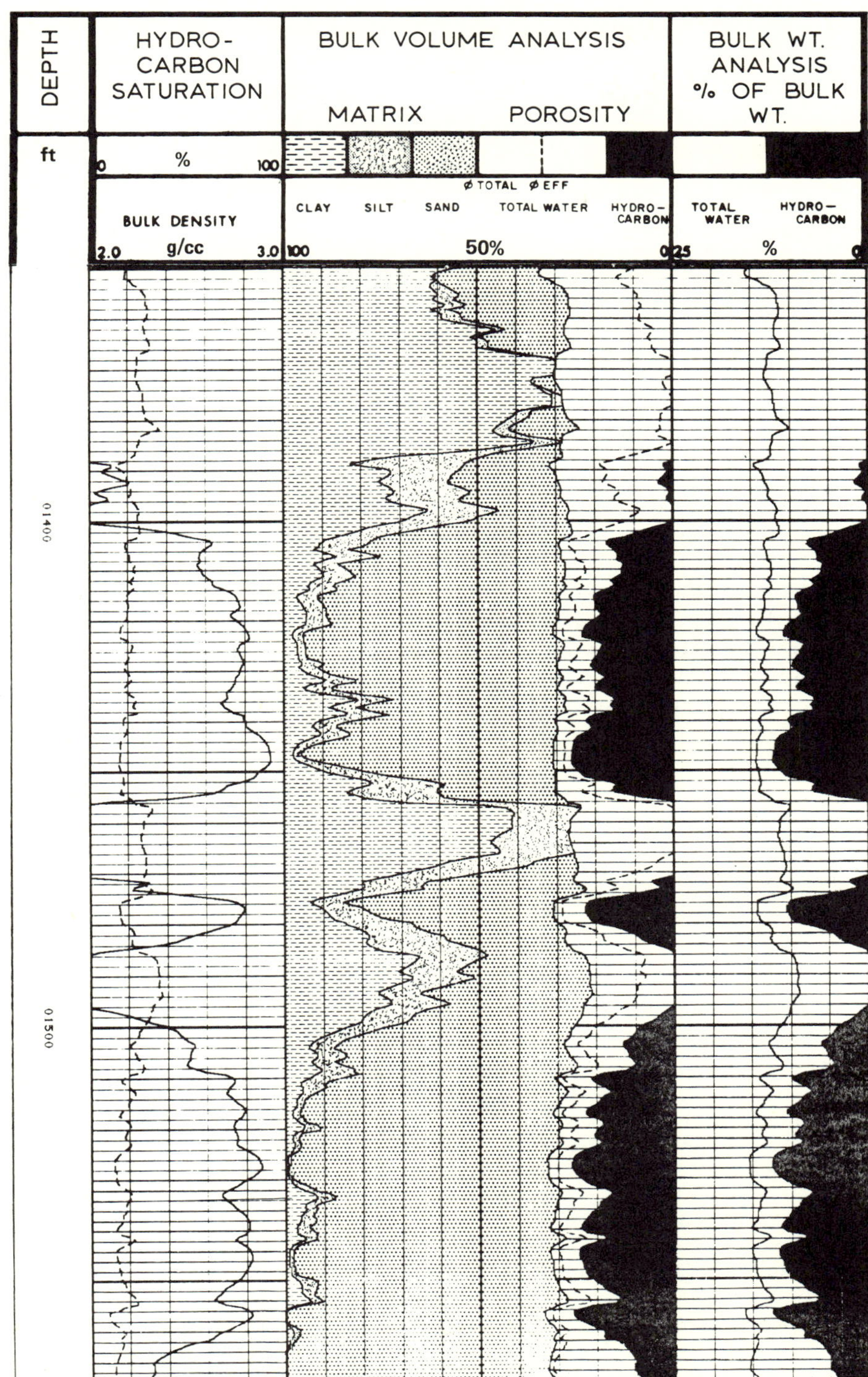

Fig. 11-11. Example of computerized Athabasca tar-sand evaluation.

Fig. 11-10 shows a comparison of the bulk-weight tar content (BW_t) determined from the core analysis and that computed from log data.

Presentation of digital tar-sand evaluation using well logs

Several types of analog and graphical presentation of log-analysis results have been discussed in the literature [8,19,32,41].

The field example presented in Fig. 11-11 illustrates some of these basic concepts which have been incorporated into Dresser Atlas' EPILOG programs to achieve realistic digital tar-sand evaluation. The test well discussed is located in the Athabasca tar-sand deposit of northeastern Alberta, Canada.

Conclusions

Tar-sand deposits can be evaluated by realistic and reliable concepts using geophysical well-logging data. Digital log analysis gives a continuous in-situ determination of the reservoir characteristics.

Application of such log-analysis concepts in tar-sand deposits is feasible from both the technical and economical viewpoints. Selective coring and specific core analysis, however, will still be required on a limited basis in newly discovered tar-sand deposits to establish a new set of interpretation parameters, which are characteristic for the specific tar-sand deposit under evaluation.

Acknowledgment

The authors wish to express their appreciation to M. Hallstrom of Dresser Atlas, Houston, Texas, for his valuable suggestions in the preparation of this manuscript.

References and Bibliography

1 R.P. Alger, L.L. Raymer, W.R. Hoyle and M.P. Tixier, "Formation Density Log Applications in Liquid-Filled Holes", *J. Pet. Technol.*, *15*, 321—333 (1963).

2 R.W. Baltosser and H.W. Lawrence, "Application of Well Logging Techniques in Metallic Mineral Mining", *Geophysics*, *35(1)*, 143—152 (1970).

3 D.W. Bennett and A. King, "Drilling and Coring in Athabasca Oil Sands Combines Oilfield and Mining Techniques", *North. Min.*, p. 42 (June 12, 1975).

4 R.P. Bond, R.P. Alger and A.W. Schmidt, "Well Log Applications in Coal Mining and Rock Mechanics", *Trans. Soc. Min. Eng.*, *250(12)*, 355—362 (1971).

5 R.A. Broding, C.W. Zimmerman, E.V. Somers, E.A. Wilhelm and A.A. Stripling, "Magnetic Well Logging", *Geophysics*, *17(1)*, 1—26 (1952).

6 M.A. Carrigy, "Geology of the McMurray Formation, Part III. General Geology of the McMurray Area", *Res. Counc. Alberta, Geol. Div., Mem.*, *I*, 130 pp. (1959).

7 K.A. Clark, "Permeabilities of the Athabasca Oil Sands", *Trans. Can. Inst. Min. Metall.*, *63*, 151—156 (1959).

8 H.N. Collins, "Log—Core Correlations in the Athabasca Oil Sands", SPE 5037, *Soc. Pet. Eng. AIME, 49th Annu. Fall Meet.*, Houston, Texas (Oct., 1974).

9 J.A. Czubec, "Recent Russian and European Developments in Nuclear Geophysics Applied to Mineral Exploration and Mining", *Log Anal.*, pp. 20—34 (Nov.—Dec., 1971).

10 J.A. Czubec, "Pulsed Neutron Method for Uranium Well Logging", *Geophysics, 37(1)*, 160—173 (1972).

11 V.N. Dakhnov, *Geophysical Well Logging*, Gostoptekhizdat, Moscow (1959). (In Russian, translated into English by G.V. Keller, Colorado School of Mines, Golden, Colo., 1962.)

12 C.J. Decker and M. Martin, "The Laterolog and Salt Mud Logging in Kansas", *Oil Gas J.*, *50(41)*, 1191—129 (1952).

13 R. Desbrandes, *Theorie et Interpretation des Diagraphies*, Editions Technip, Paris, 545 pp. (1968).

14 H.G. Doll, "The SP Log: Theoretical Analysis and Principles of Interpretation", *Trans. AIME*, *11*, 146—185 (1948).

15 H.G. Doll, "The Laterolog — A New Resistivity Logging Method with Electrodes Using an Automatic Focusing System", *Trans. AIME*, *192*, 305—316 (1951).

16 J.R. Eade, "Round Robin Study of Analytical Procedures of Various Laboratories on Assay Analysis of Athabasca Tar Sands", *5th Formation Evaluation Symposium of CWLS*, Calgary, Alta., Paper Q, (May, 1975).

17 W.H. Fertl, "Status of Shaly Sand Evaluation", *4th Formation Evaluation Symposium of CWLS*, Calgary, Alta., Paper F, (May, 1972).

18 W.H. Fertl, P.E. Pilkington and J.B. Scott, "A Look at Cement Bond Logs", *J. Pet. Technol.*, *26*, 607—617 (1974).

19 R.W. Fetzner, W.L. Hensen and F.J. Feigl, "Athabasca Oil Sand Evaluation Using Core and Log Analysis", *Log Anal.*, *6(4)*, 21—29 (Nov.—Dec., 1965).

20 M. Gondouin and C. Scala, "Streaming Potential and the SP Log", *J. Pet. Technol.*, *10(8)*, 170—179 (1958).

21 W.G. Green and R.E. Fearon, "Well Logging by Radioactivity", *Geophysics, 5(3)*, 272—283 (1940).

22 M. Holgate, "The Microlog as a Porosity Datum for the Neutron Log in Swanhill Field, Alberta", *Trans. Annu. Meet. Petrol., Natl. Gas Div.*, Calgary, Alta. (May, 1960).

23 J.C. Jenkings, "Practical Application of Well Logging to Mine Design", *Annu. Meet. AIME*, SME-Prepr., 69-F-73, Washington, D.C. (1969).

24 F.P. Kokesh, "Gamma Ray Logging", *Oil Gas J.*, *50(12)*, 284—298 (1951).

25 F.P. Kokesh, R.J. Schwartz, W.B. Wall and R.L. Morris, "A New Approach to Sonic Logging and Other Acoustic Measurements", *J. Pet. Technol.*, *17(3)*, 282—287 (1965).

26 E.J. Lynch, *Formation Evaluation*, Harper and Row, New York, N.Y., 422 pp. (1962).

27 R. Meinhold, *Geophysikalische Messverfahren in Bohrungen*, Geest und Portig, Leipzig, 237 pp. (1965).

28 G.R. Pickett, "Acoustic Character Logs and Their Applications in Formation Evaluations", *J. Pet. Technol.*, *15*, 659—668 (1963).

29 S.J. Pirson, *Handbook of Well Log Analysis for Oil and Gas Formation Evaluation*, Prentice-Hall, Englewood Cliffs, N.J., 326 pp. (1963).

30 S.J. Pirson, *Geologic Well Logging Analysis*, Gulf, Houston, Texas, 370 pp. (1970).

31 A. Poupon, C. Clavier, J. Dumanoir, R. Gaymard and A. Misk, "Log Analysis of Sand—Shale Sequences — A Systematic Approach", *J. Pet. Technol.*, *22(7)* (1970).

32 R.C. Sah, A.E. Chase and L.E. Wells, "Evaluation of the Alberta Tar Sands", *49th Annu. Fall Meet.*, Houston, Texas, SPE 5034 (Oct., 1974).

33 F. Segesman and M.P. Tixier, "Some Effects of Invasion on the SP Curve", *J. Pet. Technol., 11(6)*, 138—146 (1959).
34 C.W. Tittle, H. Faul and C. Godoman, "Neutron Logging of Drill Holes: The Neutron—Neutron Method", *Geophysics, 16(4)*, 626—658 (1951).
35 J. Tittman and J.S. Wahl, "The Physical Foundations of Formation Density Logging (Gamma—Gamma)", *Geophysics, 30(2)*, 284—296 (1965).
36 J. Tittman, H. Sherman, W.A. Nagel and R.P. Alger, "The Sidewall Epithermal Neutron Porosity Log", *J. Pet. Technol., 18*, 1351—1363 (1966).
37 M.P. Tixier, R.P. Alger and C.A. Doh, "Sonic Logging", *J. Pet. Technol., 11*, 106—114 (1959).
38 M.P. Tixier and R.P. Alger, "Log Evaluation of Non-Metallic Mineral Deposits", *Trans. SPWLA*, Paper R, 1—22 (1967).
39 J.S. Wahl, J. Tittman and C.W. Johnstone, "The Dual Spacing Formation Density Log", *J. Pet. Technol., 16*, 1411—1417 (1964).
40 S.H. Ward and K.A. Clark, *Res. Counc. Alberta, Rep., 57*, 136 pp. (1950).
41 R. Woodhouse, "Athabasca Tar Sand Reservoir Properties Derived from Cores and Logs", *Trans. SPWLA*, Paper T (1976).
42 A.E. Worthington and R.F. Meldon, "Departure Curves for the Self-Potential Log", *J. Pet. Technol., 10(1)*, 11—16 (1958).
43 M.R.J. Wyllie, "A Quantitative Analysis of the Electrochemical Component of the SP Curve", *Trans. AIME, 186*, 17—26 (1949).
44 M.R.J. Wyllie, A.R. Gregory and G.H.F. Gardner, "An Experimental Investigation of Factors Affecting Elastic Wave Velocities in Porous Media", *Geophysics, 23*, 459—494 (1958).
45 C.J. Zablocki, "Some Applications of Geophysical Well Logging Methods in Mineral Exploration Drill Holes", *Trans. SPWLA*, Paper U, 13 pp. (1966).

Appendix I *

Government Policy Statement with respect to Oil-Sand Development — Part 1 (October 1962)

Recently the Government authorized the first commercial development of the oil sands and other applications currently are pending.

The Government has an obvious responsibility to regulate the timing and the extent of oil-sand production to protect the interests of the public as the owners of this resource and to ensure that the position of conventional oil production in Alberta is not jeopardized by the loss of already limited markets to a new source of supply. No economic advantage to consumers of oil products will accrue through the development of the oil sands since synthetic crude oil from the sands and conventional crude oil will, for at least some time, be competitive.

The conventional oil industry has invested nearly $ 4 billion in exploration and development in Alberta and the impact of its operations is a major factor in the buoyancy of Alberta's economy. In addition, it generates over 40% of total provincial public revenues. Obviously it would be detrimental to the public interest to permit unregulated development of an alternative source of supply to impair the economic soundness of the conventional oil industry by further reducing its already limited market. This situation is aggravated by the fact that conventional oil is prorated to available market while oil from the sands cannot be so prorated because a constant plant throughput is essential to make such development economically feasible. Having regard to these circumstances, the policy of the Government will be to so regulate oil-sand production that it will supplement but not displace conventional oil. At the same time, an opportunity will be provided for the orderly development of the oil sands within the limits dictated by the Government's responsibility to the public interest in preserving the stability of conventional oil development and the necessary incentive to ensure its continued growth.

For such production from the oil sands as may be able to reach markets clearly beyond present or foreseeable reach of Alberta's conventional industry, there is no need to restrict the rate of production from the oil sands and, provided the development program meets with the approval of the Oil and Gas Conservation Board, the Government will authorize it.

On the other hand, for such oil-sand production as would be in competition with present or foreseeable markets for conventionally produced Alberta crude oil, the impact on the conventional industry will be carefully considered. In this instance, the Government's judgement is that the best interests of the province will be served:

(1) in the initial stages of oil-sand development, by restricting production to some 5% of the total demand for Alberta oil, i.e., at a level of the order of that recently approved for Great Canadian;

(2) as market growth enables the conventional industry to produce at a greater proportion of its productive capacity, by permitting increments in oil-sand production as recommended by the Oil and Gas Conservation Board, and on a scale, and so timed, as to retain incentive for the continued growth of the conventional industry; and

---

* Canada, see Chapter 5 prepared by F.K. Spragins.

(3) by relating the scale and timing of increments of oil-sand production also to the life index of proved reserves of conventional oil allowing the index to decline gradually from present levels but ensuring that it does not drop below twelve to thirteen years.

This policy will afford flexibility in application and will ensure that the orderly development of the oil sands will proceed as rapidly as their production can be integrated into the over-all oil economy of the Province.

As the Government now sees the situation, total oil-sand production probably will not exceed 200,000 bbl/day by 1975 and, depending upon the total oil demand and the capacity of the conventional industry, it could be less.

All plans for oil-sand development require the approval of the Oil and Gas Conservation Board. The Government will look to the Board for assistance in implementing this general policy and for guidance should it become necessary to select among competing development proposals.

In short, it is the Government's intention to assure to the conventional oil industry a rate of production and a share of available markets in excess of what currently prevails and also to ensure that a reasonable share of future increased markets will be available to conventionally produced oil. This will still give scope for the orderly development of the oil sands under a regulated program that will protect the public interest by preventing detrimental dislocations in the Provincial economy.

Appendix II *

Government Policy Statement with respect to Oil-Sand Development — Part 2 (February 1968)

In October 1962, the Government announced its policy for the development of the Athabasca and related oil sands. The essential features of this policy are as follows:

(1) applications for oil-sand production "able to reach markets clearly beyond present or foreseeable reach of Alberta's conventional industry" would be approved providing the development program meets the conservation and related requirements of the Oil and Gas Conservation Board and

(2) applications involving the marketing of oil-sand production "within reach of Alberta's conventional industry" would

(a) for initial development be restricted to a volume in the order of 5% of the total market for Alberta crude oil and

(b) for subsequent development be restricted to ensure that "market growth enables the conventional industry to produce at a greater proportion of its productive capacity . . ." and, with respect to the scale and time of incremental oil-sand production, by relating such production to "the life index of proven reserves of conventional oil . . .".

The Government's intent in the policy was to provide for the orderly development of the oil sands in such a manner as to supplement but not displace production from the conventional industry.

There have been several developments since 1962 which have had an impact on the effectiveness and the interpretation of this policy.

In 1964 the Oil and Gas Conservation Board announced the adoption of a new proration plan which has had a significant effect on the development of reserves in the con-

* Canada, see Chapter 5 prepared by F.K. Spragins.

ventional industry. As a result of the new plan a detailed reappraisal of the crude-oil reserves of the Province was carried out. This resulted in increases in the reserves of a number of fields, especially Pembina. The plan also has resulted in an increase in the incentive for the institution of enhanced recovery operations leading to higher crude-oil recoveries.

Relatively few crude-oil discoveries were made during the period immediately preceding the formulation of the Government's policy in 1962. In the subsequent years industry's exploratory efforts were more successful. In 1964 and 1965 discoveries were made in the Mitsue and Nipisi Gilwood Sand. In 1965 also the discovery in Rainbow brought about the Keg River oil play in northwest Alberta. These latter discoveries added materially to the Province's crude-oil reserves and it is expected that the Rainbow play will continue for several years.

The growth of the market for Alberta crude oil has been substantial but somewhat less than had been anticipated at the time of its appraisal in 1962. This has been due primarily to a higher than expected level of production in Saskatchewan. During the last year increased exports to the United States have brought the total market for Alberta crude oil very close to the level forecasted by the Board in 1962.

The impact of the reassessment of older reserves, the institution of numerous enhanced-recovery schemes, the new discoveries and the market circumstances have increased the life index for conventional crude oil from the 22 years of 1962 to a current level of some 31 years, rather than to the 21 years previously expected.

In addition to the above developments substantial reserves of heavy hydrocarbons that have many similarities with the Athabasca-type oil sands have been discovered in the general Cold Lake area. While the Oil and Gas Conservation Board recently has found these reserves to fall within the definition of "oil sands" in The Oil and Gas Conservation Act, the definition appears to require clarification. Moreover, the different definitions in various Provincial statutes require standardization.

During the last year the Government has received representations from some companies requesting both a clarification and modification of the existing policy. The Government asked the Oil and Gas Conservation Board to consider these matters. The Board did so and advised the Government that it believed certain aspects of the present policy should be clarified and that the policy should be amended in a manner which would encourage greater market growth than would otherwise occur and by this means enable further oil-sand development without prejudice to the conventional industry. It considered these important in the long-term interest of the Province in the development of its natural resources and to enhance its position as a major and growing source of petroleum supply on the North American continent. Further development would ensure that the Province would be able to take full advantage of market opportunities expected with the growing supply deficiencies in the United States, and enable it to maintain its technological position as a source of synthetic crude oil having regard to potential developments elsewhere — especially in the oil-from-coal and the oil-from-shale programs in the United States. The Board suggested that certain clarifications and one amendment could be made in the policy without change in its broad intent and presented some preliminary proposals to achieve the objective.

The Government then asked the Board to discuss its preliminary proposals with representatives from the Alberta Division of the Canadian Petroleum Association and the Independent Petroleum Association of Canada. Following this discussion the Board advised the Government of the views of the various parties and in the light of them suggested certain amendments to its original proposals. The Government reviewed the amended Board proposals and on June 16, 1967, the Honourable Mr. Patrick and I discussed them in detail with an industry group representing the Alberta Division of the Canadian Petroleum Association, the Independent Petroleum Association of Canada and a number of individual companies interested in oil-sand development. The Government has considered the views expressed at this meeting and the submissions which it received following the meeting and has now come to its decision.

The Government agrees with most of the industry and the Board that there are certain features of the present oil-sand development policy that require clarification. Additionally, the Government believes that the policy should be amended to encourage further growth in the total crude-oil market and thereby permit further oil-sand development.

The clarifications are as follows:

(1) The Government believes that the heavy oils of the Cold Lake type must be subject to the same policy as the Athabasca-type oil sands. It takes this position because of the magnitude of the Cold Lake reserves in relation to conventional reserves, the similarity of the crude hydrocarbons themselves, and the probable similarity of in-situ recovery methods for Cold Lake-type heavy oils and Athabasca-type oil sands. Moreover, the Government believes that regulation of the rate of production of the Cold Lake-type heavy oils by the "approval" system used with oil sands is more practical than by prorating as is done with the light and medium crude oils. The lack of ready interchangeability among the heavy oils and the fact that by upgrading processes they, like the oil-sand oil, could compete in the market for light and medium crude oil suggests serious problems if the regulation of production were based on proration to market demand. Consistent with this the Government has decided that the definition of oil sands should be amended in order to remove ambiguities. Furthermore, it believes that common definitions need to be adopted in all Provincial statutes and regulations, thus ensuring a consistent mineral acquisition and development policy. The Government recognizes that because of the gradation in characteristics of the heavy oils it will be difficult to arrive at satisfactory definitions and that some arbitrariness will probably be necessary.

The Government believes that the best way of developing satisfactory definitions would be through the advice of a special committee composed of representatives from the Alberta Division of the Canadian Petroleum Association, the Independent Petroleum Association of Canada, the Department of Mines and Minerals and the Oil and Gas Conservation Board. The Board will be asked to convene such a committee.

(2) There needs to be clarification of what markets would be considered "beyond present and foreseeable reach of Alberta's conventional industry". The Government believes that the distinction between "within reach" and "beyond reach" markets should not be confined to a geographical one but that "beyond reach markets" should include any markets, including specialty markets, which Alberta's conventional industry is not now serving nor can reasonably be expected to serve in the foreseeable future because of price, quality specification or other reasons. Decision in an actual case might be based on the recommendation of the Board following the public hearing of an application.

(3) With respect to an application proposing the marketing of oil-sand production within reach of the conventional industry, but not in "new" markets as defined later, the Government believes that, as at present, the application should be approved only when indicated to be desirable on the basis of the trend in the life-index of the conventional industry. However, the criterion of percent utilization of productive capacity referred to in the present policy is no longer useful and will be discarded.

(4) With respect to an application proposing the marketing of oil-sand production in markets that are beyond reach of the conventional industry, the present policy is satisfactory and will be continued with such production being unrestricted so long as the development program meets the conservation and related requirements of the Oil and Gas Conservation Board.

(5) Under the present policy experimental operations in the oil sands, not involving commercial production, are encouraged and authorized by the Board and the Government without public hearing.

The Government believes it desirable that this be continued and, for clarification, points out that such operations may involve temporary production and marketing of oil-sand products at levels considered subcommercial by the Board.

One amendment is made in the present policy. This is discussed in the following:

(1) The Government believes that in order to encourage greater growth in the total crude-oil market than would otherwise occur and thereby permit further oil-sand development, the present policy requires amendment with respect to the treatment of applications that provide for marketing a product from oil sands "within reach" of the conventional industry. Where it can be demonstrated that the applicant's marketing proposal would provide such additional growth by the development of a "new" market the Government is prepared to authorize further production of oil-sand product at volumes equal to 50% of the new market. A "new" market would be one not being served today; one over and above the normal growth in existing markets; and one representing a net increase in total market.

The Government believes that applications approved under this modification of the policy would provide the conventional industry with an immediate share of markets which if otherwise obtained at all would have been obtained several years later. The modification, therefore, is unlikely to have any significant adverse effect upon the conventional industry.

It is recognized that during the next few years it is particularly difficult to estimate market growth. In view of this the Government believes it desirable to establish specific limitations on the additional volume of oil-sand production that would be approved under this amendment of the 1962 policy. Accordingly, the total volume of commercial oil-sand production, including the presently authorized production, that will be permitted to enter new markets within reach of the conventional industry will be restricted to 150,000 bbl/day. Unless some wholly unforeseen set of circumstances should develop, this limit will remain in effect for five years. During this period the limit will be reviewed and, if conditions warrant, it may be increased for a succeeding period.

In addition to these matters relating to the circumstances under which additional oil-sand production would be authorized, the Government has also given serious consideration to the question of the royalty payable to the Crown on products derived from bituminous sands or oil sands owned by the Province. Such royalties are prescribed by regulation made under The Mines and Minerals Act, 1962. The Act authorizes the establishment of rates of royalty either of general application or with respect to any specified operation.

In January of 1963, Bituminous Sands Royalty Regulation No. 1 was established fixing the royalty payable until March 31, 1972, on the products recovered in the operation of Great Canadian Oil Sands, Ltd. (45,000 bbl of synthetic crude oil and some 300 long tons of sulfur per day). The present royalty is based upon the value at the plant site of these products and the rate averages out at about 12% on the synthetic crude oil and at 16.66% on the sulfur. (The total royalty is equivalent to some 20% of the value of the raw bitumen from which the synthetic crude oil and sulfur are derived.)

The Government has decided that when the present royalty arrangement with Great Canadian is reviewed, and in the case of any other commercial development of oil sands, it will express the royalty as one applicable to the raw bitumen recovered, at its value at the recovery site. This change in basis will result in comparable royalty treatment regardless of the extent of upgrading and will ensure that there will be no royalty incentive against extensive upgrading of the bitumen in Alberta.

Whether the future royalty rate on the raw bitumen will be altered from one which would yield the same return as under the present arrangements with Great Canadian will depend on future circumstances and whether any changes are found necessary in royalty rates as they apply to the production of Provincially owned oil and gas generally.

In considering the royalty rate expressed on the raw bitumen basis during the first term of a Crown lease, the Government would bear in mind the provision of the lease and The Mines and Minerals Act, 1962, relating to the maximum royalty rates applicable, during the first term of the lease, to the products derived from bituminous sands or oil sands. The total royalty would not exceed that which could be fixed under these limits.

Crown royalties applicable to crude oil produced from wells have in the past been set for periods of ten years, the last regulation coming into effect on April 1, 1962. Accordingly, the next general review of royalty rates will be in 1972.

Appendix III *

Order-in-Council No. 244/72

Edmonton, Alberta February 23, 1972

The Executive Council has had under consideration the report of the Honourable the Minister of Mines and Minerals, dated February 18, 1972, stating that:

Whereas the Energy Resources Conservation Board has inquired into and heard an application by Atlantic Richfield Canada Ltd., Canada-Cities Service, Ltd., Gulf Oil Canada Limited and Imperial Oil Limited (hereinafter referred to as the "applicants") for an amendment of Approval No. 1223 under the Oil and Gas Conservation Act:

Therefore, upon the recommendation of the Honourable the Minister of Mines and Minerals, the Executive Council advises that, pursuant to the Oil and Gas Conservation Act and The Energy Resources Conservation Act, the Lieutenant Governor in Council hereby approves the amendment of Approval No. 1223 in the form hereto attached, subject to the following conditions:

Firstly, that the applicants will grant to Canadian citizens who are residents of the Province of Alberta an opportunity to purchase an equity in the Syncrude project to be operated by Syncrude Canada Ltd., the nature, method of allocation and distribution of the equity to be subject to the approval of the Government of Alberta;

Secondly, that there will be a director of the company incorporated to issue the said equity, a Canadian citizen resident in Alberta whose appointment shall be effective only upon the prior approval of the Government of Alberta;

Thirdly, that the applicants shall give in writing prior to August 31, 1973, to the Government of Alberta a firm commitment to proceed with the Syncrude project;

Fourthly, that the applicants undertake in connection with the Syncrude project wherever practical and reasonable

(a) to use the engineering services of firms or companies whose personnel are residents of Alberta;

(b) to use construction firms owned by residents of Alberta;

(c) to purchase equipment and supplies manufactured in Alberta;

(d) to employ residents of Alberta;

and the applicants will, from time to time, but not less than once every three months, submit to the Government of Alberta a report with respect to items (a), (b), (c) and (d) above, with such explanation as the Government of Alberta may require;

Fifthly, that insofar as it is reasonable to do so, the applicants will ensure that the production, processing and manufacture of by-products developed from the operation of the project will be carried out in the Province of Alberta.

---

* Canada, see Chapter 5 prepared by F.K. Spragins.

Appendix IV *

MINES AND MINERALS

Office of the Minister

407 Legislative Building
Edmonton, Alberta, Canada T5K 2B6
September 14, 1973

Imperial Oil Limited
500 Sixth Avenue S.W.
CALGARY, Alberta
T2P 0S1

Atlantic Richfield Canada Ltd.
650 Guinness House
CALGARY, Alberta
T2P 0Z6

Canada-Cities Service, Ltd.
1100, 550 — 6th Avenue S.W.
CALGARY, Alberta
T2P 2M7

Gulf Oil Canada Limited
707 — 7th Avenue S.W.
CALGARY, Alberta
T2P 2H7

Dear Sirs:

Re: *Syncrude Project*

Reference is made to the recent discussions concerning the above project between representatives of Her Majesty the Queen in the Right of the Province of Alberta ("Her Majesty"), Imperial Oil Limited ("Imperial"), Canada-Cities Service, Ltd. ("Cities"), Atlantic Richfield Canada Ltd. ("ArCan") and Gulf Oil Canada Limited ("Gulf") (Imperial, Cities, ArCan and Gulf and herein jointly called the "Lessees"). As a result of these discussions agreement in principle has been reached concerning certain matters respecting the Syncrude project, including the initial royalty to be paid unto Her Majesty, rights of residents of Alberta to acquire interests in the Syncrude project and related facilities, conditions concerning renewals of leases and the Lessees' commitment to proceed with the Syncrude project.

Her Majesty understands that the Lessees have agreed to proceed with the Syncrude project on the following understandings:

(A) Interpretation

(1) In this letter:

(a) "*Accounting Manual*" shall mean the Accounting Manual to be annexed to the Definitive Agreement;

(b) "*capital costs*" shall mean the capital costs determined in accordance with the provisions of the Accounting Manual incurred by the Lessees on the Syncrude project (other than working capital employed therein) after February 23, 1972;

* Literal text of Letter Agreement; see Chapter 5 prepared by F.K. Spragins.

(c) *"Definitive Agreement"* shall mean the agreement to be hereafter prepared and executed between the parties to this letter setting out in definitive terms the general understandings contained in this letter;

(d) *"Syncrude"* shall mean Syncrude Canada Ltd., a company incorporated under the laws of the Province of Alberta;

(e) *"Syncrude project"* shall mean the project approved in Approval No. 1725 of the Energy Resources Conservation Board, as such project may be amended from time to time by any approval hereafter issued in substitution therefor or amendment thereof under The Oil and Gas Conservation Act with the approval of the Lieutenant Governor in Council; excluding however the utilities plant and the synthetic crude pipeline both as hereinafter defined;

(f) *"synthetic-crude pipeline"* shall mean the pipeline or pipelines hereafter constructed, together with all land and facilities acquired or constructed in connection therewith, primarily for the purpose of transporting crude oil from the Syncrude project to Edmonton, Alberta;

(g) *"the leased substances"* shall mean all substances which the Lessees have the right to recover pursuant to the leases, or any of such substances;

(h) *"the leases"* shall mean Government of Alberta Bituminous Sands Lease No. 17, together with any other documents of title issued in substitution therefore;

(i) *"utilities plant"* shall mean the utilities plant approved in draft Approval No. HE 7313 of the Energy Resources Conservation Board, as such Approval may be hereafter amended from time to time under The Hydro and Electric Energy Act with the approval of the Lieutenant Governor in Council;

(j) *"date of start of production"* shall mean the first date after the Syncrude project has produced an aggregate of 5,000,000 barrels of synthetic crude oil;

(k) *"year"* shall mean a calendar year.

(B) Approval of the Syncrude project

(2) The Lessees acknowledge that Order in Council 244/72, as amended by Order in Council 1337/73, approved the Energy Resources Conservation Board's amendment of its approval of the Syncrude project subject to the following conditions:

(a) that Lessees will grant to Canadian citizens who are residents of the Province of Alberta an opportunity to purchase an equity in the Syncrude project to be operated by Syncrude, the nature, method of allocation and distribution of the equity to be subject to the approval of the Government of Alberta;

(b) that there will be a director of the company incorporated to issue the said equity, a Canadian citizen resident in Alberta whose appointment shall be effective only upon the prior approval of the Government of Alberta;

(c) that the Lessees shall give in writing prior to September 17, 1973, to the Government of Alberta a firm commitment to proceed with the Syncrude project;

(d) that the Lessees undertake in connection with the Syncrude project wherever practical and reasonable:

(I) to use the engineering services of firms or companies whose personnel are residents of Alberta;

(II) to use construction firms owned by residents of Alberta;

(III) to purchase equipment and supplies manufactured in Alberta;

(IV) to employ residents of Alberta; and the Lessees will, from time to time, but not less than once every three months, submit to the Government of Alberta a report with respect to items (I), (II), (III) and (IV) above, with such explanation as the Government of Alberta may require;

(e) that insofar as it is reasonable to do so, the Lessees will ensure that the production, processing and manufacture of by-products developed from the operation of the Syncrude project will be carried out in the Province of Alberta.

(3) The Lessees hereby commit to proceed with the Syncrude project, subject to the provisions hereof, and agree to comply with the other conditions of Order in Council 244/72, as so amended. Her Majesty agrees that this commitment to proceed satisfies the third condition of the said Order in Council, as so amended.

(4) Each of the Lessees hereby covenants and agrees that it shall not mine, produce or process the leased substances except in accordance with the provision of Approval No. 1725 of the Energy Resources Conservation Board, as such Approval may be hereafter amended from time to time under The Oil and Gas Conservation Act with the approval of the Lieutenant Governor in Council.

(C) Provisions concerning royalty

(5) The royalty to be rendered and paid to Her Majesty pursuant to the leases during the term of these provisions shall be a portion of all leased substances derived from the leased lands calculated as hereinafter provided. It is agreed that such royalty portion is and always has been reserved to Her Majesty. The amount of the royalty shall be determined each year after the date of start of production and shall be that percentage of all leased substances recovered in the Syncrude project which are sold or otherwise disposed of during such year (excluding operating and utility fuel requirements and losses of the Syncrude project and the utility plant) and which have an aggregate value equal to 50% of the excess of gross revenue over expenses of the Syncrude project for that year. That excess shall be determined each year after the date of start of production by deducting from the gross revenue of the Syncrude project from all sales or other dispositions (valued at the plant gate at the time of sale or other disposition) during such year the following costs and expenses:

(a) allowed operating costs incurred during such year (which costs shall not include income tax);

(b) amortization of all capital costs (except working capital) of the Syncrude project incurred after February 23, 1972, on a straight-line basis as follows:

(I) all capital costs incurred on or before the beginning of the fifth anniversary of the date of start of production shall be amortized over an assumed life of 20 years commencing on the fifth anniversary of the date of start of production;

(II) all capital costs incurred after the fifth anniversary of the date of start of production shall be amortized over an assumed life commencing on the date such costs were incurred and ending on the 25th anniversary of the date of start of production;

except that if the date of start of production or the date a capital cost was incurred (as the case may be) is not the first day of a year then:

(A) the capital costs which may be amortized during the balance of the year after the fifth anniversary of the date of start of production or the date such capital cost was incurred, shall only be that fraction of the capital costs which could be amortized during such year, if it were a full year, which is equal to the fraction which the number of days in such year after the fifth anniversary of the date of start of production, or the date such capital cost was incurred, bears to 365, and

(B) during the final year in which capital costs may be amortized, capital costs may be amortized for that number of days in such final year up to the 25th anniversary of the date of start of production;

and provided further that in computing cost simple interest at the rate of 8% per annum may be charged up to the date of start of production on costs of construction incurred between September 1, 1973, and the date of start of production providing that the total amount of such interest charges shall not exceed $ 90,000,000;

(c) deemed interest expense (in lieu of actual interest charges) at the rate of 8% per annum of 75% of the average capital (including working capital but excluding any interest on costs of construction permitted pursuant to sub-paragraph (b) above) employed during such year reducing such capital (other than working capital) on a straight-line basis as follows:

(I) capital employed at the date of start of production shall be amortized over an assumed life of 25 years from the date of start of production;

(II) additional capital employed after the date of start of production shall be amortized over an assumed life commencing on the date of employment of such capital and ending on the 25th anniversary of the date of start of production.

Gross revenue, the costs and expenses which may be deducted therefrom, and capital costs shall be calculated in accordance with the provisions of the Accounting Manual. In computing such gross revenue, costs and expenses, no revenue shall be included from the sale or other disposition of technology or assets developed or acquired by the Lessees prior to February 23, 1972, and no charge shall be made for the use of such technology or assets. Any income earned from the Syncrude project prior to the date of start of production shall be deducted in computing costs incurred prior to the date of start of production. Any amount received from the sale or other disposition of material and equipment in the Syncrude project shall be deducted in computing capital costs.

Leased substances recovered in the Syncrude project which are stored on the project site shall not be included in the above calculations until they are sold or otherwise disposed of.

If the sum of the deductions permitted by subparagraphs (a), (b) and (c) hereof during any year exceeds the gross revenue for such year the amount of such excess shall be carried forward cumulatively and allowed as an operating cost in computing the excess of gross revenue over expenses of the Syncrude project during succeeding years.

(6) Within 15 days after the end of each month of each year Syncrude shall make a good faith estimate in accordance with the Accounting Manual of the amount of royalty payable with respect to the Syncrude project for such year. The Lessees shall cause Her Majesty to be advised in writing of the amount of such estimated royalty within such 15-day period and at the same time shall pay Her Majesty a royalty instalment in an amount equal to 1/12th of such estimated royalty. The total of the royalty instalments paid in respect of such year shall be adjusted within 90 days after the end of the year in accordance with the actual royalty payable with respect to the Syncrude project for the year as verified by audited financial statements (which shall be supplied to Her Majesty) of the royalty payable for such year. Such financial statements shall include details of the calculation of such royalty and an auditor's certificate as to such calculation.

(7a) Representatives of Her Majesty shall have access at all reasonable times to all information, data, contracts and agreements relating to the Syncrude project, the utilities plant and the synthetic crude oil pipeline that the Lessees are entitled to disclose, including, without limitation, information, data, contracts and agreements relating to:

(I) the design, engineering, construction or operation of the Syncrude project, the utilities plant and the synthetic crude pipeline, or any of them,

(II) the purchase or other acquisition of materials and supplies for the Syncrude project, the utilities plant and the synthetic-crude pipeline, or any of them, and

(III) the sale or other disposition of leased substances from the Syncrude project.

All such information, data, contracts and agreements shall at all times be kept secret and confidential and the Definitive Agreement shall set out appropriate confidentiality requirements for the representatives of Her Majesty.

(7b) From and after the date of this letter representatives of Her Majesty,

(I) shall have the right to meet with the Lessees, or their representatives, once each month to be advised of such matters relating to the Syncrude project as Her Majesty may reasonably request, and shall receive notice of and have the right to attend all meetings of directors of Syncrude;

(II) shall have the right from time to time to audit, at Her Majesty's expense, the past and current costs and expenses incurred or committed to be incurred with respect to the Syncrude project; and

(III) shall be entitled to currently receive all data and information concerning the Syncrude project in order to keep Her Majesty fully informed of all matters relating to the Syncrude project.

(c) Her Majesty shall designate such representatives by notice in writing to Syncrude within 90 days after the date hereof, and shall have the right from time to time to change such representatives by a similar notice in writing.

(d) Neither Her Majesty, nor any of Her representatives, shall have any right to vote at any such meeting and shall not have the right to approve or disapprove any action or proposed action by any of the participants in the Syncrude project and shall not be liable for any costs, expenses or liabilities thereunder.

(8) The Lessees shall cause Syncrude to give Her Majesty written notice of the date of start of production within 15 days after such date.

(9) Notwithstanding the above, Her Majesty shall have the right at any one time after the fifth anniversary of the date of start of production to take Her royalty in an amount equal to $7\frac{1}{2}\%$ of the total annual production of any and all of the leased substances recovered from the Syncrude project, excluding however leased substances (other than synthetic crude oil) consumed in the operation of, or lost in, the Syncrude project or the utility plant. Her Majesty may exercise this right by delivering a notice in writing to that effect to Syncrude, whereupon this right shall be exercised as of the effective date specified in the notice (which shall not be earlier than the date of delivery of the notice) or the fifth anniversary of the date of start of production, whichever is the later. If Her Majesty exercises this right, the royalty payable pursuant to these royalty provisions shall, after the effective date of such exercise, be calculated pursuant to this clause instead of pursuant to clause 5 hereof.

(10) Any sale of the leased substances, or any of them, until otherwise ordered by the Minister of Mines and Minerals, shall include the royalty share thereof belonging to Her Majesty. Her Majesty reserves the right to take Her royalty in kind at the plant gate by taking an amount of any and all of the leased substances recovered in the Syncrude project:

(a) having an aggregate fair market value for each year equivalent to the value of the royalty calculated in accordance with clause 5 hereof for such year, or

(b) which during each year shall be equal to the royalty calculated in accordance with clause 9 hereof for such year,

whichever clause is the applicable.

If Her Majesty exercises the right to take Her royalty in kind pursuant to this clause 10, She shall be entitled to take in kind each month 1/12th of the total amount She would be entitled to take during such year, based where applicable upon the estimates referred to in clause 6, and subject to adjustment within 90 days after the end of such fiscal year in accordance with the actual royalty for such year as verified by the audited financial statements referred to in clause 6. If Her Majesty exercises the right to take in kind set forth in this clause, Her Majesty shall provide all tanks and other facilities required to take the production in kind.

(11) In the event of the sale by the Lessees of the royalty share of the leased substances belonging to Her Majesty, the deductions that may be allowed for charges incurred in transporting such royalty share of the leased substances shall:

(I) in the case of synthetic crude oil transported in the synthetic crude pipeline, be the tariffs charged by such pipeline, and

(II) in the case of other leased substances, be the reasonable deductions specified by the Minister of Mines and Minerals.

(12) These provisions concerning royalty shall come into effect on the date hereof and shall remain in full force and effect until changed as hereinafter provided. Her Majesty shall have the right to change these royalty provisions at any time after the tenth anniversary of the date of start of production. In addition, should substantial changes in circumstances occur which are not now reasonably within the contemplation of the parties hereto:

(a) Syncrude may request a review and change of these royalty provisions at any time thereafter, and

(b) Her Majesty may review and change these royalty provisions at any time after the fifth anniversary of the date of start of production.

Without restricting the generality of the foregoing, the parties hereto agree they anticipate that, within five years after the date of start of production, the Syncrude project will, over a six-month period, be producing synthetic crude oil at an average rate of at least 100,000 bbl/d.

(13) Notwithstanding anything herein contained or implied to the contrary, these royalty provisions will be subject to review and revision in the event of changes in Federal Government policy or laws which could materially affect the position of Her Majesty or any of the other parties hereto.

(14) The Minister of Mines and Minerals shall recommend to the Lieutenant Governor in Council that regulations be passed pursuant to section 174 of The Mines and Minerals Act prescribing the above royalty with respect to the leases.

(D) Environmental considerations

(15) Reference is made to the letter dated July 13, 1973, from the Minister of the Environment to the President of Syncrude concerning environmental matters relating to the Syncrude project. The Lessees agree to comply with the provisions of the said letter in their construction and operation of the Syncrude project. Her Majesty agrees that the commitment contained in paragraph 6 of the said letter is a commitment binding upon Her Majesty.

(16) As required by sub-paragraph 5 (c) of the said letter, the Lessees agree to deposit with Her Majesty funds, or a bond acceptable to Her Majesty, in an amount sufficient to ensure to the satisfaction of Her Majesty proper reclamation of the lands involved in the Syncrude project. If, upon termination of the Syncrude project, the lands are reclaimed to the satisfaction of Her Majesty the said funds or bond shall be returned to, or to the order of, the Lessees. Otherwise, the said funds or bond shall be forfeited.

(E) Public participation

(17a) The Lessees hereby grant Her Majesty an irrevocable option to acquire an interest in the Syncrude project, including the project site, the leases and rights granted thereby, and all facilities acquired or constructed as part of the Syncrude project, which interest may be for an undivided percentage interest of not less than 5% and up to and including 20%.

(b) The above option may be exercised at any time during the period from the date hereof and up to and including that date which is six months after the date of start of production, or the 31st day of December, 1982, whichever is the earlier. If the option is exercised prior to the date of start of production the interest for which the option is exercised will be deemed to have been acquired as and from the date of delivery to Syncrude of the notice exercising the option. If the option is exercised on or after the date of start of production the interest for which the option is exercised will be deemed to have been acquired as and from the date of start of production.

(c) The option may be exercised by notice in writing to Syncrude delivered within the said period setting out the interest for which the option is exercised and within 60 days after delivery of such written notice the optionee will pay to the Lessees the cost of acquiring such interest, computed as hereinafter stated.

(d) If the option is exercised the optionee will pay, and the Lessees will receive as proceeds of disposition, an undivided percentage share of all costs (net of income) incurred in the Syncrude project after February 23, 1972, and up to the date from which the interest is acquired under option equal to the percentage interest which the optionee elects to acquire, and the optionee shall become a full joint venture participant in the Syncrude project as and from such date. Such costs shall be computed in accordance with the Accounting Manual and shall include interest (compounded annually) on such costs at 8% per annum. No portion of such costs shall be attributed to reserves of leased substances.

(e) Upon exercise of the said option the Lessees shall assign to the optionee, in proportion to their respective interests therein, the interest which the optionee has elected to acquire in the Syncrude project, including the project site, the leases and all rights granted thereby, and all facilities constructed in connection therewith, and an equal percentage share of all outstanding shares of Syncrude and the Lessees and the optionee shall execute a joint venture agreement, and the Lessees, the optionee and Syncrude shall execute an operating agreement (each such agreement to be in a form to be annexed to the Definitive Agreement) for the operation of the Syncrude project. Syncrude shall be the operator under such operating agreement. The optionee shall not acquire any ownership in the technology and assets developed or acquired by the Lessees prior to February 23, 1972.

(f) Her Majesty may assign all or any part of the option granted in this clause 17 to an entity or entities to be formed for the purpose of permitting the people of Alberta to participate, inter alia, in the Syncrude project. The word "optionee" herein shall mean Her Majesty and any entity or entities to which this option, or any part thereof, is assigned.

(18a) The synthetic crude pipeline shall be constructed and owned as to an undivided 80% thereof by Her Majesty and/or an entity or entities hereafter formed by Her Majesty for the purpose of permitting the people of Alberta to participate, inter alia, in the synthetic crude pipeline, and as to the remaining 20% thereof by the Lessees or their assignees in the respective undivided percentage interests in which the Lessees presently own the leases.

(b) Her Majesty and/or the entity or entities formed by Her Majesty and the Lessees shall execute an agreement (in a form to be annexed to the Definitive Agreement) providing for the design, construction and operation of the synthetic crude pipeline.

(c) Notwithstanding anything herein or in such agreement contained or implied to the contrary:

(I) each owner of an interest in the synthetic crude pipeline shall be obligated, but only to the extent of its equity interest therein, to execute such guarantees, stock subscription agreements or similar documents in aid of financing as may be required by lending institutions in any financing arrangements for the synthetic crude pipeline;

(II) the Lessees and Her Majesty and their respective assignees of any interest in the Syncrude project will dedicate their respective shares of the synthetic crude oil recovered from the Syncrude project to the synthetic crude pipeline;

(III) the pipeline tariff for synthetic crude oil from the Syncrude project shall be sufficient to cover operating costs, interest on debt, recapture of capital (over the projected life of the Syncrude project) and a reasonable return, provided that if the capacity of the synthetic crude pipeline exceeds 125,000 bbl/d, the tariff charged for synthetic crude oil from the Syncrude project shall not exceed the amount which could reasonably be charged for the pipeline if its capacity were only 125,000 bbl/d.

(19) Her Majesty, and/or an entity or entities hereinafter formed by Her Majesty for the purpose of permitting the people of Alberta to participate therein and other operations, shall have an undivided 50% interest in the utilities plant. The Lessees, or another corporation designated by them (if they desire to have another corporation construct and operate the utilities plant) shall own the remaining 50% interest in the utilities plant. The respective owners of the utilities plant, the Lessees and Syncrude, shall execute an agreement (in a form to be annexed to the Definitive Agreement) for the construction and operation of the utilities plant; which agreement shall, inter alia, ensure the owners of the utilities plant a reasonable return calculated on the basis of 75% debt and 25% equity and will provide for the utilities plant to be operated for the owners of the Syncrude project on a cost of service basis.

(F) Lease Renewal

(20) The terms of those portions of the following Alberta Bituminous Sands Leases owned by the Lessees, namely Leases 22, 29, 31, 32, 40, 41 and 78, containing rights which in the opinion of the Minister lie within the mineable portion of the bituminous sands area shall be for an initial term of 21 years renewable as hereinafter prescribed:

(a) such portions of the said leases may be renewed at the end of the initial 21-year term in accordance with Alberta Regulation 130/69 provided that the lessees thereof have entered into exploration commitments satisfactory to the Minister designed to evaluate the economically extractable deposits of bituminous sands underlying such portions and the thickness and composition of associated overburden materials;

(b) work carried out prior to the date hereof shall at the discretion of such Minister be credited towards the lessee's exploration commitment for lease renewal;

(c) the exploration commitment shall be carried out in conformity of time schedules related to the expiry of the initial terms of the said leases, subject to the discretion of such Minister:

(i) at least six months prior to renewal of those leases expiring in 1975 to 1979 inclusive; and

(ii) by January 1, 1980, for those leases expiring in 1980 and later;

(d) the renewal shall otherwise be subject to the provisions of Alberta Regulation 130/69.

(G) Conditions

(21) The agreement of Her Majesty and the Lessees is subject to the following conditions, namely:

(a) that such Syncrude contractors as Syncrude may request shall enter into a site agreement or agreements with labor organizations in a form which will have the effect of bringing the various trade components under one set of working conditions and which will achieve labor stability through to the completion of the project;

(b) that a Federal income tax ruling or other advice, satisfactory to her Majesty and the Lessees, be obtained on or before November 16, 1973, to the effect that either the royalty reserved by Her Majesty pursuant to the royalty provisions referred to herein will not be included in computing the income of the Lessees for Canadian income tax purposes, or payments of such royalty to Her Majesty will be a deductible expense of the Lessees in computing their income for Canadian income tax purposes; and, in any event, such royalty payments will not be considered to be taxes on income from mining operations which, after 1976, may not be deduced in computing income under the Canadian Income Tax Act;

(c) that the Federal Government does not regulate directly or indirectly the prices of synthetic crude oil below the levels attainable in a free international market.

(H) Definitive agreement

(22) Forthwith after the Lessees' approval of this letter a Definitive Agreement shall be prepared and executed between Her Majesty and the Lessees. The Definitive Agreement shall incorporate the general understandings contained herein and shall have annexed thereto the Accounting Manual and the forms of agreements referred to in clauses 17 (e), 18 (b) and 19 hereof for the construction and operation of the Syncrude project, the synthetic crude pipeline and the utilities plant.

If the Lessees agree with the above understandings, will each of the Lessees please execute and deliver all five copies of this letter so that Her Majesty and each Lessee may have one fully executed copy thereof.

Yours very truly,
Bill Dickie, Q.C.
Minister of Mines and Minerals

Agreed to this 14th day of September, 1973

IMPERIAL OIL LIMITED
Per

ATLANTIC RICHFIELD CANADA LTD.
Per

CANADA-CITIES SERVICE, LTD.
Per

GULF OIL CANADA LIMITED
Per

Appendix V *

MINES AND MINERALS

Office of the Minister

407 Legislative Building
Edmonton, Alberta, Canada T5K2B6
December 13, 1973

Imperial Oil Limited
500 Sixth Avenue S.W.
CALGARY, Alberta T2P 0S1

Atlantic Richfield Canada Ltd.
650 Guinness House
CALGARY, Alberta T2P 0Z6

Canada-Cities Service, Ltd.
1100, 550 — 6th Avenue S.W.
CALGARY, Alberta T2P 2M7

Gulf Oil Canada Limited
707 — 7th Avenue S.W.
CALGARY, Alberta T2P 2H7

* Literal text of Amended Letter; see Chapter 5 prepared by F.K. Spragins.

Dear Sirs:

Re: *Syncrude Project*

Reference is made to the letter dated September 14, 1973, between us (herein called the "Syncrude Letter") setting out the agreement in principle which had been reached between us concerning certain matters respecting the Syncrude Project.

This letter will confirm that you have requested, and we have agreed to, certain amendments to the Syncrude Letter as follows:

1. Division C, containing paragraphs 5 to 14 inclusive, of the Syncrude Letter is hereby deleted in its entirety and the following is substituted therefor:

(C) Provisions concerning joint venture

(5a) Her Majesty and the Lessees shall form a joint venture for the conduct of the Syncrude project, and shall enter into an agreement (in a form to be annexed to the Definitive Agreement) providing for the operation of such joint venture on the terms and conditions herein set out. Her Majesty shall commit Her Lessor's interest in the leases and in the leased substances, and each of the Lessees shall commit its lessee's interest in the leases and the leased substances, to such joint venture.

(5b) Her Majesty shall have a participating interest in such joint venture equal to 50% of the net profits therefrom, computed as hereinafter set out. Her Majesty agrees that, so long as She is a party to such joint venture, She shall not exercise any right She may have under the leases or the laws applicable thereto to prescribe a royalty pursuant to the leases. The remaining participating interests in the joint venture shall be owned by the Lessees in proportion to their respective interests in the leases, subject to the right of public participation referred to in paragraph 17 hereof.

(5c) The joint venture agreement shall provide for the joint operation of the leases, but shall not provide for any transfer of interests in the leases and shall not grant Her Majesty any property interest in the plant or equipment acquired for such operations. Syncrude shall operate the joint venture for the joint venture participants but shall not be paid any fee, or be entitled to any profit from the joint venture, for such operation.

(6) All costs, expenses and losses of the joint venture shall be paid and borne by the Lessees, in proportion to their respective participating interests therein and the Lessees shall, in proportion to their respective participating interests indemnify and save harmless Her Majesty from any and all actions, suits, proceedings, claims, costs, demands and expenses which may be brought against or suffered by Her Majesty, or which Her Majesty may sustain, pay or incur, by reason of any matter or thing arising out of or in any way attributable to the joint venture or the operations conducted by the Lessees, or any of them, or by Syncrude, or their respective servants, agents, employees or contractors with respect thereto.

(7) The net profit from the joint venture shall be determined each year after the date of start of production by deducting from the gross revenue of the joint venture from all sales or other dispositions (valued at the plant gate at the time of sale or other disposition) during such year the following costs and expenses:

(a) allowed operating costs incurred during such year (which costs shall not include income tax);

(b) amortization of all capital costs (except working capital) of the Syncrude project incurred after February 23, 1972, on a straight-line basis as follows:

(I) all capital costs incurred on or before the beginning of the fifth anniversary of the date of start of production shall be amortized over an assumed life of 20 years commencing on the fifth anniversary of the date of start of production;

(II) all capital costs incurred after the fifth anniversary of the date of start of production shall be amortized over an assumed life commencing on the date such costs were incurred and ending on the 25th anniversary of the date of start of production;

except that if the date of start of production or the date a capital cost was incurred (as the case may be) is not the first day of a year then:

(A) the capital costs which may be amortized during the balance of the year after the fifth anniversary of the date of start of production, or the date such capital cost was incurred, shall only be that fraction of the capital costs which could be amortized during such year, if it were a full year, which is equal to the fraction which the number of days in such year after the fifth anniversary of the date of start of production, or the date such capital cost was incurred, bears to 365, and

(B) during the final year in which capital costs may be amortized, capital costs may be amortized for that number of days in such final year up to the 25th anniversary of the date of start of production;

and provided further that in computing capital costs simple interest at the rate of 8% per annum may be charged up to the date of start of production on costs of construction incurred between September 1, 1973, and the date of start of production providing that the total amount of such interest charges shall not exceed $ 90,000,000;

(c) deemed interest expense (in lieu of actual interest charges) at the rate of 8% per annum of 75% of the average capital (including working capital but excluding any interest on costs of construction permitted pursuant to subparagraph (b) above) employed during such year reducing such capital (other than working capital) on a straight-line basis as follows:

(i) capital employed at the date of start of production shall be amortized over an assumed life of 25 years from the date of start of production;

(ii) additional capital employed after the date of start of production shall be amortized over an assumed life commencing on the date of employment of such capital and ending on the 25th anniversary of the date of start of production.

Gross revenue, the costs and expenses which may be deducted therefrom, and capital costs shall be calculated in accordance with the provisions of the Accounting Manual. In computing such gross revenue, costs and expenses, no revenue shall be included from the sale or other disposition of technology or assets developed or acquired by the Lessees prior to February 23, 1972, and no charge shall be made for the use of such technology or assets. Any income earned from the Syncrude project prior to the date of start of production shall be deducted in computing costs incurred prior to the date of start of production. Any amount received from the sale or other disposition of material and equipment in the Syncrude project shall be deducted in computing capital costs.

Leased substances recovered in the Syncrude project which are stored on the project site shall not be included in the above calculations until they are sold or otherwise disposed of.

If the sum of the deductions permitted by subparagraphs (a), (b) and (c) hereof during any year exceeds the gross revenue for such year the amount of such excess shall be carried forward cumulatively and allowed as an operating cost in computing the net profit of the Syncrude project during succeeding years.

(8a) Her Majesty shall be entitled to, and shall receive, in satisfaction of Her participating interest share of the net profits of the joint venture during each year after the date of start of production, a share of all leased substances recovered in the Syncrude project by the joint venture which are sold or otherwise disposed of during such year (excluding operating and utility fuel requirements and losses of the Syncrude project and the utility plant) and which have an aggregate value equal to 50% of the net profit of the joint venture for that year. Syncrude shall make a good faith estimate, in accordance with the Accounting Manual, each month of each such year of the net profit of the joint venture, and of Her Majesty's share thereof, for that year, and shall deliver to, or to the order of, Her Majesty at the plant gate during such month leased substances recovered in the Syncrude project having an aggregate value equal to 1/12th of Her Majesty's share of the estimated net profit for the year. The total amount of leased substances so delivered to Her Majesty each year shall be adjusted within 90 days after the end of the year in accordance with the actual net profit of the joint venture for the year as verified by audited financial statements (which shall be supplied to Her Majesty). Such financial statements shall include details of the calculation of Her Majesty's share of such net profit and an auditor's certificate as to such calculation.

(8b) Whenever and so long as Her Majesty shall fail to take in kind Her participating interest share of the leased substances recovered in the Syncrude project each of the Lessees shall dispose of that percentage of such share which is equal to its respective participating interest on behalf of Her Majesty and shall account to Her Majesty for all proceeds therefrom, less the reasonable charges which Her Majesty may allow in connection with such disposition, within 15 days following the end of the month in which such disposition is made; provided that each of the Lessees may not enter into any contract for such disposition having a term longer than one year and shall not make any such disposition which is contrary to any instructions in writing they may have received from Her Majesty.

(8c) If Her Majesty takes Her participating interest share of the leased substances in kind, Her Majesty shall provide all tanks and other facilities required to take such share in kind.

(9) In the event of any disposition by any of the Lessees of its proportionate share of the participating interest share of the leased substances belonging to Her Majesty, the deductions that may be allowed for charges incurred in transporting such share of the leased substances shall:

(I) in the case of synthetic crude oil transported in the synthetic crude pipeline, be the tariffs charged by such pipeline, and

(II) in the case of other leased substances, be the reasonable deductions specified by the Minister of Mines and Minerals.

(10) Her Majesty shall have the right at any one time after the fifth anniversary of the date of start of production to terminate the joint venture and to elect to take a gross production royalty in an amount equal to $7\frac{1}{2}\%$ of the total annual production of any and all of the leased substances recovered from the Syncrude project, excluding however leased substances (other than synthetic crude oil) consumed in the operation of, or lost in, the Syncrude project or the utility plant. Her Majesty may exercise this right by delivering a notice in writing to that effect to Syncrude, whereupon this right shall be exercised as of the effective date specified in the notice (which shall not be earlier than the date of delivery of the notice or the fifth anniversary of the date of start of production, whichever is the later).

(11a) Her Majesty shall have the right to require a review and change of the joint venture provisions at any time after the tenth anniversary of the date of start of production. In addition, should substantial changes in circumstances occur which are not now reasonably within the contemplation of the parties hereto:

(I) Syncrude may request a review and change of the joint venture provisions at any time thereafter, and

(II) Her Majesty may request a review and change of the joint venture provisions at any time after the fifth anniversary of the date of start of production.
Without restricting the generality of the foregoing, the parties hereto agree they anticipate that, within five years after the date of start of production, the Syncrude project will over a six-month period, be producing synthetic crude oil at an average rate of at least 100,000 bbl/d.

In the event the parties are unable to agree on any change of the joint venture provisions requested pursuant to this paragraph, Her Majesty shall have the right at any time thereafter to terminate the joint venture by delivering a notice in writing to that effect to Syncrude.

(11b) Notwithstanding anything herein contained or implied to the contrary, any party hereto shall have the right to terminate the joint venture provisions at any time, by delivering notice in writing to that effect to Syncrude, in the event of changes in Federal Government policy or laws which could materially affect the position of Her Majesty or any of the other parties hereto.

(11c) In the event of the termination of the joint venture provisions as herein provided Her Majesty shall have the right thereafter to prescribe a royalty to be paid under the leases pursuant to the provisions of The Mines and Minerals Act. Her Majesty shall endeavour to ensure that this royalty will be of such a kind that it will either not be included in computing the income of the Lessees, or if so included will be deductible by the Lessees, under the provisions of the Income Tax Act of Canada as presently worded and under the proposed amendments thereto referred to in paragraph 6.028 of the Corporate Tax Guide published by the Department of National Revenue. Her Majesty understands that these proposed amendments will provide that mining taxes (including income royalties) will not be deductible in computing income for 1977 and subsequent taxation years.

(12) Representatives of Her Majesty shall have access at all reasonable times to all information, data, contracts and agreements relating to the Syncrude Project, the utilities plant and the synthetic crude oil pipeline that the Lessees are entitled to disclose, including, without limitation, information, data, contracts and agreements relating to:
(I) the design, engineering, construction or operation of the Syncrude Project, the utilities plant and the synthetic crude pipeline, or any of them,
(II) the purchase or other acquisition of materials and supplies for the Syncrude Project, the utilities plant and the synthetic crude pipeline, or any of them, and
(III) the sale or other disposition of leased substances from the Syncrude Project.
All such information, data, contracts and agreements shall at all times be kept secret and confidential and the Definitive Agreement shall set out appropriate confidentiality requirements for the representatives of Her Majesty.

(13a) From and after the date of this letter representatives of Her Majesty,
(I) shall have the right to meet with the lessees, or their representatives, once each month to be advised of such matters relating to the Syncrude project as Her Majesty may reasonably request, and shall receive notice of and have the right to attend all meetings of directors of Syncrude;
(II) shall have the right from time to time to audit, at Her Majesty's expense, the past and current costs and expenses incurred or committed to be incurred with respect to the Syncrude project; and
(III) shall be entitled to currently receive all data and information concerning the Syncrude project in order to keep Her Majesty fully informed of all matters relating to the Syncrude Project.

(13b) Her Majesty shall designate such representatives by notice in writing to Syncrude within 90 days after the date hereof, and shall have the right from time to time to change such representatives by a similar notice in writing.

(13c) Neither Her Majesty, nor any of Her representatives, shall have any right to vote at any such meeting and shall not have the right to approve or disapprove any action or proposed action by any of the participants in the Syncrude Project and shall not be liable for any costs, expenses or liabilities thereunder.

(14) The Lessees shall cause Syncrude to give Her Majesty written notice of the date of start of production within 15 days after such date."

2. Paragraph 17 (a) of the Syncrude Letter shall be amended by deleting the word "an" in the second line thereof and inserting in place thereof the words "a share of the Lessees'"; by deleting the word "interest" in the fifth line thereof and substituting therefor the word "share"; by inserting at the end of the fifth line thereof after the word "percentage" the words "of the Lessees"; and by adding at the end thereof after the figures "20%" the following, "after deduction of any joint venture interest or royalty interest to which Her Majesty is entitled pursuant to Division C hereof".

3. Paragraph 21 of the Syncrude Letter shall be amended as follows:

(a) by deleting from the second line of subparagraph (a) thereof the words "a site agreement or";

(b) by deleting subparagraphs (b) and (c) thereof in their entirety.

If you agree with the foregoing, will each of you please execute and deliver all five copies of this letter so that Her Majesty and each of the Lessees may have one fully executed copy thereof. Upon execution and delivery of all copies the Syncrude Letter shall be amended as set out above.

Yours very truly,
Bill Dickie, Q.C.
Minister of Mines and Minerals

Agreed to this 13th day of December, 1973.

IMPERIAL OIL LIMITED
Per:

ATLANTIC RICHFIELD CANADA LTD.
Per:

CANADA-CITIES SERVICE, LTD.
Per:

GULF OIL CANADA LIMITED
Per:

Appendix VI

Fracturing of Reservoir Rocks

WALTER H. FERTL and GEORGE V. CHILINGARIAN

Introduction

Several basic objectives are accomplished by fracturing a reservoir rock. These objectives include (1) to overcome the reservoir damage around the borehole *, (2) to

* Due to drilling fluid invasion, swelling of clays by the mud filtrate, and precipitation of various chemical compounds.

create deep penetrating fractures to increase formation permeability and thus well productivity, (3) to assist in secondary and/or tertiary recovery projects in tight reservoirs and heavy-oil or tar-sand deposits, and (4) facilitate the subsurface injection or disposal of brine and industrial waste material. Induced fractures, which radiate from the wellbore, act as gathering lines connecting porous and permeable zones.

Basic rock mechanics and fracturing concepts are equally important in present-day drilling and completion operations, which aim at maximum well control combined with minimum cost. Properly engineered drilling operations attempt to avoid, or at least minimize, the danger of well kicks, stuck drill pipe, lost circulation of drilling muds, etc. Primary and squeeze cementing operations, fluid injection, sand consolidation, and hydraulic fracturing all require a basic understanding of two key formation parameters, namely, formation *pore pressure* and *fracture pressure*.

Pertinent literature related to (1) the theory and practical implications of consolidation and the state of stress in compacting sediments [20] and (2) the importance of formation pressures to the exploration, drilling and production of oil and gas resources [11] has been recently summarized in detail. An excellent basic theoretical analysis of fracturing has been presented by Craft et al. [9].

Inasmuch as special subsurface recovery schemes for hydrocarbon production from tar sand may require an artificial increase of reservoir permeability, such as due to fracturing, the present discussion attempts to introduce the reader to some of the basic concepts and theories involved. A brief look is taken at some of the concepts of both *hydraulic* and *nuclear* fracturing.

Hydraulic fracturing

Basic considerations

Hydraulic fracturing is the process of creating an artificial fracture or fracture system in a subsurface formation by injecting a fluid, i.e., the *frac-fluid*, under pressure into the rock. Depending on the prevailing in-situ stress environment, either horizontal or vertical fractures can be created. Such hydraulic fracturing is accomplished as a result of overcoming the native, natural state of stresses by exceeding the failure limit of the rock.

Fig. A-1 is a schematic diagram showing relationship between fluid pressure and time. To fracture a reservoir rock, such as a tar sand, the pressure must be applied by injecting a fluid down a wellbore and into the formation. A specific pressure is required to initiate

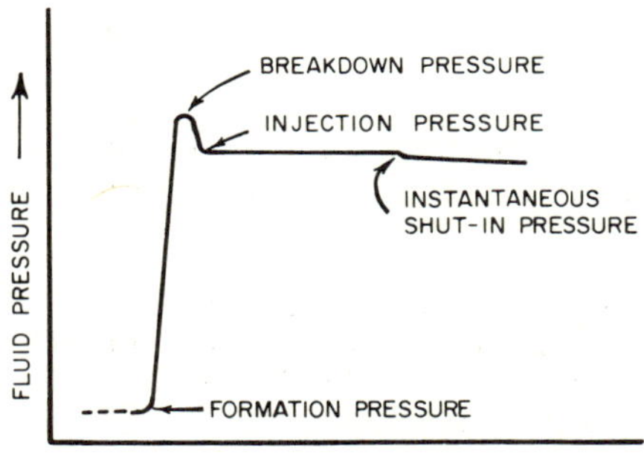

Fig. A-1. Schematic relationship between pore-fluid pressure and time, showing breakdown, injection, instantaneous shut-in, and formation pressures.

fracturing of the reservoir rock. This pressure is called the *breakdown pressure* of a formation or referred to as *fracture initiation pressure*. A slightly lower pressure will be necessary to continue fluid injection and propagation of the created fracture; this is called the *injection pressure* or *fracture propagation pressure*.

These pressures are related to formation pore pressure, lithology of the reservoir rock, age and depth of the formation, and the in-situ rock stress environments.

Reservoir rocks under high formation pressure usually are harder to fracture (i.e., require higher fracture initiation pressures) than low-pressured reservoir rocks. Decline in fracturing and reservoir pressures is commonly observed in old, depleted oil and gas fields. The opposite, i.e., a concurrent increase in both fracturing and reservoir pressures, is frequently observed in secondary recovery projects, such as water-flood operations.

The mechanics of fracture initiation and extension, *and* the resulting fracture geometry, are related to the stress condition around the borehole, the properties of the rock, the characteristics of the frac-fluid, and the manner in which the frac-fluid is injected. Besides the actual mechanics of failure of reservoir rocks (i.e., their fracturing), however, consideration also needs to be given to the amount of energy loss while pumping the fracturing fluid through the well head down the wellbore and injecting into the reservoir rock. The transmission efficiency of the energy delivered at the well head by the surface pumping equipment and transmitted to the formation underground is a function of (1) the frictional pressure drop in the wellbore and fracture itself, and (2) the pressure drop due to leakoff (i.e., flow) from the open fracture into the rock.

Successful fracturing of deep wells in excess of 10,000 ft requires a thorough understanding of treatment hydraulics and fracturing fluid behavior. It should be pointed out, however, that the tar-sand deposits of present economic interest are encountered at rather shallow depth.

Concepts of hydraulic fracturing are related to (1) mechanics of rock failure, (2) general state of stress, (3) elasticity versus plasticity, (4) fluid penetration into the formation, and (5) homogeneity and isotropy versus heterogeneity [25].

Basically the two main areas of investigation relate to (1) concepts of fracture mechanisms on a microscopic scale, and (2) concepts of strength and deformation on a macroscopic scale [6].

Numerous theories have been proposed to predict the breakdown or rupture behavior of rocks surrounding a wellbore. It is generally accepted that the Griffith failure theory [12] describes quite well the behavior of isotropic materials, whereas the behavior of anisotropic materials based on experimental data seems to be best defined by modified versions of theories developed by Griffith [12], Walsh and Brace [27] and the Coulomb-Navier Theory [6]. The latter theory is a special case of Mohr's graphical method to determine the limits of failure [18,21,24].

Mechanics of hydraulic fracturing

The main criteria of consideration in hydraulic fracturing include the stress conditions around a wellbore, the properties of the formation and frac-fluid, and the physical manner in which the latter is injected.

Fracture-gradient determinations. Hubbert and Willis [13] postulated that in geologic regions with normal faulting the least stress should be horizontal and most probably would range between 1/2 to 1/3 of the effective pressure of the overburden. Furthermore, hydraulic fracture pressures should be nearly equal to the least-principal compressive stress component in the subsurface.

Overburden pressure (P_o) equals the sum of formation pressure (P_f) plus the vertical matrix stress (σ_v). Basically then, fracture propagation pressure (FP) has to overcome the pore pressure (P_f) and the horizontal rock matrix stress (σ_H):

$$FP = P_f + \sigma_H \tag{A-1}$$

Assuming then, as Hubbert and Willis postulated, that

$$\sigma_H = (1/3 \text{ to } 1/2)\, \sigma_v = (1/3 \text{ to } 1/2)\, (P_o - P_f) \tag{A-2}$$

one can mathematically express the fracture pressure gradient (*FPG*) as follows:

$$FPG = \frac{P_f}{D} + (1/3 \text{ to } 1/2) \frac{P_o - P_f}{D} \tag{A-3}$$

where D is depth in ft, if *FPG* is reported in psi/ft.

This fracture pressure gradient is dependent on the overburden, pore pressure gradient, and rock frame stress. Hence, for normal hydrostatic pressure conditions and an overburden gradient of 1.0 psi/ft, the minimum and maximum *FPG* are equal to 0.64 psi/ft and 0.73 psi/ft, respectively.

Figure A-2 graphically presents both minimum and maximum fracture pressure gradients. For all normally-pressured formations, the pressure gradients remain constant with increasing depth. The difference (i.e., tolerance) between maximum and minimum fracture pressure gradients, however, decreases with increasing pore pressure gradient. Hence surge pressures, which occur while running pipe, starting mud pumps, etc., become more and more critical with higher pore pressure gradients (Fertl, 1976, [11]).

Several other fracture pressure gradient prediction methods have been developed by the oil industry (Fig. A-3). Techniques particularly related to drilling operations include the methods proposed by Kelly and Matthews (1967, [16]), Eaton (1969, [10]), Taylor and Smith (1970, [23]), MacPherson and Berry (1972, [15]), Anderson et al. (1973, [1]), and Christman (1973, [8]). Regardless of the prediction method used, formation pore pressure and fracture pressure gradient curves (plotted versus true vertical depth) are the two basic parameters for any properly engineered well drilling and completion plan.

Fracture orientation. Theoretically, hydraulic fracturing occurs in a plane perpendicular to the direction of the *least* principal stress. Hence there is always a relationship between the relative magnitude of principal stresses and fracture orientation.

As shown in Fig. A-4, (1) if $\sigma_{H_2} > \sigma_{H_1} > \sigma_v$, fracture is horizontal and normal to σ_v; (2) if $\sigma_{H_1} > \sigma_v > \sigma_{H_2}$, fracture is vertical and normal to σ_{H_2}; and (3) if $\sigma_{H_2} > \sigma_v > \sigma_{H_1}$, fracture is vertical and normal to σ_{H_1}.

In areas with normal faulting, the least principal stress is horizontal and thus vertical fractures result. If horizontal principal stresses are not equal, then the vertical fracture will have a preferred azimuth perpendicular to the direction of minimum stress. In very shallow formations and thrust-faulted areas, the formations may be under considerable horizontal compressive stress. The vertical overburden stress, therefore, may be the least principal stress, resulting in horizontal fractures. At the present time, however, many engineers claim that formation of horizontal fractures occurs only rarely.

Productivity increase. Production increases from artificially fractured reservoir rocks are due to (1) increasing conductivity through damaged reservoir zones of reduced permeability; (2) exposure of new, otherwise tight zones; and (3) artificial change in the flow pattern in a reservoir. In the case of properly designed fracturing operation, one is able to evaluate and predict the expected productivity increase resulting from the above factors.

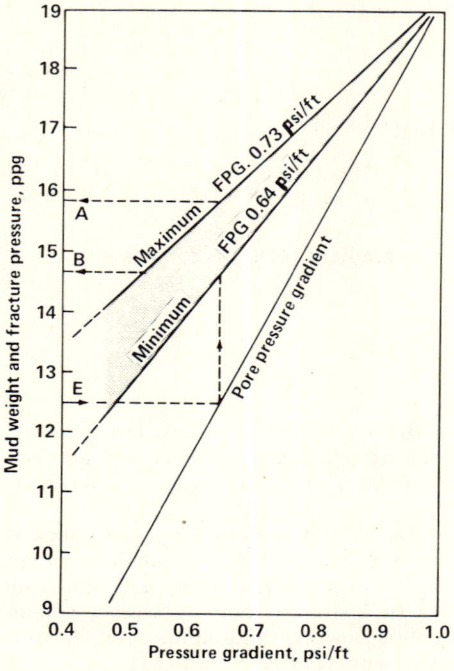

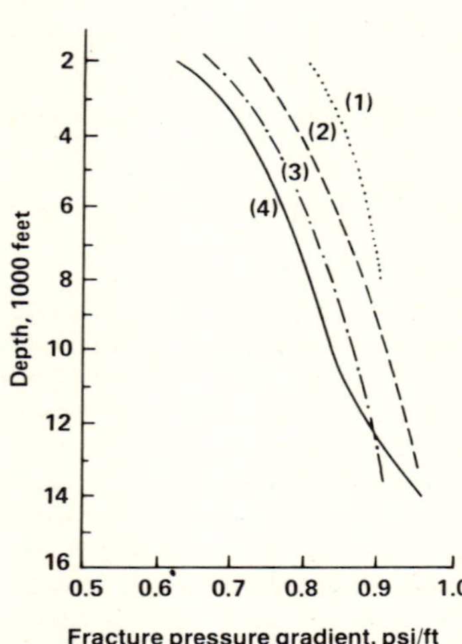

Fig. A-2. Graphical presentation of maximum and minimum fracture pressure gradients (*FPG*) as proposed by Hubbert and Willis [13], in: Fertl [11].

Fig. A-3. Comparison of several fracture pressure gradient predictions in U.S. Gulf Coast areas. (After Fertl [11].) *1* = East Louisiana average trend (Taylor and Smith [23]); *2* = South Texas (Matthews and Kelly [16]); *3* = Louisiana Gulf Coast (Matthews and Kelly [16]); *4* = average empirical trend for ten offshore fields in West Louisiana (McPherson and Berry [15]).

Bypassing of most damaged zones frequently requires only a short fracture. It is, however, extremely important that the fracture is propped around the wellbore (Wood and Junkins, 1970, [28]). The expected production increase can be estimated (1) from appropriate nomograms as a function of the depth of the damaged zone and the ratio of the permeabilities in the damaged and undamaged zones, (2) from production logging, and (3) from transient pressure tests. McGuire and Sikora (1960, [17]) found that productivity increase in vertical fractures primarily depends upon the permeability contrast of the fracture and the formation. As illustrated in Fig. A-5, the length of the fracture is of small significance unless extremely high-conductive fractures can be generated. This chart has been developed for wells on a 40-acre spacing and wellbores of 6 in. in diameter. Table A-1 lists scaling factors for other well spacings.

Fracture width and permeability. Permeability of natural and artificial unpropped fractures is a function of the width of the fracture. An approximate relationship between fracture permeability (k, in darcys) and the width of a fracture (W, in inches) can be

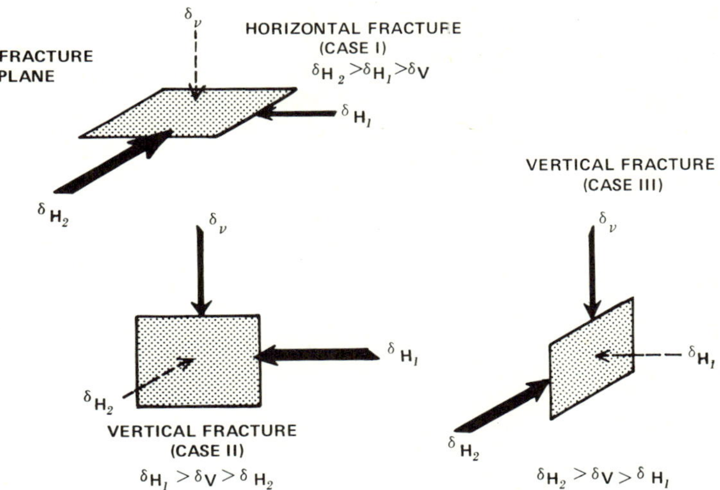

Fig. A-4. Effect of relative magnitude of principal stresses on fracture orientation.

shown as follows:

$k = 54 \times W^2 \times 10^6$

For example, a fracture having a width of 0.001 inches has a permeability of 54 darcys. This very high permeability created by a rather narrow fracture demonstrates why

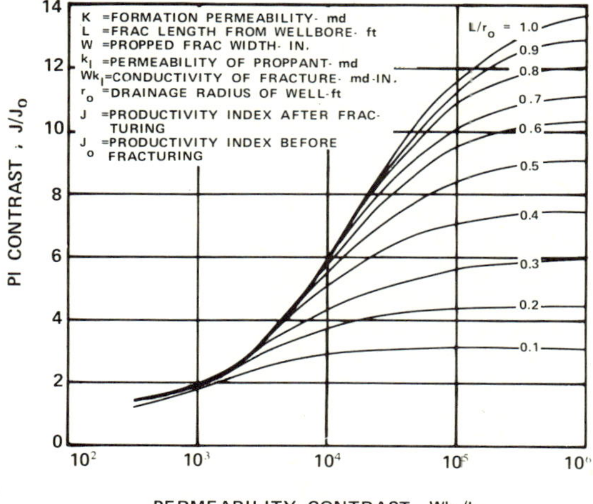

Fig. A-5. Effect of (1) permeability contrast between fracture and formation, and (2) fracturing length on the expected increase of the productivity index in a hydraulically fractured well. Curves are based on 40-acre spacing and 3-inch wellbore radius. Scaling factors must be used for other well spacings (see Table A-1). (After McGuire and Sikora [17].)

TABLE A-I

Scaling factors for productivity estimates (After McGuire and Sikora [17]; courtesy of SPE of AIME)

Well spacing (acres)	Drainage radii (ft)	Scaling factors for	
		abscissa (Wk_1/k)*	ordinate (J/J_0)*
10	330	2.00	1.11
20	467	1.42	1.05
40	660	1.00	1.00
80	933	0.71	0.95
160	1320	0.50	0.91
320	1867	0.35	0.87
640	2640	0.25	0.84

* Where W = propped fracture width (inches); k_1 = permeability of proppant (mD); k = formation permeability (mD); J = productivity index after fracturing; and J_0 = productivity index before fracturing.

fractures significantly affect the production capabilities of reservoirs. Even small fractures act as "pipeline" from the fractured area into the wellbore, the result being increased production.

If a formation is fractured while drilling, the high fracture permeability may result in high drilling fluid (mud) losses into the formation.

It is generally assumed that the induced fractures have widths of around 0.1 inch. Fracture width is a function of rock elasticity, injection rate, properties of fracturing fluid, and fracture size (Smith, 1965, [22]). This has been mathematically described by Perkins and Kern (1961, [19]).

Propping agents for fracturing. Hydraulically created fractures have to be propped * to maintain the desired fracture conductivity. This fracture conductivity depends on the type, size, uniformity, amount (density), and the manner of placement of fractures in the formation. Other important factors include the proppant's strength, degree of its embedment into the formation and its deformation, its chemical and temperature stability, density, and the availability in large quantities at reasonable cost.

Regardless of the type of proppant used (sand, glass beads, aluminum, nut shells, etc.) the propping of the fractures immediately around the wellbore is of utmost importance. Ideally this region should be packed by achieving a "sand-out" in the fracture.

Appropriate models are available to calculate the rate of proppant's advance and the placement efficiency in horizontal fracture systems (Wahl and Campbell, 1963, [26]). The horizontal extent and vertical height of proppant-fill in a vertical fracture can also be calculated (Kern et al., 1959, [14]).

Fracturing fluids. A large variety of fracturing fluids are available. Factors which should be considered in the selection of these fluids include the rheological properties, avail-

* Propping agents hold fractures in place after pressure release and may consist of carefully sized sand (20—24 mesh = 0.0328—0.0164 in.), for example, having a packed permeability of about 300 D.

ability, and cost of the frac-fluid, and the properties of the reservoir rock and reservoir fluids. These fluids are generally classed as *water-base* or *oil-base*. The latter are used less frequently, but do find application in low-pressure and preferentially oil-wet reservoirs. The chemical nature of the reservoir rock determines if either acid-base or non-acid-base fluids should be used. For example, in rocks exhibiting high solubility in acids (such as carbonates) an acid-base fracturing fluid is usually more effective.

An optimum fluid can be selected for a given reservoir if detailed information based on proper laboratory tests is available as to the rock and reservoir fluid properties.

Impact of hydraulic fracturing. The commercial introduction of hydraulic fracturing started in 1948 and has been quickly accepted by the oil industry. By 1968, more than 500,000 fracturing jobs had been completed. Numerous areas previously thought to be non-commercial due to the low permeability of reservoir rock are now being developed all over the world.

Development of novel concepts for the in-situ recovery of hydrocarbon resources from tar sand and oil shale deposits will also have to rely heavily on special fracturing concepts.

Nuclear fracturing

Over the last two decades considerable research by both governmental agencies and several major U.S.A. oil companies has focused on the potential application of nuclear energy for facturing tight reservoir rocks containing vast hydrocarbon reserves. The worldwide abundance of nonproductive or low-productive oil and gas reservoirs, and extensive tar sand and oil shale deposits provide the necessary incentive to develop technical and economically feasible methods for increasing the ultimate recovery of these valuable hydrocarbon resources based on in-situ recovery processes.

The feasibility of nuclear fracturing, has been demonstrated in several field tests in the U.S.A. (Gnome test and Gasbuggy, Rulison, and Rio Blanco projects) and other countries. Factors frequently involved are of technical, economical, political, environmental, and emotional nature.

From a technical stand point, the major difference between hydraulic and nuclear fracturing lies in the fact that whereas a hydraulic fracturing treatment normally creates a simple fracture, nuclear fracturing will from a cavity of several hundred feet in diameter with a multitude of fractures radiating from it into the otherwise tight reservoir rock. The result is a very large *effective* wellbore radius (Atkinson and Lekas, 1963, [2]) which should allow high production rates. Oil shales and tar sands could be retorted in situ and the hydrocarbon resources removed in gaseous or liquid form, thereby eliminating the need for mining operations. If in-situ thermal recovery processes, however, are considered which are to be conducted through a fracture system, the hydraulic fracturing techniques appear to be more advantageous than the nuclear mass rubbling process.

Furthermore, in areas where the effectiveness of hydraulic fracturing has been well established, nuclear fracturing cannot compete because of its high cost. Guidelines for reservoir characteristics desired in the case of nuclear fracturing include low reservoir permeability, massive zones bounded by thick impermeable formations, gas-bearing reservoirs, and low-viscosity oil deposits.

Basically, a nuclear detonation in a wellbore creates a cavity resulting from the vaporization of the rock and its saturating fluids. Fracture system radiates from this cavity into the formation. Rock collapse into the cavity forms a chimney-rubble zone, with most of the molten material and radioactive fission products concentrated in the bottom of these zones (Atkinson, 1964, [3]). A schematic presentation of a typical post-shot environment resulting from a contained underground nuclear explosion is illustrated in Fig. A-6.

Studies of several contained underground nuclear explosions showed that the nuclear

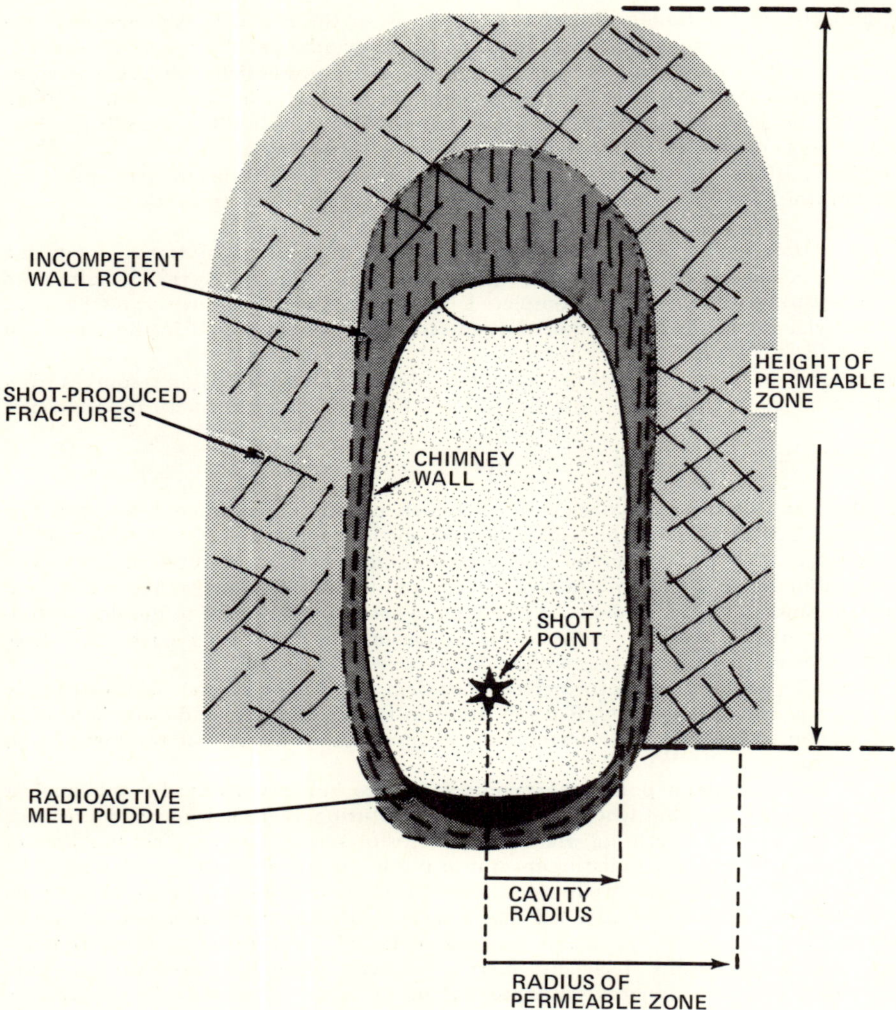

Fig. A-6. Schematic presentation of a typical post-shot environment using nuclear fracturing.

devices have yielded a rather consistent model of the geometric features. Mathematical relationships have been developed for calculating specific characteristics of the post-shot geometry as a function of the yield of the nuclear device, depth of burial, and the type of formation (Boardman et al., 1964, [4]; Bray et al., 1965, [5]).

Table A-11 shows (1) the variation in cavity radius; (2) the volume of rock (in acre-feet) in which there is a drastic permeability increase; and (3) the height of the created permeable zone as a function of yield at the depth of burial of 450 and 800 ft. Although these data are only qualitative, they do show how these parameters are expected to vary.

TABLE A-II

Estimated volume of increased permeability for nuclear explosions in reservoir rock (After Coffer et al. [8]; courtesy of USAEC)

Yield (in Kt) at scaled depth of burial (ft)		Cavity radius (ft)	Depth of burial (ft)	Height of permeable zone (ft)	Volume of gross permeability increase (acre-ft)
Kt	ft				
10	800	96	1720	540	$2.1 \cdot 10^3$
	450	110	970	620	$3.2 \cdot 10^3$
100	800	167	3720	940	$1.1 \cdot 10^4$
	450	196	2100	1110	$1.8 \cdot 10^4$
500	800	252	6350	1420	$3.8 \cdot 10^4$
	450	290	3600	1640	$5.9 \cdot 10^4$
1000	800	299	8000	1690	$6.4 \cdot 10^4$
	450	346	4500	1950	$1.0 \cdot 10^5$

The greatest changes in permeability occur in the vertical direction, which will tend to restrict the size (yield) of nuclear devices used for shallow reservoir stimulation.

References

1 R.A. Anderson, D.S. Ingram and A.M. Zanier, "Determining Fracture Pressure Gradients from Well Logs", *J. Pet. Technol., 25,* 1259—1268 (1973).

2 C.H. Atkinson and M.A. Lekas, "Atomic Age Fracturing May Soon Open Up Stubborn Reservoirs", *Oil Gas J., 61(48),* 154—156 (1963).

3 C.H. Atkinson, *Subsurface Fracturing From Shoal Nuclear Detonation,* Report PNE 3001, USAEC-USMB, 18 pp. (Nov., 1964).

4 C.R. Boardman, D.D. Rabb and R.D. McArthur, "Contained Nuclear Detonations in Porous Media — Geologic Factors in Cavity and Chimney Formation, Engineering with Nuclear Explosives", *USAEC TID 7695,* 109—126 (1964).

5 B.G. Bray, C.F. Knutson, H.A. Wahl and J.N. Dew, "Economics of Contained Nuclear Explosions Applied to Petroleum Reservoir Stimulations", *J. Pet. Technol., 17,* 1145—1152 (1965).

6 D.C. Card, *Review of Fracturing Theories, UCRL 13040,* Colorado School of Mines Research Foundation, Golden, Colo., pp. 14—20 (Apr. 16, 1962).

7 S.A. Christman, "Offshore Fracture Gradients", *J. Pet. Technol., 25,* 910—914 (1973).

8 H.F. Coffer, B.G. Bray and C.F. Knutson, "Application of Nuclear Explosives to Increase Effective Well Diameters", *Eng. Nucl. Explosives, USAEC TID 7695,* 269—288 (1964).

9 B.C. Craft, W.R. Holden and E.D. Graves, *Well Design, Drilling and Production,* Prentice-Hall, 325 pp. (1962).

10 B.A. Eaton, "Fracture Gradient Prediction and Its Application in Oil Field Operations", *J. Pet. Technol., 21,* 1343—1360 (1969).

11 W.H. Fertl, *Abnormal Formation Pressures,* Elsevier, Amsterdam, 385 pp. (1976).

12 A.A. Griffith, "The Phenomena of Rupture and Flow of Solids", *Philos. Trans. Royal Soc., Ser. A, 221,* 163—166 (1929).

13 M.K. Hubbert and D.G. Willis, "Mechanics of Hydraulic Fracturing", *Trans. AIME, 210,* 153—168 (1957).
14 L.R. Kern, T.K. Perkins and R.E. Wyant, "The Mechanics of Sand Movement in Fracturing", *Trans. AIME, 216,* 403—405 (1959).
15 L.A. MacPherson and L.N. Berry, "Prediction of Fracture Gradients from Log-Derived Elastic Moduli", *Log Analyst, 13(5),* 12—29 (1972).
16 W.R. Matthews and J. Kelly, "How To Predict Formation Pressure and Fracture Gradient", *Oil Gas J., 65(8),* 92—106 (1967).
17 W.J. McGuire and V.J. Sikora, "The Effects of Vertical Fractures on Well Productivity", *Trans. AIME, 219,* 401—403 (1960).
18 A. Nadai, *Theory of Flow and Fracture of Solids,* McGraw-Hill, New York, N.Y., 208 pp. (1950).
19 T.K. Perkins and L.R. Kern, *"Widths of Hydraulic Fractures", J. Pet. Technol., 13,* 937—949 (1961).
20 H.H. Rieke, III and G.V. Chilingarian, *Compactions of Argillaceous Sediments,* Elsevier, Amsterdam, 424 pp. (1974).
21 F.B. Seely, *Resistance of Materials,* Wiley, New York, N.Y., 272 pp. (1947).
22 J.E. Smith, "Design of Hydraulic Fracture Treatments", *SPE 1286, AIME Fall Meet.,* Denver, Colo. (Oct., 1965).
23 D.B. Taylor and R.K. Smith, "Improving Fracture Gradient Estimates in Offshore Drilling", *Oil Gas J., 68(15),* 67—72 (1970).
24 S. Timoshenko, *Strength of Materials,* Van Nostrand, New York, N.Y., 480 pp. (1941).
25 H.K. Van Poollen, "Theories of Hydraulic Fracturing", *Quart. Colo. School Mines,* Golden, *52(3),* 113—131 (July, 1957).
26 H.A. Wahl and J.M. Campbell, "Sand Movement in Horizontal Fractures", *SPE* 564, *SPE Prod. Res. Sym.,* Norman, Oklahoma (Apr., 1963).
27 J.B. Waslh and W.F. Brace, "A Fracture Criterion for Brittle Anisotropic Rocks", *J. Geophys. Res., 69(16),* 3449—3456 (1964).
28 D.B. Wood and G. Junkins, "Stresses and Displacement Around Hydraulically Fractured Wells", *SPE 3030, AIME Fall Meet.,* Houston, Texas (Oct., 1970).

REFERENCES INDEX *

* Prepared by Keith Manasco.

SUBJECT INDEX *

* Prepared by Keith Manasco.